HARD CHOICES, SOFT LAW

Global Environmental Governance Series

Series Editors: John J. Kirton and Konrad von Moltke

Global Environmental Governance addresses the new generation of twenty-first century environmental problems and the challenges they pose for management and governance at the local, national, and global levels. Centred on the relationships among environmental change, economic forces, and political governance, the series explores the role of international institutions and instruments, national and sub-federal governments, private sector firms, scientists, and civil society, and provides a comprehensive body of progressive analyses on one of the world's most contentious international issues.

Also in the series

The Politics of Irrigation Reform:
Contested Policy Formulation and Implementation in Asia, Africa and Latin America
Edited by Peter P. Mollinga and Alex Bolding
ISBN 0 7546 3515 5

Agricultural Policy Reform:
Politics and Process in the EU and US in the 1990s
Wayne Moyer and Tim Josling
ISBN 0 7546 3050 1

Governing Global Biodiversity:
The Evolution and Implementation of the Convention on Biological Diversity
Edited by Philippe G. Le Prestre
ISBN 0 7546 1744 0

Linking Trade, Environment, and Social Cohesion:
NAFTA Experiences, Global Challenges
Edited by John J. Kirton and Virginia W. Maclaren
ISBN 0 7546 1934 6

Hard Choices, Soft Law

Voluntary Standards in Global Trade, Environment and Social Governance

Edited by
JOHN J. KIRTON and MICHAEL J. TREBILCOCK
University of Toronto, Canada

ASHGATE

Published by
Ashgate Publishing Limited
Gower House
Croft Road
Aldershot
Hants GU11 3HR
England

Ashgate Publishing Company
Suite 420
101 Cherry Street
Burlington, VT 05401-4405
USA

Ashgate website: http://www.ashgate.com

British Library Cataloguing in Publication Data
Hard choices, soft law : voluntary standards in global trade, environment and social governance. - (Global environmental governance)
1. International economic relations 2. Industries - Self-regulation 3. Sustainable development
I. Kirton, John J. II. Trebilcock, Michael J. (Michael John), 1941-
337

Library of Congress Cataloging-in-Publication Data
Hard choices, soft law : voluntary standards in global trade, environment and social governance / edited by John J. Kirton, Michael J. Trebilcock.
p. cm. -- (Global environmental governance)
Includes bibliographical references and index.
ISBN 0-7546-0966-9
1. Sustainable development--International cooperation. 2. Standardization. 3. Commercial policy--International cooperation. 4. International trade--Environmental aspects. 5. International trade--Social aspects. 6. Social responsibility of business. 7. International agencies. 8. Public-private sector cooperation. I. Kirton, John J. II. Trebilcock, M. J. III. Global environmental governance series

HC79.E5H3525 2004
343'.07--dc22 2004046269

ISBN 0 7546 0966 9

Printed and bound in Great Britain by MPG Books Ltd, Bodmin, Cornwall

Contents

List of Tables

List of Figures

List of Contributors

Tasso Rezende de Azevedo is a forester with the Escola Superior de Agricultura Luiz ded Queiroz at the University of São Paulo in Brazil and Director of the National Forest Programme at the Ministry of Environment in Brazil.

Sir Nicholas Bayne, KCMG, is a Fellow at the International Trade Policy Unit of the London School of Economics and Political Science.

Steven Bernstein is Assistant Professor of International Relations in the Department of Political Science at the University of Toronto.

Adelle Blackett is Assistant Professor at the Faculty of Law of McGill University. Formerly, she was a labour law and labour relations specialist at the International Labour Office in Geneva.

Benjamin Cashore is Assistant Professor of Sustainable Forest Policy and Chair of the Program on Forest Certification at the Yale School of Forestry and Environmental Studies.

Michael E. Cloghesy is President of the Conseil patronal de l'environnement du Québec (CPEQ), known in English as the Business Council on the Environment. The CPEQ is one of the social partners of 'Strengthening Canada's Environmental Community through International Regime Reform: Exploring Social Cohesion in a Globalizing Era' (the EnviReform project).

Wesley Cragg is the George R. Gardiner Professor of Business Ethics in York University's Schulich School of Business.

Hevina S. Dashwood is an Assistant Professor of International Relations at Brock University.

John W. Foster is Principal Researcher, Civil Society, for the North-South Institute.

The Honourable Thomas A. Hockin, P.C., is President and Chief Executive Officer of the Investment Funds Institute of Canada.

John J. Kirton is an Associate Professor of Political Science and Director of the G8 Research Group at the University of Toronto, and Principal Investigator of

'Strengthening Canada's Environmental Community through International Regime Reform: Exploring Social Cohesion in a Globalizing Era' (the EnviReform project).

The Honourable Roy MacLaren, P.C., has led a distinguished career in the Canadian foreign service, business, and Parliament.

Robert O. Matthews is a Professor Emeritus of Political Science at the University of Toronto.

Lisa N. Mills is an Assistant Professor in the School of Public Policy and Administration at Carleton University.

Sylvia Ostry is Distinguished Research Fellow at the Centre for International Studies of the University of Toronto, and a co-investigator in 'Strengthening Canada's Environmental Community through International Regime Reform' (the EnviReform project).

C. S. Venkata Ratnam is director of the GITAM Institute of Foreign Trade in Visakhapatnam, India.

Chris Tollefson is a Professor of Law at the University of Victoria's Environmental Law Centre and a member of 'Strengthening Canada's Environmental Community through International Regime Reform: Exploring Social Cohesion in a Globalizing Era' (the EnviReform project) at the University of Toronto.

Michael J. Trebilcock is Professor of Law at the University of Toronto and a co-investigator in 'Strengthening Canada's Environmental Community through International Regime Reform' (the EnviReform project) at the University of Toronto.

Anil Verma is Professor of Industrial Relations and Human Resource Management at the University of Toronto, where he holds a joint appointment at the Faculty of Management and the Centre for Industrial Relations.

Leah F. Vosko is Canada Research Chair in Feminist Political Economy at the School of Social Sciences in the Joseph E. Atkinson Faculty of Liberal and Professional Studies at York University.

Christopher Wilkie is Deputy Director of the Investment Trade Policy Division of the Department of International Trade in Ottawa.

Preface and Acknowledgements

This book, the sixth in Ashgate Publishing's Global Environmental Governance series, establishes the tradition of using the annual conference of the University of Toronto's 'EnviReform project' as the foundation for an edited volume exploring the central issues and contemporary challenges of integrating trade liberalisation, environmental protection, and social cohesion in a globalising age. The first such conference served as the basis for the second book in the series, *Linking Trade, Environment, and Social Cohesion: NAFTA Experiences, Global Challenges*, edited by John J. Kirton and Virginia M. Maclaren and published in 2002. That volume focussed on a pioneering regional innovation in hard law — the legalised organisation of the 1994 regime of the North American Free Trade Agreement (NAFTA) — and its implications for global trade, environmental, and social governance. The current work shifts the focus to the soft law world of voluntary standards and their contribution to the global community as a whole. In doing so, it examines how soft law processes and regimes relate to their hard law equivalents; it also considers their contribution to sectors such as forestry, where few hard law global regimes exist, as well as their use by and for the labour and business communities and their role in conflict-plagued regions such as Africa. The book's insights come at a particularly timely moment, as the world embarks on the second decade of constructing an environmentally and socially sustainable global economy, built on the foundation of the World Summit on Sustainable Development in 2002.

This volume is based in the first instance on the work of the University of Toronto project on 'Strengthening Canada's Environmental Community Through International Regime Reform' (the EnviReform project) and, in particular, its second annual conference, held on 8–9 November 2001 at the Munk Centre for International Studies, on 'Hard Choices, Soft Law: Voluntary Standards in Global Trade, Environment and Social Governance'. The conference brought together experts from the scholarly, nongovernmental, business, labour, and government communities, as well as from many major global regions — North and South America, Europe and Asia — to examine the evolution, use, impact, and future of soft law and voluntary standards in the specific sector of forestry and in the broad domain of trade and labour standards. They explored codes of corporate responsibility as a particular form of soft law, and how various international institutions, notably the World Trade Organization (WTO), the Organisation for Economic Co-operation and Development (OECD), the G7/G8, and NAFTA, rely on soft law in their work. They drew on a wealth of disciplinary traditions and fields: political science and political economy, law and regulation, economics, environmental studies, labour studies, management studies, and ethics. They also benefited from a similarly wide range of professional experience, including trade ministers, senior officials responsible for political leaders' participation in the G7/G8 summit, and the General Agreement on Tariffs and Trade (GATT), currently serving officials dealing with corporate responsibility, and major figures from the business, labour, and environmental communities.

This volume unites the contributions of the EnviReform project's participating scholars and its several social partners. Of the project's ten scholars, represented in this volume are John Kirton, Michael Trebilcock, Sylvia Ostry, and Chris Tollefson, and the project's former postdoctoral research fellow, Hevina Dashwood. Contributions from the project's social partners come from Michael Cloghesy of the Conseil patronal de l'environnement du Québec, Chris Tollefson in his capacity as a representative of the Sierra Legal Defence Fund of Canada, and Tasso Azevedo as part of the broader, global Leadership for Environment and Development (LEAD) network.

Acknowledgements

In producing this volume, we have enjoyed the exceptional support of those who have contributed in many different ways. We are grateful to the Social Sciences and Humanities Research Council of Canada, under its strategic grant programme on 'Social Cohesion in a Globalizing Age', for its award to the University of Toronto's EnviReform project. It was this award that financed the conference and research on which this volume is based, and the editorial work required to bring it to publication.

We are further grateful to the University of Toronto's Centre for International Studies at the Munk Centre, and its director, Professor Louis Pauly, who oversees the EnviReform project, serves as a member of its scholarly project team, and imparts wisdom and collegiality that have been a constant and indispensable source of sound advice and support. We also appreciate the support of the chairs and deans of the departments and faculties whose members participate as scholars in the project, and of Professor Carl Amrhein, former Dean of the Faculty of Arts and Science where most of those departments have their home. Beyond the University of Toronto, we recognise the valuable and vital part played in the EnviReform project by our social partners: the Commission for Environmental Cooperation, the Conseil patronal de l'environnement du Québec, LEAD Canada, the Canadian Auto Workers, Pollution Probe, Foodshare, and the Sierra Legal Defence Fund. We are also pleased and proud to acknowledge the critical contribution of our international partner, the Woodrow Wilson Center for International Studies in Washington DC. Under the leadership of its Director of Canadian Studies, David Biette, we were able to conduct our conference as a fully interactive videoconference with experts in Washington, to the intellectual enrichment of all, with the participation of Susan Aaronson, Robin Broad, Kimberly

Ann Elliott, Steven Tullberg, Les Simon, Geoffrey Dabelko, Bama Athreya, and Ross Gorte. The videoconference itself would not have been possible without the efforts of many people at Woodrow Wilson, among them Ken Nelson, John Tyler, Kent Hughes, and Emily Heard, and the Center's director, Lee Hamilton. Our colleague and EnviReform project team member Stephen Clarkson played an essential part in recognising this opportunity for a wider dialogue and forging the partnership that brought it to fruition.

Many others worked to make EnviReform's November 2001 conference — and thus this book — a success. At the Munk Centre for International Studies, Mary Lynne Bratti, Diane Granato, Nancy Fortin, and Tina Lagopoulous worked tirelessly and efficiently in support of our work. Scott Bohaker ensured that our videoconference and webcast — the first ever from the new Vivian and David Campbell Room in the Munk Centre — came off as we wished and as the inaugural event this state-of-the-art facility deserved. In addition, we were well served by many exceptional research assistants and student volunteers including Anju Aggarwal, Dan Ben Aron, Brendan Heath, Petra Kukacka, Marilena Liguori, Melanie Martin, Klaudyna Osika, Bob Papanikolaou, Kristiana Powell, and Zaria Shaw.

We owe a particular word of thanks to two individuals. The first is Madeline Koch, who as Managing Editor of the EnviReform project and the Global Environmental Governance series worked with her usual superb skill and dedication to produce our conference, and to transform the extensively revised and updated papers from it into this book. The second is Helen Walsh, who as President of ThinkContent, managed both the videoconference and webcast dimensions of our conference.

We are also most grateful to our colleague Professor Peter Hajnal, who once again assumed and skilfully oversaw the process of having this manuscript anonymously refereed. We are further grateful to the dedication, thoroughness, and insight of our anonymous referees, whose important comments have been taken fully into account.

As always we reserve a special word of thanks for Kirstin Howgate and her colleagues at Ashgate, for recognising the value of producing this volume and for working so effectively, and with appropriate understanding, to ensure the smooth adoption and publication of the manuscript.

We acknowledge, as always, the patience and support of our families as we laboured to organise our conference, produce our own initial papers, and convert rough drafts into finished chapters. Above all, we thank our students for their endless enthusiasm, constructive criticism, intellectual contributions and challenges, and continuing commitment to the causes dealt with in this book. It is to them that we dedicate this work.

John J. Kirton and Michael J. Trebilcock
April 2004

List of Abbreviations

ACP	African, Caribbean, and Pacific group of states
ADI	acceptable daily intake
AFL-CIO	American Federation of Labor and Congress of Industrial Organizations
APEC	Asia-Pacific Economic Cooperation
ARET	Accelerated Reduction or Elimination of Toxics
ASCM	Agreement on Subsidies and Countervailing Measures
BC	British Columbia
BIAC	Business and Industry Advisory Committee (of the Organisation for Economic Co-operation and Development)
BRACELPA	Associação Brasileira de Celulose e Papel (Brazilian Association for Pulp and Paper)
CalPERS	California Public Employees Retirement System
CAP	Common Agricultural Policy
CBD	Convention on Biological Diversity
CDCAC	Canadian Democracy and Corporate Accountability Commission
CEAWC	Committee on the Eradication of Abductions of Women and Children
CEC	Commission for Environmental Cooperation
CEPC	Carpet Export Promotion Council
CERFLOR	Programme for Forest Certification (Brazil)
CFM	community forest management
CIDA	Canadian International Development Agency
CIME	Committee on International Investment and Multinational Enterprises (of the Organisation for Economic Co-operation and Development)
CITES	Convention on International Trade in Endangered Species
CLAC	Anti-Capitalist Convergence of Montreal
CLC	Commission for Labor Cooperation
CoC	chain of custody certification
CONAMA	Comisión Nacional del Medio Ambiente (Brazilian National Environment Council)
CPEQ	Conseil patronal de l'environnement du Québec
CSA	Canadian Standards Association
CSR	corporate social responsibility

CTE	Committee on Trade and Environment (of the World Trade Organization)
DAN	Direct Action Network
DFAIT	Department of Foreign Affairs and International Trade (Canada)
DTZ	Deutsche Gesellschaft für technische Zusammenarbeit GmbH
ECOSOC	United Nations Economic and Social Council
EPZ	export processing zone
ETC Group	Action Group on Erosion, Technology, and Concentration (formerly RAFI)
FAO	Food and Agriculture Organization
FATF	Financial Action Task Force
FDA	United States Food and Drug Administration
FDI	foreign direct investment
FIDH	International Federation for Human Rights (Fédération Internationale des Ligues des Droits Humaines)
FOE	Friends of the Earth
FOGS	functioning of the GATT system
FSC	Forest Stewardship Council
FSC-BC	Forest Stewardship Council of British Columbia
FTA	Canada–U.S. Free Trade Agreement
FTAA	Free Trade Agreement of the Americas
G7	Group of Seven (G8 without Russia)
G77	Group of 77
G8	Group of Eight (France, United States, United Kingdom, Russia, Germany, Japan, Italy, Canada)
GATS	General Agreement on Trade in Services
GATT	General Agreement on Tariffs and Trade
GDP	gross domestic product
GMO	genetically modified organism
GNP	gross national product
GNPOC	Greater Nile Petroleum Operating Company
GRAIN	Genetic Resources Action International
GSP	Generalised System of Preferences
HIPCs	heavily indebted poor countries
IBAMA	Instituto Brasileiro do Meio Ambiente e dos Recursos Naturais Renováveis (Brazilian Institute for the Environment)
ICAFF	Inter-Church Coalition of Africa
ICC	International Chamber of Commerce
ICFTU	International Confederation of Free Trade Unions
ICRC	International Committee of the Red Cross
IFF	Intergovernmental Forum on Forests
IFG	International Forum on Globalization

IFIR	International Forest Industry Roundtable
IGAD	Intergovernmental Authority on Development
IGF-I	insulin-like growth factor I
IKPC	Indústrias Klabin de Papel e Celulose
ILO	International Labour Organization
Imaflora	Instituto de Manejo e Certificação Florestal e Agrícola (Brazilian Institute for Agricultural and Forest Management and Certification)
Imazon	Instituto do Homem e Meio Ambiente da Amazônia (Amazon Institute for Man and the Environment)
IMF	International Monetary Fund
INCHRITI	International NGO Committee for Human Rights in Trade and Investment
INCRA	Instituto Nacional de Colonização e Reforma Agrária (Brazilian National Institute for Colonization and Agrarian Reform)
IOE	International Organisation of Employers
IPF	IGAD Partners Forum or Friends of the Intergovernmental Authority on Development
ISO	International Organization for Standardization
ITP	International Third Position
ITTA	International Tropical Timber Agreement
JECFA	Joint Expert Committee on Food Additives
JPAC	Joint Public Advisory Committee
LMO	living modified organism
MAI	Multilateral Agreement on Investment
MNC	multinational corporation
MRL	maximum residue limit
NAAEC	North American Agreement on Environmental Cooperation
NAALC	North American Agreement on Labor Cooperation
NAFTA	North American Free Trade Agreement
NATO	North American Treaty Organization
NCP	National Contact Point
NGO	nongovernmental organisation
NIEO	New International Economic Order
OECD	Organisation for Economic Co-operation and Development
OHCHR	Office of the High Commissioner for Human Rights
OMERS	Ontario Municipal Employees Retirement System
OPEC	Organization of the Petroleum Exporting Countries
PEFC	Pan European Forest Certification
PGA	Peoples' Global Action
PPMs	production and processing methods

PVP	plant variety protection
RAFI	Rural Advancement Fund International (now ETC Group)
rbGH	recombinant bovine growth hormone
RLS	ratcheting labour standards
RPPN	Reservas Particulares do Patrimônio Natural (Privately Owned Natural Conservation Areas)
RTRs	resources and tenure rights
SFI	Sustainable Forestry Initiative of the American Forest and Paper Association
SPLA	Sudan People's Liberation Army
SPLM	Sudan People's Liberation Movement
SPS	sanitary and phytosanitary measures
TBT	technical barriers to trade
TPRB	Trade Policy Review Board
TPRM	Trade Policy Review Mechanism
TRIMS Agreement	Agreement on Trade-Related Investment Measures
TRIPs Agreement	Agreement on Trade-Related Aspects of Intellectual Property Rights
UNCED	United Nations Conference on Environment and Development
UNCTAD	United Nations Conference on Trade and Development
UNEP	United Nations Environment Programme
UNESCO	United Nations Educational, Social, and Cultural Organization
UNFCCC	United Nations Framework Convention on Climate Change
UNFF	United Nations Forum on Forests
UNICEF	United Nations Children's Fund
UNIDO	United Nations Industrial Development Organization
UNIFEM	United Nations Development Fund for Women
UPOV	International Union for the Protection of New Varieties of Plants
USS	Universities Superannuation Scheme
USTR	United States Trade Representative
WARP	Woodworkers Alliance for Rainforest Protection
WBCSD	World Business Council on Sustainable Development
WHO	World Health Organization
WSSD	World Summit on Sustainable Development
WTO	World Trade Organization
WWF	World Wildlife Fund

INTRODUCTION

Chapter 1

Introduction: Hard Choices and Soft Law in Sustainable Global Governance

John J. Kirton and Michael J. Trebilcock

The Governance Challenge

As the twenty-first century unfolds, the challenge of generating global governance to support sustainability has become ever more acute. The precarious state of so many ecosystems and communities around the world, and the interconnections among them, have now become a well-recognised reality for many who were once sceptical of the urgency and integrated nature of the threat. In the developed world, it is becoming routine for the leading countries to embark on trade liberalisation only after environmental and social consequences have been assessed and taken into account. In the developing world, a succession of financial crises in an era of rapid globalisation has demonstrated that poor people can have their life chances — and the economic, ecological, and social resources on which they depend for survival — destroyed in a single stroke. Among interested citizens throughout the world, there is a growing consensus that further trade and financial liberalisation must take place only if it protects and promotes ecological and social values as well. The leaders of the world's most powerful countries, meeting at their G8 summits from Cologne in 1999 to Evian in 2003, have recognised and reinforced the call. While a rich debate flourishes about just how best to make the difficult choices required to integrate equitably economic, environmental, and social values, all agree that such hard choices must be made.

Yet despite this emerging consensus, the intergovernmental organisations created by the global community over the past half century to address these issues remain fragile, fragmented, fixed in old frameworks, and thus often ineffective in response. The World Trade Organization (WTO), meeting at the ministerial level in November 2001 to launch the Doha Development Agenda, incorporated environmental considerations into its mandate for new trade liberalisation negotiations to an unprecedented degree. But these values remained partial, applied to a particular negotiation with an uncertain outcome rather than to the WTO's overall mandate and *modus operandi*, dependant on the overriding value ascribed to economic development and largely divorced from social concerns. The September 2003 failure of the mid-term ministerial review of the Doha round at Cancun, due in part to deep divisions between the developed North and developing South, further eroded confidence that any swift solutions will come from within the WTO. The United Nations environmental community, at the World Summit on Sustainable Development (WSSD) in Johannesburg in August 2002, for the first time acknowledged the important role of trade in securing environmental and social progress. But unlike its 1992 predecessor — the United Nations Conference on Environment and Development (UNCED) — the WSSD produced no landmark legal conventions backed by new intergovernmental organisations to govern critical areas, such as forestry, where trade, environment, and social processes and values intersect. And the social side of the UN family, from the venerable International Labour Organization (ILO) to the United Nations Educational, Social, and Cultural Organization (UNESCO), has remained as ineffective as specialised agencies such as the Food and Agriculture Organization (FAO) in modernising to meet the needs for governance in a now rapidly integrating world. Although the UN recognised the need for integrated global governance for a new era by holding its Millennium Summit in 2000, it produced there only ambitious goals in individual areas to be met during the twenty-first century. It left the achievement of these goals largely to the fragmented, incomplete, formal intergovernmental legal and institutional architecture created in the mid twentieth century and conceived in the nineteenth century before.

There is real doubt whether this nineteenth-century invention and inheritance of heavily legalised, bureaucratically heavy, hard law intergovernmental organisations, centred on the United Nations system, can adequately address the pressing needs of the new age. The UN still is grounded in, and in some ways imprisoned by, a fixed, formal, constitution-like legal charter that lacks proper recognition of the value, or even the existence, of the natural environment, not to speak of non-state actors. At the same time, the need to find a solution is becoming ever more urgent. This is particularly true with two new major rounds of trade liberalisation — the WTO's Doha Development Agenda and the Free Trade Agreement of the Americas (FTAA) — both scheduled to conclude by 2005, and with a plethora of new regional and bilateral free trade agreements being born in all global regions every year.

It is thus hardly surprising that the world has increasingly turned to soft law solutions to the hard choices its confronts. Such soft law solutions come in a richly kaleidescopic variety of forms. Yet at their centre stand two primary instruments. The first instrument is the voluntary standards that serve as equivalent to formally legislated and ratified government law and regulation. The second instrument consists of the informal institutions at the international, transnational, and national levels that depend on the voluntarily supplied participation, resources, and consensual actions of their members, rather than on the formally mandated participation and regularly assessed obligatory contributions, organisation, resources, and sanctions of the institution itself. Building on pioneering efforts such as the chemical sector's Responsible Care programme and, before it, the economy-wide work of the International Organization of Standardization (ISO), the past decade of intense globalisation has seen an explosion of such voluntary standards. Indeed, most major industry sectors in the intensely internationally exchanged global economy have now turned to the power and potential of soft law.

Moreover, starting in the 1970s, governments have turned to new informal international institutions, often dominated and delivered by leaders themselves, to bring economic, environmental, and social values together in a more integrated and equitable way. On a regional or plurilateral plane, such institutionalised 'soft law summitry' has emerged in the 1990s with the Asia-Pacific Economic Cooperation (APEC) forum, and the Summit of the Americas process. Both offer a firm commitment by a fixed date to full free trade among their members, a group that includes many of the most consequential developed and developing countries in the world. At the apex of this institutionalised system of soft law summitry stands the G8 major market democracies, which, since its creation as the G7 in 1975, have long pioneered effective new principles, norms, and concrete commitments to strengthen, balance, and integrate the trade, environment, and social realms (Kirton 2002).

With such dynamism in the soft law sphere, it is understandable that the hard law intergovernmental organisations themselves have increasingly turned to the role that can be played by more informal, soft law standards and institutions to produce real, sustainable results. Indeed, some of the more effective hard law bodies have relied on soft law instruments from the start. Selective, democratically composed organisations such as the Organisation for Economic Co-operation and Development (OECD) have long successfully used an analytically based, consensus-oriented approach that has come to integrate economic, environmental, and social values in a more coherent and equitable way. Here governments, firms, and the labour movement have participated directly as tripartite partners in the institution and task. At the global level, they have also done so in the ILO, which continues to depend on the consensus-oriented tripartism that the soft law spirit brings. In the trade field, the WTO also relies heavily — as did its predecessor, the General Agreement on Tariffs and Trade (GATT) — on the more voluntary codes of the Codex Alimentarius and the ISO. Such codes have assumed a growing relevance as new conflicts have arisen over food safety and the environmental stewardship of the world's endangered forests. Moreover, at the WSSD, the leaders and their civil society partners placed a premium on public-private partnerships as the essential centrepiece for making progress in the years ahead. At the highest level, the United Nations is moving rapidly in the direction of soft law. It is led here by Secretary General Kofi Annan's Global Compact to mobilise the power of the world's major multinational corporations and their sense of corporate responsibility to help meet the UN's environmental, social, and other goals (Ruggie 2003).

The experience thus far suggests that the soft law approach offers many advantages: timely action when governments are stalemated; bottom-up initiatives that bring additional legitimacy, expertise, and other resources for making and enforcing new norms and standards; and an effective means for direct civil society participation in global governance. These benefits are particularly important at a time when the demands of intensifying globalisation may outstrip the capacity of even the most powerful, but now often deficit-ridden, national governments to respond.

Nonetheless, the soft law approach comes with its own challenges. It may lack the legitimacy and strong surveillance and enforcement mechanisms offered by hard law. With a broader array of stakeholders, soft law may promote compromise, or even compromised, standards, less stringent than those delivered by governments acting with their full authority all alone. And soft law can lead to uncertainty, as competing sets of voluntary standards struggle for dominance, and as actors remain unclear about the costs of compliance, or its absence, and about when governments might intervene to impose a potentially different, mandatory regime.

Amidst such uncertainty and complexity, it is hardly surprising that the hard law–soft law struggle stands at the heart of many of the central economic, environmental, and social debates and issues of the day. The contest within the United States, and between the U.S. and all its other G8 partners, over the Kyoto Protocol on Climate Change centres on whether voluntary, soft law approaches are enough until science and technology develop further, or whether ratification of, and compliance with, the hard law protocol, replete with precise targets and a timetable for action, is needed now. The tension between North America, on the one hand, and Europe and Japan, on the other, over genetically modified organisms (GMOs) and food safety revolves around whether voluntary standard setting and labelling are sufficient, or whether sovereign regulation nationally or regionally by the European Union should be allowed, supported rather than constrained by the formal trade law system. In the field of health, the open question is whether the innovative public-private partnership of the Global Fund to Fight AIDS, Tuberculosis, and Malaria created at the 2001 G8 Genoa Summit will be sufficient to combat these deadly infectious diseases, which are devastating Africa, or whether the resources and authority of governments and their international organisations, such as the World Health Organization (WHO) or even the entire UN system, will be needed in the end.

Thus the twenty-first century challenge for sustainable global governance is clear. Integrating trade, environment, and social values in international regimes that promote sustainable development requires the international community to make hard choices, both in moving beyond the traditionally separated set of international institutions dealing with each individual field, and in making tradeoffs within them when priorities conflict. At a time when many doubt the probability, or even the wisdom, of the WTO embracing environmental, labour, or broader social values such as cultural diversity more fully, and when the world's environmental and social institutions remain separated, weak, and reluctant to defer to, or depend on, trade processes and institutions, it is important to explore how much and how soft law approaches offer a way ahead.

Purpose and Approach

To address this issue, this volume explores how and why soft law in general, and voluntary standards and informal institutions in particular, have worked — and can work — in support of economically, ecologically, and socially sustainable global governance. This volume has three central purposes. The first is to explore how, how much, and where voluntary standards and informal institutions have worked, by themselves and in tandem with highly legalised and organised hard law organisations, to produce more effective global governance promoting economic openness, environmental enhancement, and social cohesion in an integrated, equitable and mutually supportive way. The second purpose is to identify why soft law standards and institutions have worked for this purpose, with such explanations serving as an analytical and empirical foundation for consideration of how they can be made to work better amidst real world pressures and political configurations. The third purpose is to point to how voluntary standards and informal institutions might and should be used, in ways that work to deliver the emerging consensus for sustainable development shared by so many citizens of the globe.

To meet these objectives, this volume assembles the contributions of leading experts from North and South America, Europe, and Asia, and from the disciplines of political science, philosophy, law, economics, environmental science, labour studies, and management studies. The contributors include the leaders of several professional communities, including those who have been international civil servants, ministers of a major national government who negotiated environmentally and socially consequential trade agreements, and civil society activists directly designing and delivering sustainable global governance on their own.

This work further explores its subject across a wide range of institutions, sectors, and instruments. It does so at three different levels: first, the largely nongovernmental processes producing ecologically and socially sustainable practices in the key sector of forestry, where no real intergovernmental hard law or institutions of global reach exist; second, the way governments with their hard law and civil society with its voluntary standards and soft law have come together in the critical area of setting labour standards and developing codes of corporate responsibility; and third, in the central intergovernmental institutions for integrating trade with environmental and social values, notably the multilateral WTO, the regional North American Free Trade Agreement (NAFTA), and the softer law bodies of the OECD and G8. The volume moves analytically from the specific to the general, from the sector of forestry, through the crosscutting dimension of labour standards, to the instruments of corporate responsibility, and the culminating concern with how the world's major economic and environmental institutions combine hard and soft law approaches in their work. Substantively, this work thus combines the often separately explored issues of forestry practices, trade-labour relationships, and codes of corporate responsibility, for treatment in an integrated fashion, under the embracing concept of soft law. In its exploration, the volume deals with a broad variety of sectors, from food safety to the oil industry, and with a wide array of regions, from the G8 and OECD world, through the developing countries as a whole, to critical regions such as South America, South Asia, and Africa.

The Analytical Framework

Despite their proliferating presence and potential importance, soft law instruments, and their hard law complements, have attracted inadequate analysis and generated little analytically based agreement on their nature, advantages and disadvantages, operation, and causes and effects. This is particularly when their inter-relationship is considered. Indeed, there is a wide range of disagreement about the content, the causes, and the consequences of soft law in general, and voluntary standards and informal institutions in particular, as they operate in the contemporary world.

The existing, theoretically grounded, empirical literature, the policy-oriented debate, and the analyses in this volume, revolve around the five basic, interrelated questions that require further exploration: What is hard and soft law? Where has each flourished? What are the advantages and disadvantage of each, and how do they coexist? What are the causes that allow each to arise and flourish? What consequences do each have?

Hard and Soft Law Defined

The first basic question concerns the core content of the concepts of hard and soft law. To be sure, each term can be seen as an inclusive, expansive, and flexible category. Both terms are used with a great variety of meanings in the existing literature. In this volume, however, they are applied with a more bounded degree of definitional freedom. Each author has been left to employ the definition most suitable for the particular subject at hand, yet within a restricted range that allows for comparison and coherence across the work as a whole. It is thus important to begin with a common understanding of the core, essential characteristics of each conceptual category.

Contemporary students of international politics, in the mainstream liberal institutionalist tradition, take as their analytical and, often, normative ideal the hard, formally authorised, codified law and regulation comparable to that which exists at the domestic level within the hierarchies of strong nation-states. Yet even amidst the 'anarchy' of the inter-state system, such hard law is not a single thing. Rather, in the now dominant 'legalisation' phase of the liberal institutionalist school, hard law is a collage composed of three dimensions — precision, obligation, and delegation (Abbott et al. 2000). Each of these dimensions varies a great deal. Hard law is thus an international, or by extension an inter-actor, arrangement that has a very high value on each of those three dimensions.

Such a formulation allows soft law — as the other category in a conceptually dichotomous world — to cover a vast array of the remaining terrain, and a wide variety of the levels and forms of legalisation that exist within. Empirically, in this non-highly legalised category, soft law extends downward from the commanding, highly legalised heights of hard law to embrace UN specialised agencies, non-binding forestry principles, technical standards, and even the G7 spheres of influence and balance of power itself.

For some purposes, it may indeed be useful to conceive of the hard law–soft law duality as a continuum rather than a dichotomy, and to explore how the different dimensions of the common concept of law vary with full freedom in the heuristic and empirical world. However, for this volume, some clear — if generous — definitional boundaries should be set. This analysis thus begins by taking law to mean regimes and by using the standard conception of regimes as explicit or implicit principles, norms, rules, and decision-making procedures on which actors expectations converge in a given issue, subject to the sole qualification that regimes may refer not just to individual issue areas but to global governance as a whole (Krasner 1983). Hard law is defined as a regime relying primarily on the authority and power of the state — ultimately its legitimate monopoly on the means of coercion — in the construction, operation, and implementation, including enforcement, of arrangements at international, national, or subnational level.

Soft law, in contrast, refers to regimes that rely primarily on the participation and resources of nongovernmental actors in the construction, operation, and implementation of a governance arrangement. Although the parameter of 'primarily' permits some ambiguity (as well as flexibility), the key characteristics of a soft law arrangement are clear, in contrast with a hard law arrangement. First, in a soft law regime, the formal legal, regulatory authority of governments is not relied upon and may not even be contained in the institutional design and operation. Second, there is voluntary participation in the construction, operation, and continuation. Any participant is free to leave at any time, and to adhere to the regime or not, without invoking the sanctioning power of state authority. Third, there is a strong reliance on consensus-based decision making for action and, more broadly, as a source of institutional binding and legitimacy (Ikenberry 2001). Fourth, and flowing from the third, there is an absence of the authoritative, material sanctioning power of the state — police power as a way to induce consent and compliance.

It follows that soft law regimes can take many forms, including those in which governments and their resources are present but do not dominate the regime. Indeed, governments participate in the soft law regimes of the Forest Stewardship Council (FSC). Intergovernmental organisations such as the ILO and OECD can employ soft law practices when they rely on the tripartite participation of government, business, and labour to devise and manage regimes, and when they produce voluntary codes that their members — or more often constituents within them — are to varying degrees free to adopt or not. More broadly still, soft law can embrace customary international law and state practice, whether codified or not, thus adding public international law to the many schemes that flourish, such as private commercial arbitration, under private international law. Indeed, there is often a blend, with public authority in the form of intergovernmental institutions using more private authority–inspired practices, as with NAFTA's Chapter 11 investment regime, which depends on the systems based on private commercial arbitration embodied in the World Bank Group's International Center for the Settlement of Investment Disputes and the United Nations Commission on International Trade Law (Kirton and Maclaren 2002).

National governments can and do use soft law, such as voluntary codes on the behaviour of their home-based multinationals abroad, as part of their array of statecraft instruments. Indeed, the defining features of soft law are dominance by actors and by sources of authority other than governments and an inter-subjective consciousness on the part of the agents involved of the principles, norms, rules, and decision-making procedures involved. Strongly patterned or self-absorbed behaviour or the economists' 'revealed preferences' do not count.

Soft law, so defined, comes in many forms, from private and voluntary codes and certification and labelling systems to transparency obligations on the part of governments. Yet two central categories constitute the foundation for this book. The first is voluntary standards — the principles and norms that depend on consent, consensus, and resources other than governmental authority for their work. The second is informal institutions — those transnational and intergovernmental bodies, including those exclusively among governments, that similarly rely on consent and consensus, rather than on codified and ratified charters, and intergovernmental bureaucracies with resources of their own. Informal institutions with voluntary standards or commitments and compliance systems are thus the hard core of soft law. Yet soft law includes formal intergovernmental organisations using voluntary codes. Excluded are the exclusively intergovernmental bodies when employing formal international law, and the fourth, entirely theoretical, cell of informal institutions with formal law.

The Fields Where Soft Law Has Flourished

In its many forms, soft law has long existed in many fields of public and international life. Indeed, since the great three-stage cadence of international balancing to institutionalisation to constitutionalisation was interrupted by the crises of the 1970s and the challenges of globalisation that came in their wake, soft law arrangements have flourished all across the global stage (Ikenberry 2001). Traditionally, at the international level, hard law has emerged only in a very circumscribed form, as with the International Court of Justice, to govern the behaviour of states across a wide array of activities. Elsewhere, from the mid nineteenth century onward, it has grown in stronger forms in particular institutions, from the ILO through the Bretton Woods bodies to the UN galaxy, where a particular, narrowly defined subject area or value is involved. The new challenge of integrating, in an equitable and mutually reinforcing way, the values of economic growth through liberalisation, environmental protection and enhancement, and social cohesion has thus placed an enormous strain on the inherited intergovernmental regime.

It is thus the soft law institutions that have taken the lead. They have thrived among governments, where the institutions take form infrequently as observed collectivities, and where there are few detailed codified rules to restrict those engaged. They often directly involve and are driven by the government leaders, or ministers meeting together, for they have the flexibility to go beyond narrow codified bureaucratically entrenched rules and to mobilise the resources required to put new principles and norms into effect. They have arisen particularly strongly within open, democratic societies. Here governments play a restricted role and command relatively fewer resources, the popular demand for ecologically and socially sustainable economic growth and liberalisation is strongest, and civil society actors — beginning with business and labour — are sufficiently strong and legitimate to demand and secure a direct governance role. Although no single soft law institution yet meets all these characteristics, in terms of governance that is truly global in the sense of embracing all issues and regions in the world, the G8 system comes the closest, and may serve as a concrete example when it finds a way to involve civil society effectively and directly.

In its form as voluntary standards, soft law has emerged earlier and flourished more broadly, often over a broader span of public life. Businesses and civil society actors have long voluntarily arrived at arrangements to guide their shared concerns. What is new in recent decades is how firms have begun, at the level of the individual firm, business sector, or the whole economy, to design and adopt voluntary codes that extend from their core commercial interests, and closely related issues of public safety, to embrace broader environmental and social concerns, including issues of gender, minority rights, and human rights. Also new is the willingness of civil society actors, beyond the labour movement, such as nongovernmental organisations (NGOs), to focus on core economic issues and become involved with business in producing and managing voluntary codes. Thus voluntary codes have arisen, starting in the developed world, to embrace most industry sectors, and most socially and economically crosscutting values in public life. Indeed, the world may be approaching the point where there are too many, often competing codes, rather than too few.

The Coexistence and Convergence of Hard and Soft Law

In both theory and, over time, in practice, both hard law and soft law have distinct, if not always unique, advantages and disadvantages of their own. Hard law offers the legitimacy, the strong surveillance and enforcement mechanisms, and the guaranteed resources that soft law often lacks. Governments acting alone with their full authority can produce high standards and clear and durable solutions — even ones that presciently address future problems in a timely fashion and prevent them cost-effectively. In contrast, with a broader array of stakeholders and interests, soft law arrangements outside the traditional single industry confines can deliver standards less stringent than those required to meet current and future demands. While the 'conflict of laws' challenge exists in the realm of hard law as well, the lack of authority associated with soft law can lead to a proliferation of competing standards that generates duplication, confusion, and uncertainty about which will prevail and which governments themselves might ultimately adopt.

To be sure, hard law and soft law are each the conceptual antithesis of the other, and each brings distinctive advantages and disadvantages from its separate identity. Yet in the real world, hard law and soft law approaches might be more complementary than competitive. Each can have its own empirical domain where it is destined to flourish, and does. Yet the two might also blend and overlap in a single regime, or combine separately to reinforce each other and fill the gaps in an overall tableau of global governance as a whole.

Of even greater interest are the process and pathways by which a regime in one sphere leads to its cognate in the other. A hard law or a soft law might well continue to operate, without acquiring the characteristics or inspiring the creation of a cognate regime in the other realm. But soft law arrangements can lead to hard law versions, beginning, at the national level, with the apparently innocuous steps of the courts reinforcing them in decisions or regulations, or governments employing them in their procurements, financing, or other activities. Indeed, much of the value of soft law might come from its role as a pioneer or stimulus of hard law. Reciprocally, hard law institutions and even regulations might contain soft law components, and increasingly rely or add to them to wield influence in a fast-moving world. It could further be noted that the speed of contemporary communications in furthering the exchange of ideas has speeded up these processes considerably.

Thus the question of how one might lead to the other is a central issue in the hard law–soft law debate. This volume accordingly addresses several question about the static co-existence and dynamic interrelationship of hard and soft law. Under what conditions can soft law replace or reinforce the hard law approach of formal legally entrenched commitments and governmentally grounded regulation? What are the alternative sequences and mixed resulting forms? What are the advantages and disadvantages or each approach and particular blends?

The Causes of Hard and Soft Law

What factors cause hard and soft law to emerge, survive, dominate, or induce the antithetical form of law to acquire its features as the years unfold? To date, a wide array of central causal drivers have been suggested, from those who variously regard soft law as second-best substitutes for failed hard law or as strategic choices of nongovernmental actors driven by self-interest and efficiency and market concerns. A broad array of potential causes have thus been offered as relevant, from the macroscopic, structural, international distribution of power and character of the international political economy to the microscopic rational calculations and strategies of the individual agents participating in the regimes.

Consistent with the liberal institutionalist literature on legislation, recent, and rapidly developing innovative scholarship on private authorities suggests a valuable array of three sets of causal factors (Cutler, Haufler, and Porter 1999; Cutler 2003; Haufler 2001). These are structural, largely static, power-based considerations; dynamic, historical trends in regulation; and calculations of efficiency and market contexts, grounded in transaction costs and strategic calculations, on the part of the actors involved. In the analyses in this volume, this trilogy offers a broad guide, enriched by a second dichotomy — the crosscutting concern of whether soft law behaviour in particular is shaped by rationally calculated interests or by inter-subjectively shared, separately existing norms.

The Consequences of Soft Law

The ongoing debate over the causes of soft and hard law is accompanied by an equally lively one over their consequences. Here the basic immediate question is about regime effectiveness — whether soft law is as effective as its hard law equivalent in meeting the need or solving the problem for which it was designed and introduced. The broader question is who benefits, with the leading candidates ranging from the more powerful states who can afford it to the modern capitalist class (who in an age of flexible accumulation needs to be able to walk away from agreements when the pattern of international competition changes (Cutler 1999; Scheuerman 1999). The latter argument could well explain why soft law regulation has thrived in the postmodern, late capitalist age (Cutler 2003).

In considering consequences, this volume recognises these guideposts by considering, narrowly, the instrumental effectiveness question of whether hard and soft law actually deliver their alleged and intended benefits and, more broadly, the distributional question of who benefits and who pays the cost in each case. Of particular concern in the latter case are two fundamental issues. The first is whether soft law approaches benefit primarily the rich developed North and its most powerful members, and whether they can be made to meet the needs of the developing South. The second is the large normative issue of whether the private allocation of values, through private, non-binding regulatory mechanisms, are in various senses 'good' for societies as a whole (Abbott and Snidal 2001). This is a critical question for those concerned about the effect of globalisation on social cohesion within countries around the world.

Beyond these six central questions lie two additional issues taken up in this book. The first is what solutions the analyses based on this framework might yield for policy makers confronting the challenges of how to combine trade, the environment, and social values in a more coherent, effective, and sustainable whole. And the second, amidst the ongoing need for further analysis and experimentation, is the emerging consensus about the power and potential brought by soft law solutions.

The Authors' Arguments

To address these questions, the book opens, in Part I, 'Setting Standards for Sustainable Forestry', by exploring the role of soft law and standards setting for sustainability in a single sector — that of forestry. This is vitally important and controversial sector, where formal international law remains largely absent and where competing voluntary standards vie for dominance.

In Chapter 2, 'Non-state Global Governance: Is Forest Certification a Legitimate Alternative to a Global Forest Convention?', Steven Bernstein and Benjamin Cashore note that the failure at the state-to-state level to negotiate an international convention on forestry practices has resulted in the emergence of voluntary forest certification and eco-labelling schemes sponsored by environmental NGOs or forest company and land owner associations. Bernstein and Cashore focus on the certification scheme developed by the FSC, which has largely been an NGO initiative and competes in many countries with other certification schemes promoted by forest companies and land owner groups. The authors evaluate such initiatives from the perspective of both international and domestic legitimacy, recognising that 'governance without government' poses particular challenges for traditional notions of political legitimacy. However, they argue for a very broad concept of legitimacy, rooted in the organisational sociological literature, that views a rule or institution as legitimate if relevant audiences accept it as appropriate. At the international level, Bernstein and Cashore argue that the FSC certification scheme appears to be consistent with various international rules and norms (such as the WTO's Agreement on Technical Barriers to Trade). However, at the domestic level, non-state, market-driven governance regimes tend to be more vigorously contested, as reflected by the existence of competing certification regimes in the forestry sector in many countries. Whether, in the end, such regimes can persist, flourish, and prove effective in the absence of direct or indirect government support remains an open question.

In Chapter 3, 'The Forest Stewardship Council: A Developing Country Perspective', Tasso Rezende de Azevedo reviews the evolution of the FSC forest certification regime in Brazil. Prior to the establishment of the regime, Brazilian exports of tropical forest products faced boycotts by consumer groups in Europe and the United States. These, however, had little or no impact on forestry practices in Brazil, in large part because most tropical timber is used locally. While the FSC certification programme currently applies to quite a limited percentage of Brazilian forests, it has had a significant impact in improving the working conditions in forest activities, forest management practices through the formulation of the detailed management plan required to acquire certification, and community relations with indigenous people in areas where there are forest activities. FSC certification has substantially raised and reinforced awareness in the Brazilian population at large of sustainable forestry practices and changed buying practices. One of the limitations of the programme is that smaller forestry companies find the terms of certification too costly and complex to comply with, which in part explain the relatively small coverage of the programme. Nevertheless, its impact on forest practices in Brazil is, on balance, positive. More generally, the FSC has achieved an impressive record — almost 40 million hectares of certified forests in 58 countries and more than 3000 companies having the 'chain of custody' certification (although 81 percent of areas certified according to FSC rules are located in North America and Europe, and only 19 percent in other regions in the world).

In Chapter 4, 'Indigenous Rights and Forest Certification in British Columbia', Chris Tollefson describes the development of the FSC certification regime in British Columbia.

He focusses in particular on the elaboration of Principle 3 in the FSC's Ten Broad Principles of Forest Stewardship. This requires FSC-certified forest managers to recognise and respect the legal and customary rights of indigenous people to control their lands, territories, and resources. In British Columbia, hard law protections of this right have faltered and proved incomplete, although recent decisions by the Supreme Court of Canada provide indigenous people in British Columbia and elsewhere in Canada somewhat greater prospects for the protection of indigenous land and related rights. However, even this case law does not go nearly as far as the FSC-BC certification regime in elaborating the protection of indigenous rights related to forests. The involvement of representatives of indigenous people in the governance of the FSC-BC certification regime has made it an appealing alternative vehicle to pursuing costly and protracted litigation. Tollefson takes an optimistic view of the prospects of the certification regime in British Columbia and considers that soft law of this kind can often render the hard choices required by hard law more attractive over the longer run. In this sense, soft law may be an indispensable precursor to hard law.

Part II on 'Setting Standards for Labour' explores the broad, crosscutting realm of trade and labour standards, where the hard law of the ILO has long operated, but voluntary standards are now seen as having an essential role. The debate revolves around the role of voluntary standard setting with respect to labour practices and the relative strengths and weaknesses of such a strategy relative to more hard law options.

In Chapter 5, 'Codes of Corporate Conduct and the Labour Regulatory State in Developing Countries', Adelle Blackett evaluates corporate codes of conduct in the context of labour standards. More specifically, she critiques a recent proposal by Charles Sabel, Dara O'Rourke, and Archon Fung for a competition-based variant, which they call 'ratcheting labour standards' (RLS) and which they advance as a form of a race to the top. In Blackett's view, these forms of corporate self-regulation are likely to supplant rather state-based regulatory initiatives in the labour domain, rather than coexist with them. She cites in particular the experience with export processing zones. Moreover, they discount the importance of democratic participation by workers in the formulation and enforcement of the norms applicable to the workplace through collective bargaining and other forms of tripartitism. Thus, in her view, corporate codes of conduct should at best be viewed as complements to traditional forms of hard law labour regulation, rather than as substitutes for them.

In Chapter 6, 'Standard Setting at the International Labour Organization: The Case of Precarious Employment', Leah Vosko considers the limitations of the ILO's standard-setting process by exploring its treatment of nonstandard and contingent, or precarious, forms of employment. From 1980 to 1996, the ILO, in evolving standards in this area, proceeded from the premise of parity of treatment in order to mitigate precarity — that in most material respects workers employed under nonstandard employment arrangements should be treated similarly to workers with full-time standard forms of employment. However, beginning in 1997 the consensus-based tripartite decision-making processes of the ILO reached an impasse on these issues, with worker and employer representatives simply negating the others' positions, and it began to shift course. In Vosko's view, this case reveals the limits of international standard setting in the labour context within the ILO, and in particular consensus-based decision-making regimes.

In Chapter 7, 'Hard Law or Soft Law: India and International Labour Standards', C. S. Venkata Ratnam and Anil Verma focus on the specific case of India as a developing country. They describe the scepticism of developing countries generally, including India, toward efforts by developed countries to promote a link between international labour standards and trade sanctions. Developing countries view such a link as a form of disguised protectionism directed at imports from developing countries. Ratnam and Verma argue that developing countries are not opposed to international labour standards. But these countries view these standards as appropriately the domain of the ILO where compliance with conventions that have been ratified by member states fall short of hard law, in the sense that sanctions for non-compliance typically involve investigation, public reporting, technical assistance, and the like. In the case of India, a number of the major ILO conventions have not been ratified. Full compliance with such agreements, such as those relating to freedom of association and child labour, would present considerable immediate difficulties for India. Ratnam and Verma describe various social labelling programmes that have been developed within India in industrial sectors such as carpets, garments, leather products, and tea. They generally take the view that these are useful intermediate steps toward more hard law commitments. They also describe a number of domestic tripartite initiatives designed to enhance labour standards in India. In general, the authors argue that soft law strategies are to be preferred from a developing country perspective to hard law international obligations in the form of trade sanctions attaching to violations of international labour standards.

In Chapter 8, 'Trade Policy and Labour Standards: Objectives, Instruments, and Institutions', Michael Trebilcock traces the history of the links among trade, trade policy, and labour standards back to the middle of the nineteenth century. He notes that even the Havana Charter, intended to create the multilateral trading system after World War II, allowed that unfair labour practices would be constrained by multilateral rules as an impediment to international trade. Three rationales that have commonly been advanced for a link between trade policy and labour standards: unfair competition, a race to the bottom, and labour standards as international human rights. Trebilcock largely rejects the first two rationales, but finds the third compelling. He argues, however, that this does not provide a case for privileging core international labour standards over other forms of international or universal human rights. Viewed in this broader context, core international labour standards such as freedom of association, freedom from forced labour, abolition of child labour, and freedom from discrimination in the workplace might, under appropriate circumstances, justify trade sanctions for persistent and systematic abuses thereof. That said, complex issues relating to institutional competence and legitimacy remain to be resolved, in particular the relative roles and responsibilities of the WTO and ILO. Nevertheless, it is difficult, if not impossible, to reject out-of-hand some sort of link between trade policy and core international labour standards. In this respect, Trebilcock is doubtful that certification, labelling, or corporate codes of conduct regimes are likely to be an adequate alternative, in all contexts, to hard law international obligations that at the limit may attract trade sanctions for violations. This is in part because of information problems that are likely to afflict these voluntary regimes, collective action problems on the part of the consumers, and the fact that violations of core international labour standards often occur in non-traded sectors rendering consumer responses in importing countries irrelevant.

Part III, 'Creating Codes of Corporate Responsibility', focusses on a particular soft law instrument of increasing attractiveness and use. In an age of globalisation and the rolling back of the state in many aspects of social and economic policy, large multinational corporations (MNCs) have assumed a greater role and importance both internationally and in the countries in which they operate. The pressures for MNCs to adopt policies on socially responsible behaviour raises a number of difficult questions about their role in the communities in which they invest. Pressure is exerted on them from above, in the form of global initiatives establishing codes of conduct, and from below, through the efforts of NGOs and transnational social movements. Consensus is lacking on the extent of a corporation's social responsibility, and corporations operate in vastly different regulatory environments, and even in conditions of civil war. Dilemmas emerge as tension builds between the appropriate weight that should be accorded to voluntary initiatives, on the one hand, and regulatory initiatives enacted by governments in regard to corporate responsibility, on the other. Although there can be no easy resolution to these dilemmas and tensions, efforts at international standard setting will be critical to the evolution of hard and soft law respecting corporate responsibility.

In Chapter 9, 'Corporate Social Responsibility and the Evolution of International Norms', Hevina Dashwood argues that MNCs are increasingly becoming attuned to corporate social responsibility (CSR), adopting codes of conduct, and recognising how their corporate reputation affects their profitability. This trend is driven by a globalisation that has brought CSR to the forefront of the global agenda, shifted norms about acceptable behaviour for MNCs, and altered the way MNCs calculate their interests. Some MNCs have created roles and identities in which they view CSR as consistent with their corporate objectives. Both interests and norms thus account for the growing attention and acceptance of MNCs to CSR. In developing this argument, Dashwood considers how CSR should be conceptualised. She concludes that CSR cannot be solely left to the voluntary or private sector and that an international regulatory framework is vital for the consistent adoption of minimal standards of acceptable behaviour.

In Chapter 10, 'The Role of Nongovernmental Organisations and Social Movements in Developing Countries', John Foster goes beyond a consideration of incremental change of individual or sectoral corporate actions or debates in regard to voluntary codes of various descriptions to reflect more broadly on the recent history of civil society engagement with issues of corporate accountability and their shift in emphasis toward more fundamental issues. He argues that while the balance of gross power has shifted over the last 30 years in favour of corporate actors, NGOs and social movements are shifting the field of engagement, seeking to reassert democratic citizen power, and calling into question the posture and actions of state actors, both nationally and multilaterally. Yet his brief review of individual campaigns, codes, and sectoral structures, and defensive battles against privatisation, investor guarantees, and intellectual property rights leads him to conclude that this social movement of globalisation from below is faced with a profound political challenge. The playing field will continue to be tilted in favour of corporate and private privilege and protection, unless democratic control is reasserted on a transnational or international basis as well as municipal, state, tribal, or national level.

In Chapter 11, 'Multinational Corporations, Globalisation, and the Challenge of Self-Regulation', Wesley Cragg evaluates the option of ethically guided self-regulation for corporations. He asks if, in principle, ethical codes of conduct are capable of taking on the task of regulating international commerce. The answer he finds is that there is now widespread agreement that MNCs do have responsibilities that extend beyond their shareholders to their stakeholders, about the values and principles on which regulatory frameworks should be constructed, and a supportive foundation of dialogue and debate within and across the government, private, and voluntary sectors. However, there is much less consensus on how these general principles and values should be operationalised in specific cultural and national settings. Questions arise about how the general principles should be interpreted and applied, who should be involved in the interpretive enterprise, and how responsibility for ensuring respect and compliance should be assigned. The answers currently offered in very influential circles are often fundamentally unsatisfactory. What is required is a new social contract that reallocates responsibilities for ensuring that economic activity generates adequate public benefits among government, the private sector, and civil society — the building blocks of which are already in place.

In Chapter 12, 'Canadian Corporate Responsibility in Sudan: Why Canada Backed Down', Robert Matthews offers an extended case study of the relationship between Canada and Sudan, and more specifically of the activities of a Canadian corporation, Talisman Energy, in oil operations in southern Sudan. Nearly two million people in Sudan have been killed since civil war broke out in 1983, and more than 4.5 million have been internally displaced. Criticism of Talisman's activities have focussed on its contribution to the massive displacement of people on and around the drilling site, and the millions of dollars in oil revenues going to the central government from the sale of oil that has increased the regime's capacity to wage civil war and reduced its incentive to negotiate in good faith with opposition interests. Despite Talisman's endorsement of the International Code of Ethics for Canadian Business and its commitment to issuing an annual corporate social responsibility report, Matthews finds that its interests and authority are such that its activities necessarily undermine international human rights in Sudan. Accordingly, he argues that trade or other economic sanctions imposed by the Canadian government would have been appropriate in terminating the company's activities in the region. In short, Matthews rejects the adequacy of soft law self-enforced ethical commitments in this context, and argues for the imposition by the Canadian government of harder law obligations on Talisman, enforceable through economic sanctions.

This volume concludes in Part IV, 'International Institutions and Soft Law', with an analysis of how soft and hard law can be combined in order to forge a link among trade, environment, and social cohesion in ways that promote sustainable development. It first presents analyses by two of Canada's former trade ministers, the Honourable Thomas A. Hockin and the Honourable Roy MacLaren, who were present at the 1986 birth and 1994 delivery of the Uruguay Round that created the WTO, as well as Canada's chief Uruguay Round negotiator, Sylvia Ostry, who hosted the 1988 Montreal ministerial meeting that served as the mid-term review for the round. All are leading scholarly analysts and authors in their own right (Hockin 2001; MacLaren 1981; Ostry 1997). In this section, they assess the approach taken to combining soft and hard law with trade, environment, and labour values in the initial WTO, in the contemporaneously negotiated regional NAFTA and in the Doha Development Agenda that now lies ahead. They reflect on their respective experiences as these bear on the relationship between trade policy and sustainable development. From this foundation, this section goes on to explore the link between hard and soft law through the perspective of the private sector, the work of the Codex Alimentarius, and experiences gained in bodies such as the G7/8 and OECD.

In Chapter 13, 'The World Trade Organization, the North American Free Trade Agreement, and the Challenge of Sustainable Development', Hockin argues that the environmental provisions in NAFTA itself and in the NAFTA environmental side accord suggest the possibility of linking trade and the environment in the broader multilateral system. He points out that the distinction between hard law and soft law is less of a dichotomy and more of a continuum. The NAFTA environmental provisions illustrate this continuum. A number of the provisions are largely exhortatory or aspirational in nature. Others impose obligations of transparency in the promulgation and enforcement of regulations and provide a complex and multi-stage mechanism for adjudicating complaints that one of the signatory countries is persistently favouring to enforce its environmental laws and regulations. This process presents the possibility of a significant monetary fine at the end, the proceeds of which are to be devoted to improving environmental conditions in the country complained against.

In Chapter 14, 'Integrating Environment and Labour into the World Trade Organization', MacLaren offers an opposing view. In seeing the NAFTA environmental provisions as the result of domestic political pressures in the U.S., he is doubtful of the political feasibility of transferring such provisions to the multilateral context. This is especially so, given the increasing resistance of developing countries to links between trade and the environment and trade and labour, born out of a scepticism that these links are mostly new protectionist strategies of developed countries directed against imports from developing countries.

In Chapter 15, 'The Future of the World Trading System: Beyond Doha', Ostry identifies the Uruguay Round of negotiations as a watershed in the evolution of the multilateral system. This is in large part because of the round's focus on 'within the border' measures such as intellectual property, services, and investment, in contrast to the traditional preoccupation of the GATT with border measures. This shift of focus has provoked a new and complex set of political dynamics surrounding multilateral trade negotiations, in part reflected in a growing and co-ordinated activism on the part of developing countries, exemplified in the Doha declaration on the application of the agreement on Trade-Related Aspects of Intellectual Property Rights (TRIPS), and in part in a growing international network of antiglobalisation NGOs. These NGOs, by exploiting economies of scale and scope associated with the internet (itself a form of globalisation), have become an important new political force in the international economic arena. However, Ostry notes that this 'hedgehog' coalition is extremely diverse in the perspectives and concerns that it represents. It has made little progress in articulating a coherent alternative vision of the global economic order to the classic neoliberal version that has traditionally underpinned the multilateral system. Uncertainties about where this new constellation of political forces will lead create similar uncertainties about the future of multilateral system, including links with trade policy, on the one hand, and with the environment and labour standards, on the other.

In Chapter16, 'Enhancing Global Governance: Corporate Social Responsibility and the International Trade and Investment Framework', Christopher Wilkie focusses on policy-oriented means of addressing the conundrum of the limits of soft law approaches to trade-related CSR issues within the trade and investment framework. This is a challenge, for reasons that have to do with the natural limits to disciplines and approaches in each, and a tendency to 'stove pipe' two nonetheless related disciplines in academic literature. It is helpful to review a recent successful initiative, namely the OECD Guidelines for Multinational Enterprises, as an illustration of this, and also because it offers lessons about how sensitive underlying issues can be dealt with within the rubric of international trade rules. Wilkie suggests taking into account trade-related aspects of these issues within the WTO family of agreements. Due to the complexities involved in changing existing or negotiating additional rules, he advocates doing so under existing rules, and sees scope for such action, including through the WTO's Trade Policy Review Mechanism. Whatever approach is taken, recent moves toward the acceptance of *amicus curiae* briefs as part of the WTO's Dispute Settlement Understanding, as well as enhanced sensitivities to relevant exceptions to the GATT and the General Agreement on Trade in Services (GATS), a non-adversarial approach has the potential to aid a WTO that is challenged with the seemingly impossible tasks of overseeing its international trade rules and resisting protectionism while at the same time heeding calls to take into account broader social issues in its work. In so doing, this would have the potential of assisting in the WTO's contribution to global governance as well.

In Chapter 17, 'A Corporate Perspective on Globalisation, Sustainable Development, and Soft Law', Michael Cloghesy outlines the work of a major Canadian business organisation, the Conseil patronal de l'environnement du Québec (CPEC), as a way of identifying the approach of the corporate community to sustainable development through the use of soft and hard law instruments alike. He argues that one cannot put much faith in international conventions or other means of hard law at the international level to bring about needed changes to ensure global sustainability. Rather, both internationally and nationally, gains toward sustainability can be more easily achieved through international standards, voluntary programmes such as Responsible Care, more positive and open dialogue on sustainable development among government, private sector, and civil society, and a recognition that the regulatory framework or hard law approach has its limitations. The best way to achieve results is a flexible and open approach to the use of various tools, both hard and soft. Globalisation, despite its costs, may be the sole source of improving the financial status, and thus sustainability, for a number of developing countries. If globalisation is to go forward, the WTO should become a sustainability organisation that will be transparent and open to civil society. To be sure, it will be a challenge to move trade issues ahead in this fashion. But the gains will make the effort more worthwhile.

In Chapter 18, 'Terminating Agricultural Biotechnology? Hard Law, Voluntary Measures, and the Life Sciences Industry', Lisa Mills examines the biotechnology industry, noting how its attendant health and environmental risks present challenges to both hard and soft law instruments. Hard law instruments designed to facilitate trade and investment in biotechnology have led to conflict over the role of science and demands for the explicit inclusion of non-scientific factors in international standard setting. In the absence of effective and binding international law to protect consumers and the environment, consumer pressure, and NGO action, along with European Union regulation, have resulted in costs to corporations and a number of voluntary commitments. However, soft law voluntary standards and commitments to reject particular technologies have resulted in an uneven distribution of consumer and environmental risk, with companies excluding genetically modified food from markets under high consumer pressure, but not other markets. Ongoing industry consolidation may also represent a threat to voluntary commitments by corporations to change their approach to technology development and public consultation, as companies that have made commitments are merged with others. And as distinctions among categories of GMOs blur in the third generation of biotechnology crops, existing hard law instruments may not be adequate to ensure the protection of biodiversity or the reduction of social and environmental risk. Given the requirements of the TRIPS and Sanitary and Phytosanitary agreements, only additional hard law instruments can provide universal protection and can allow states to protect their jurisdictions without fear of WTO challenges. Mills concludes that 'flawed as it may be, a hard law framework is necessary for the effective regulation of biotechnology'.

In Chapter 19, 'Hard and Soft Law in International Institutions: Complements, Not Alternatives', Nicholas Bayne broadens the vision to offer a concluding judgement on the central issue examined in this volume. He compares two major intergovernmental institutions based on hard law — the EU and the WTO — with two based on soft law, namely the OECD and the G7/8 summit. Such a comparison leads to the conclusion that hard law and soft law are not alternatives. Rather, they serve different purposes and complement one another. Hard law has the greatest potential, but soft law can do some things that hard law cannot. It is not easy to mix hard law and soft law in the same institution, but soft law stimuli are often essential to get the hard law moving elsewhere.

The Analytical Answers

Taken together, these contributions bring a rich array of insights and evidence to each of the basic questions, discussed above, that are at the heart of the debate over hard law and soft law. The analytical advances they bring are discussed in turn below.

Hard and Soft Law Defined

Most of the authors in this volume easily affirm the view that the distinction between hard law and soft law is less a dichotomy than a continuum, while maintaining the core, distinctive features of each. Hard law is widely conceived, as embodied in formal international conventions (Cloghesy) and, in keeping with the categorisation of 'legalisation', as involving obligations of substance and procedure and specified mechanisms for enforcement. To many authors, the practical essence of hard law lies in its economic sanctions, such as fines (Hockin) and, above all, restrictions on trade (Ratnam and Verma, Trebilcock, Matthews, Wilkie).

Soft law, while united by its lack of reliance on governmental authority and resources, encompasses a much wider domain. It ranges from customary rights (in the case of Tollefson in regard to indigenous peoples) to the exhortatory or aspirational nature of some provisions in formal international agreements, including the critical question of whether the innovative environmental provisions of the NAFTA preamble have any legal force and value at all (Hockin). These chapters show that soft law indeed comes in many concrete forms, including certification and eco-labelling schemes (Bernstein and Cashore, Azevedo), as well as codes of corporate responsibility and reporting mechanisms at the industry sector or economy-wide level (Ratnam and Verma). Soft law can be designed to influence international commerce as a whole (Cragg, Wilkie) or domestic practice (Ratnam and Verma). It can extend particular values such as the environment or labour, or it can encompass a broad and integrated set of social concerns (Cragg). It can involve one or more major stakeholders, or it can be limited and competitive or inclusive and co-operative. It can be regional or global in its coverage, or confined to specific industry sectors (as with the FSC in forestry or Responsible Care in the chemical sector), or it can embrace the economy as a whole (as with labour standards). In its design, it can reach from the competition-based ratcheting-up model (Blackett) to enforcement by consumer boycott (Azevedo).

Despite this diversity, soft law as explored here is united by two core features. The first is its reliance on open, consensus, multi-stakeholder dialogue, often including government, firms, and civil society, to produce the many international standards and voluntary programmes that emerge (Cloghesy). The second is its reliance, in order to foster compliance, on investigation, public reporting, and technical assistance, rather than on hard law sanctions, even in the conventions of the intergovernmental ILO (Ratnam and Verma). This suggests that the definitional essence of soft law may lie less in the governmental or nongovernmental character of their pioneering agent or ongoing institutional location, than in the non-authoritatively coercive processes pursed to create them and to ensure that they have effect.

The Fields Where Hard and Soft Law Flourish

With its focus on the forestry sector, labour standards, and the trade institutions, this volume can provide only a partial answer to the question of where soft law has flourished most. Yet a few patterns are nonetheless evident. Voluntary standards seem to have arisen in almost all internationally traded resources and goods sectors, of concern to the developed and developing worlds alike. These include forestry, chemicals, carpets, garments, leather, tea (Ratman and Vernam), and oil (Matthews). They have also thrived on an economy-wide level, with an ever broadening sets of values, to the point where there is a single global regime, ideationally embodied in an underlying consensus over common core ethical principles and norms (Cragg) and concretely expressed in the UN's Global Compact (Dashwood) and the OECD Guidelines for Multinational Enterprises (Wilkie). This suggests that soft law is becoming more expansive across several dimensions and integrating into a coherent, consensual whole. In this view, only the still contested implementing modalities, including the application of general principles to particular national and cultural circumstances, remain to be worked out (Cragg). Soft law in general has gone global, even if its particular characteristics remain subject to international dispute and more limited application in any individual form.

The Advantages and Disadvantages of Hard and Soft Law

There is also considerable agreement on the advantages and disadvantages of soft and, conversely, hard law. Soft law is more easily achieved (Cloghesy), and allows stakeholders to go further as well as faster than hard law (Tollefson). Furthermore, it is politically attractive and cost effective for governments, NGOs, other stakeholders, and, perhaps particularly, poor groups, as an alternative to costly and protracted litigation (Tollefson). In the face of fast-moving science, it is superior to the slow, rigid, inflexible hard law that is legally entrenched in the scientific methods, findings, and categories of old (Mills). It commands considerable consensus over the principles and norms relating to the responsibilities of multinational corporations and is backed by a supportive, multi-stakeholder dialogue and debate (Cragg, Wilkie). Above all, soft law is often legitimate in the sense of being consistent with international rules and norms — even if, at the domestic level, the contest goes on (Bernstein and Cashore, Cragg).

Yet equally coherent and extensive is the portrait of soft law's disadvantages that appears. Small groups find voluntary standards costly and complex to implement and comply with, leading many to stay out (Azevedo). Democratic participation is compromised when particular stakeholders are left out. Individual institutions may lack the competence and resources of their governmental equivalents, although governments offer no guarantees themselves in this respect (Trebilcock). And questions of legitimacy, especially in practical applications, remain (Bernstein and Cashore, Trebilcock, Cragg). Moreover, there is the great outstanding issue of compliance. These include the information problems likely to afflict voluntary regimes, collective action problems on the part of consumers, and the many violations in the non-trade sectors where consumer responses in importing countries will not work (Trebilcock). How to design and deliver effective compliance mechanisms in soft law systems remains the outstanding challenge in the years ahead.

The Coexistence and Convergence of Hard and Soft Law

There is less agreement among the authors about how hard law and soft law coexist and converge. To some, soft law can be inherently risky, as firms with high voluntary standards — such as in the food and oil industries — are swallowed up by those with lower standards (Mills). Moreover, competitive forms of corporate self-regulation, such as RLS, are likely to supplant rather than supplement or coexist with existing state-based regulatory initiatives (Blackett). Alternatively, soft law can lower the hard choices in making hard law, by serving as a precursor and proving ground for hard law in the long run (Tollefson). It is thus a useful intermediate steps toward hard law commitments (Ratnam and Verma).

The overall view is hard law and soft law are complements rather than competitors, as they serve different purposes (Wilkie). While it is hard to combine both in the same institution, soft law stimuli are often needed to generate hard law in other bodies (Bayne). This conclusion points to the benefits of institutional competition, broadly defined. It may also suggest that soft law has a comparative advantage in producing new regimes with innovative principles and norms, leaving it to hard law to add the effective enforcement mechanisms over time. This latter generalisation is one that warrants more analytical attention in the years ahead.

The Causes of Hard and Soft Law

The analyses in this volume point to the causal importance of power, market, and historical factors, and of both rational and social processes, in creating, expanding, and rendering effective soft law. But they demonstrate that some category of causes are more important than others, and point to the causal power of particular factors within each cluster.

There is strong support for the familiar feeling that soft law is a second-best substitute for a first-best hard law, being created when and because the relevant hard law does not exist and the intergovernmental negotiations to produce it have failed, as in the case of forestry (Bernstein and Cashore). Yet it further comes where hard law exists, but has failed or does not reach far enough to meet the demand, as in the aboriginal rights aspect of forestry (Tollefson).

More broadly, the relative power of national governments, and the historical legacy of heavy hard law institutions, with their formidable sunk costs and revered regulations, retain their relevance in producing the soft law standards of the current age. But the dominant view is that market forces, in the new form of intense globalisation, is a major driver now (Dashwood, Cloghesy). The overall portrait is of a world where the declining role, desire (in part due to competitiveness and ideological concerns), and capacity (due to fiscal deficit concerns) of the state to regulate with hard law have led private sector actors — led by MNCs more attuned to market forces — to assume a larger role and turn to soft law as their preferred instrument (Foster). In some cases, this has yielded impressive results (Wilkie). Yet some analysis also points to the rising salience and global networking reach of NGOs demanding higher corporate standards and compliance. It is not clear what co-opting, compromises, or control this coming together will bring. It is clear that the MNCs or the needs of late-stage capitalism have not already won, even if they currently seem particularly strong (Foster). The broadly multi-stakeholder, consensus-oriented dynamic of soft law, with its still open outcomes, remains the dominant dynamic that globalisation brings.

Globalisation also emerges as particularly powerful in bringing CSR to the top of the public policy and corporate agenda, leading senior executives to see how corporate reputation affects their firm's profitability, altering the way they calculate their MNC's interests, changing their norms, producing corporate standards, and, even, in some cases, having some MNCs adopt new CSR-centric corporate identities and create compatible new roles (Dashwood). Here, as with other forms, both interests and norms work together to drive the emergence and effectiveness of soft law (Wilkie).

This overall portrait is reinforced by a focus on the particular agents that produce the soft law that this volume explores. Firms matter, for it was the forest companies and land owners that offered forest certification schemes that competed with the MNCs (Bernstein and Cashore), and the ongoing industry consolidation that saw firms with high voluntary standards swallowed up those without proved relevant in the food safety case (Mills). But at least equally relevant were several other types of actors. The FSC was largely an NGO initiative (Bernstein and Cashore, Azevedo). 'Pressure from below' and civil society more broadly shape outcomes (Dashwood, Foster). The communities in which MNCs operate are relevant (Dashwood). Consumer pressure, despite the many constrains on it, also counts (Mills, Azevedo, Trebilcock). And 'pressure from above' — through the autonomous action of regional organisations such as the EU and broadly multilateral bodies such as the UN — plays its part (Mills, Dashwood). The only actors whose interests have been arguably overlooked are the MNCs' host governments (Matthews), even if the Canadian government — in this case — generally supports, as a broad policy, CSR for Canadian firms abroad.

The Consequences of Soft Law

Are soft law and hard law actually effective in delivering their alleged benefits? Who benefits most — powerful actors, the rich North or poor South, or society as a whole? In these analyses, hard law receives a balanced judgement. In addition to the standard repertoire, the costs of hard law, in these analyses, include the conflict it creates over the role of science and the ensuing demands for the explicit inclusion of non-scientific factors in international standard setting (Mills). Yet the absence of binding, hard law leads to costs to corporations as well (Mills).

But soft law seems to bring an equally balanced set of benefits and costs, too. In the case of forestry, the FSC has significantly working conditions, forest management practices, community relations with indigenous peoples, the awareness of the Brazilian people about sustainable forestry, and altered buying practices. In the case of food safety, however, it has resulted in uneven distribution of consumer and environmental risk, as firms have kept genetically modified foods out of only those markets where high consumer pressure exists (Mills).

Across the traditional North-South divide, there is stronger support for the view that soft law can benefit both sides. In forestry, the FSC has secured a broad global coverage that embraces at least 58 countries, with most of the areas certified by FSC rules lying in developed countries and less than 20 percent in the developing world (Azevedo). In regard to labour standards, developing countries much prefer soft law strategies to hard law solutions that come with trade sanctions for violations attached (Ratnam and Verma). And soft law offers flexibility for firms that operate in vastly different regulatory environments and even in conditions of civil war, especially in the developing world (Dashwood, Wilkie). There are thus some grounds for believing that soft law can serve the needs of a broad spectrum of society, if not everyone or society as a whole.

Proposed Solutions

These analysis provide a sound foundation for beginning to consider solutions, perhaps offering a ready recipe for policies to put in place right away. Most generally, the argument that there is already a wide consensus on basic principles and norms suggests the desirability and feasibility of moving to a single set of global soft law standards, building in the first instance on the UN's Global Compact or the OECD Guidelines for Multinational Enterprises, as enriched by the body of human rights standards and law. Efforts to create such an overall normative edifice, and integrate individual codes and standards into it, could well reduce the problems presented by the competitive complexity created by several standards and schemes covering a single issue area, and even the uncertainty and rivalry between hard and soft law. Cragg's conclusion that the building blocks for constructing such a meta-regime are already in place suggests that rapid action is possible here. It may mean that the next UN-sponsored summit, perhaps a five-year review of the Millennium Summit with a wide, meaningful multi-stakeholder reach, could include this task of articulating a coherent, comprehensive normative edifice, along with reviewing and advancing its hard-law-like commitments in individuals fields.

A second suggestion that flows from these analyses is the need to make the WTO (Cloghesy) — and, more generally, the hard law economic multilateral organisations — more transparent and open to civil society. There is a case to be made for taking into account trade-related CSR issues in the international trade and investment framework (as has been done at the OECD), including at the WTO (Wilkie). To be sure, organisations such as the powerful International Monetary Fund (IMF) have made real advances in this direction in recent years. The soft law legacy speaks powerfully to the importance of widely participatory, equally empowering, consensus-oriented multi-stakeholder regimes, and points to how the 'sunlight' of transparency, properly designed and delivered, can substitute for the heavy, coercive sanctions of hard law regimes. It may take some time to create effective standards through such a process, but legitimate regimes and strong, long-lasting, low-cost compliance are likely to be the reward.

The Concluding Consensus

All authors in this volume emerge from their respective analyses as partial, rather than pure, optimists or pessimists about the proven power and future potential of soft law. Of the many optimists, most see the value of soft law as a generative or at least permissive way to build the kind of hard law that the world needs, as opposed to a static, stand-alone superior substitute for hard law. Thus, although soft law can do some things hard law cannot (Bayne), it is not an across-the-board alternative (Trebilcock). At best, it is a complement (Azevedo, Blackett, Wilkie). Moreover, in the long run, it is primarily valuable as a sometimes indispensable proving ground — a precursor and a useful intermediate step to hard law, or even a stimulus for it (Azevedo, Tollefson, Ratnam and Verma, Bayne).

Other authors are more sceptical, remaining agnostic in the debate. To them, it is an open question whether soft law regimes can be effective without government support, while governments have a long list of outstanding questions about them and see current answers as unsatisfactory (Bernstein and Cashore). Indeed, in at least one case, the failure of the consensus-based, tripartite decision-making process at the ILO casts both soft and hard law into doubt (Vosko). Moreover, despite much consensus on their underlying principles and norms, difficult and contested questions remain about their interpretation, application, and ability to induce compliance (Cragg). At the moment, securing sustainability thus requires a flexible and open use of instruments both hard and soft (Cloghesy).

Yet some authors are still more pessimistic about soft law. While virtually none are hard law 'hard liners', they see hard law as having greater potential than soft (Bayne). Some, emphasising the shortcomings of soft law, are reluctant to leave corporate social responsibility to the voluntary or private sector, looking to hard law to ensure consistently applied, minimum standards (Dashwood). Others feel the *status quo* will continue to favour corporations, unless democratic control is exerted at all levels, presumably including governments' with their formal authority (Foster). And still others judge that only hard law can bring the properly targeted formal sanctions that the challenge of compliance sometimes demands (Trebilcock).

Even more pessimistic about soft law are those who highlight the harm it has done. Even MNCs equipped with CSR principles undermine human rights in distant, conflict-ridden poor countries, and only government-enforced economic sanctions can ensure their respect (Matthews). And in the case of food safety, only new hard law instruments can provide the desired universal protection from the hard law WTO itself (Mills).

Thus, the great debate over the proper role and relationship of soft law and hard law in sustainable global governance will continue. It will do so with as much passion as ever, but at a more deeper and more refined analytical level. As it does, this volume provides four particular prominent conclusions, covering all contexts explored here, to guide the way.

First, soft law has its place, but is by no means a silver bullet solution in all spheres, or even most. On the central question of whether soft law options are best seen as a substitute for — or a complement to — harder law options, the case of forestry sustains this conclusion. Here an impasse at the international level in negotiating an international convention on forest practices, as Bernstein and Cashore highlight, and a similar political impasse at the domestic level in reconciling forest practices with the rights of indigenous peoples, as outlined by Tollefson, suggest that, in some cases, soft law options may be a viable alternative or even a substitute for unattainable harder law. However, in other contexts, viewing soft law as a substitute for harder domestic or international law obligations (such as in the context of labour law) may result in more effective hard law options being supplanted and the various values that these seek to vindicate being devalued. Such values include democratic participation by workers in workplace governance, as Blackett points out.

Second, even where soft law options are reasonably viewed as useful strategies, the question arises in every context whether they can be fully effective without complementary state action or recognition or support of some kind. That is to say, soft and hard law options may in many cases be complementary, as Bernstein and Cashore indicate, and as Wilkie argues is the case to some degree with the experience of the OECD Guidelines for Multinational Enterprises.

Third, in many contexts the further question arises as to whether soft law options, through encouraging innovation, adaptability, and limited risk taking, are properly viewed as a precursor or intermediate step toward harder law options in the future. As Bayne points out, they can exist within the same institution, but, far more frequently, soft law institutions provide the leadership needed to get the hard law moving in other bodies.

Fourth, and last, all three questions cast in sharp relief the central point that hard law and soft law options are not sharp dichotomous choices. Rather, they are choices of various strategies, or combinations of strategies, arrayed along a wide and subtle spectrum of alternatives from soft to hard law. In many contexts, combinations of soft and hard law options — either at a given moment, or over time — may yield the most productive strategies in promoting international norms relating to the environment, labour standards, and social governance.

References

Abbott, Frederick M. and Duncan Snidal (2001). 'International "Standards" and International Governance'. *Journal of European Public Policy* vol. 8, no. 3, pp. 345–370.

Abbott, Kenneth W., Robert Keohane, Andrew Moravcsik, et al. (2000). 'The Concept of Legalization'. *International Organization* vol. 54, no. 3, pp. 401–420.

Cutler, A. Claire (1999). 'Public Meets Private: The International Unification and Harmonization of Private International Trade Law'. *Global Society* vol. 13, no. 1, pp. 25–48.

Cutler, A. Claire (2003). *Private Power and Global Authority: Transnational Merchant Law in the Global Political Economy*. Cambridge University Press, Cambridge.

Cutler, A. Claire, Virginia Haufler, and Tony Porter, eds. (1999). *Private Authority and International Affairs*. State University of New York Press, Albany, NY.

Haufler, Virginia (2001). *A Public Role for the Private Sector: Industry Self-Regulation in a Global Economy*. Carnegie Endowment for International Peace, Washington DC.

Hockin, Thomas A. (2001). *The American Nightmare: Trade Politics after Seattle*. Lexington Books, Lanham.

Ikenberry, John (2001). *After Victory: Institutions, Strategic Restraint, and the Rebuilding of Order after Major Wars*. Princeton University Press, Princeton.

Kirton, John J. (2002). 'Embedded Ecologism and Institutional Inequality: Linking Trade, Environment, and Social Cohesion in the G8'. In J. J. Kirton and V. W. Maclaren, eds., *Linking Trade, Environment, and Social Cohesion: NAFTA Experiences, Global Challenges*, pp. 45–72. Ashgate, Aldershot.

Kirton, John J. and Virginia W. Maclaren, eds. (2002). *Linking Trade, Environment, and Social Cohesion: NAFTA Experiences, Global Challenges*. Ashgate, Aldershot.

Krasner, Stephen D. (1983). *International Regimes*. Cornell University Press, Ithaca, NY.

MacLaren, Roy (1981). *Canadians behind Enemy Lines, 1939–1945*. University of British Columbia Press, Vancouver.

Ostry, Sylvia (1997). *The Post–Cold War Trading System: Who's on First?* University of Chicago Press, Chicago.

Ruggie, John G. (2003). 'The United Nations and Globalization: Patterns and Limits of Global Adaptation'. *Global Governance* vol. 9 (July-September), pp. 301–322.

Scheuerman, William (1999). 'Economic Globalization and the Rule of Law'. *Constellations* vol. 6, no. 1, pp. 3–25.

PART I
SETTING STANDARDS FOR SUSTAINABLE FORESTRY

Chapter 2

Non-state Global Governance: Is Forest Certification a Legitimate Alternative to a Global Forest Convention?

Steven Bernstein and Benjamin Cashore[1]

When traditional forms of multilateral governance fail to materialise, is non-state governance a viable alternative? This question besets the forestry issue area, where repeated attempts to forge a binding international convention to manage the use and preservation of forests have failed (Wang 2001; Dimitrov 2003). Most relevant states now strongly oppose a global forest convention, including some that previously voiced support. Moreover, even prominent nongovernmental organisations (NGOs) have withdrawn their support, fearing a convention might do worse at promoting sustainable forest management than no treaty at all. In the wake of this failure, voluntary forest certification and eco-labelling schemes sponsored by environmental NGOs or forest company and landowner associations have arisen to fill this governance gap. Can such schemes gain acceptance as a viable alternative to a global forest convention?

The central argument of this chapter is that as a non-state form of governance, transnational forest certification schemes require political authority beyond the state in order to succeed. In the abstract, power, legitimacy, and community constitute political authority (Bernstein 2003). In other words, authority exists when a community or audience authorises commands to which it feels bound, empowering those rules and, possibly, mechanisms to enforce them. But in the realm of global governance, especially in cases where hard law is absent, legitimacy takes on added importance, because, by definition, non-state governance schemes lack the traditional enforcement capacities associated with the sovereign state — the traditional site of authority in the international system where power, legitimacy, and political community appeared fused. Although one should always be sensitive to the importance of power for political authority, including where it resides and how it manifests, the decoupling of coercive force and legitimate rule is the most striking feature of new sites of authority found in contemporary global governance. According to James Rosenau (2003), 'the essence of [new sites of authority] is that they derive their legitimacy from the voluntary and conditional participation of individuals who can revoke their consent at any time'.[2] This chapter focusses on legitimacy for this reason. Nonetheless, it is sensitive to the interaction between legitimacy and power in two regards: first, that underlying power configurations in the international system, as well as in domestic and international marketplaces, could interact with legitimacy dynamics to produce consequences that outside observers might not necessarily view as good or just; and second, that schemes lacking in power resources risk being paper tigers, not really constituting governance in any meaningful way.

Legitimacy of any form of governance may rest theoretically on deeper notions of justice (Rawls 1971) or legitimate procedures — what Thomas Franck (1990) calls generically 'right process'. But on a practical level, non-state governance must ultimately be deemed acceptable or legitimate among its relevant audiences. Criteria of legitimacy in global governance are thus contingent on historical understandings of legitimacy at play and the shared norms of the particular community or communities granting authority, since a scheme's viability depends upon those audiences recognising its legitimacy. In the case of forestry, relevant audiences include forest landowners and forest management companies, producers of forest products and purchasers of those products further down the supply chain (Cashore, Auld, and Newsom 2004; Sasser 2003). Governments also play an indirect role, as the source of authority for the web of existing rules and institutions that form the backdrop of any global governance scheme. While legitimacy does not guarantee a certification scheme's effectiveness in promoting sustainable forest management (Cashore and Noah 2003), it is a necessary condition, especially in the absence of coercive state power.[3]

The legitimacy of transnational governance schemes must be achieved at two levels since, by definition, they occupy political space both internationally and domestically. Internationally, legitimacy requires that the basic institutional elements of the scheme — its norms and rules — fit with existing accepted norms and institutions of global environmental and economic governance. Overt conflict, either with existing hard law or with less formal but widely agreed or accepted international norms or soft international law, would militate against the legitimacy of any emerging governance scheme.[4] The particular nature of forest certification as a market-based scheme raises the related question of whether acting in the marketplace can avoid conflicts that have militated against a binding global forest convention among states. This question is especially relevant for a central theme of this book: under what conditions can non-state or voluntary forms of governance achieve desirable environmental, social, and economic outcomes, if at all?

Domestic legitimacy is required because targeted firms also operate within domestic regulatory and competitive environments, with their own implicit and explicit norms of behaviour that set the boundaries of what relevant audiences would be willing to accept as appropriate. Two interrelated components of domestic legitimacy must be achieved: first, relevant audiences must accept the notion that a non-state form of governance is an appropriate mechanism for addressing global environmental problems; second, domestic audiences evaluate which of the competing domestic forest certification programmes, with very different conceptions of non-state governance, are most legitimate (Cashore, Auld, and Newsom 2003, 2004).

The chapter proceeds in three parts. First, it discusses the most prominent transnational certification scheme, the Forest Stewardship Council (FSC), as well as competitor schemes. The FSC is identified as a non-state, market-driven form of governance and placed in the context of more traditional forms of global governance. Second, the chapter addresses international legitimacy requirements. The FSC's relative success in meeting these requirements provides one piece of the puzzle of why forest certification has gained some acceptance as a viable form of governance whereas traditional forms of interstate governance have largely failed. The third section examines domestic legitimacy requirements and strategies employed to gain legitimacy for governance schemes. The conclusion identifies limitations to this approach, and cautions against the view that transnational legitimacy automatically or easily translates into effective governance.

Forest Certification and Global Governance

Forms of Global Governance

Despite a great deal of talk about 'governance without government' in the international system, global governance usually brings to mind large and arguably powerful intergovernmental institutions such as the World Trade Organization (WTO), the International Monetary Fund (IMF), or even the United Nations and its affiliated agencies. These organisations, and the broader regimes or institutional arrangements in the areas they help manage, for the most part rest ultimately on the authority of their state members. They are organisations and institutions established by states and for states in order to solve collective action problems and manage interdependence.

As recent writing on global governance makes clear, however, non-state actors — a term that refers to any actor other than governmental actors — increasingly participate in global governance if defined broadly as the method or means of realising shared values, interests, and goals that may or may not derive from a formal centralised political power or authority. Non-state — or soft law — mechanisms, as defined in this volume, meet governance criteria if they purposely steer actors toward collective goals or values, and are recognised as authoritatively doing so (Rosenau 1995). In other words, they qualify as governance as long as what they do rests on the acceptance of shared rule.

The role of non-state actors in global governance ranges from sources of information, influence, or monitoring of traditional intergovernmental institutions to governance schemes that manage in the absence of government control or even blessing. In between are various forms of hybrid governance where governments and nongovernmental groups — be they firms, nongovernmental political or activist groups, individuals, or groups of experts — jointly participate. Table 2-1 identifies ideal forms of global governance depending on the level of governmental authority and control.[5] The key difference relevant to this discussion is in the location of authority. What separates 'non-state market-driven' governance, as described by Benjamin Cashore (2002), from both traditional international governance and hybrid governance is that authority is diffuse and located in the marketplace. Specifically, producers and consumers along the supply chain grant authority as products move through the market from (in the case of forests) extraction to end users. The institutions or governance schemes themselves are empowered ultimately by acceptance by these players, but also by drawing legitimacy from existing webs of authoritative international norms that define and regulate appropriate behaviour and practices in the issue area. Wider processes and audiences, including environmental groups and governments, may influence both supply chain dynamics as well as these broader norms.

International governance, in contrast, ultimately rests on the sovereign authority of states, either directly or because governments have delegated that authority to an institution or to private actors. A non-state market-driven governance scheme does not rely on states and may explicitly exclude them, leaving them in a secondary role of interested actor, albeit a powerful actor and one that may potentially challenge the scheme's legitimacy. While non-state market-driven forms of governance in generic terms might occur within a single state's territory, the focus here on global governance specifically concerns schemes that transcend state borders.

Table 2-1 Alternative Authorities in Global Governance

Features	Non-state Market-Driven Governance	Shared Private/Public Governance	Traditional International Governance
Location of authority	Diffuse: producers and consumers along the supply chain (audience/market players); non-state institution as location, interpreter, implementer of rules.	Some delegation possible (e.g., *de facto* granting authority to technical experts), but sovereign governments remain ultimate authority (explicit or implicit). Transfer of authority is rare.	Sovereign governments. Some delegation to institutions is possible, e.g., to interpret rules. Transfer of authority is rare.
Source of authority	Shifting international norms enabling markets, economic incentives, acceptance of programme by supply and demand side audiences.	State sovereignty and consent (deep structure of international system). Possibly legalisation or constitutionalisation.	State sovereignty and consent (deep structure of international system). Possibly legalisation or constitutionalisation.
Role of government	Interested player, potential facilitator or debilitator.	Shares policy-making authority.	Has policy-making authority.

Source: Adapted from Cashore (2002).

The reasons for the emergence of these new forms of governance vary and are open to debate. Private governance among corporations may result from firms seeking to co-ordinate standards, form associations, provide services for their own benefit, or avoid state regulation. Similarly, states may enable or encourage private or hybrid governance for reasons ranging from the need to co-operate with firms when attempting transnational regulation to attempts to create regulatory environments consistent with their home firms' interest, to an inability or unwillingness to reach agreement at an intergovernmental level, whether owing to competing interests or structural impediments to co-operation among states. Furthermore, private or hybrid governance may form owing to an ideological or political desire to allow the market to operate with the least amount of political interference. From the perspective of non-state actors, private or hybrid governance may stem from either an attempt to avoid state regulation or to create regulation or incentives for changes in behaviour when governments are unable or unwilling to act. While it is beyond the scope of this chapter to explore fully the reasons for the emergence of non-state governance, some of the factors that led global forestry governance to move in this direction are considered.

Forest Certification and the FSC

Forest certification and labelling programmes are designed to officially recognise those companies and landowners who voluntarily operate well-managed or sustainable forestlands according to predefined criteria. Forest company or landowner associations and environmental NGOs have established the schemes, setting up processes and procedures for developing and administering rules, and monitoring compliance. Governments may be involved indirectly if the scheme falls under larger regulatory or voluntary systems of standards where governments play a role, such as the International Organization for Standardization's (ISO) 14000 series of environmental standards. However, governments can also take on a new role as an interested — but entirely outside — actor, such as in the case of the NGO-led FSC, the main focus of this chapter. The FSC is a primary example of the new breed of governance schemes that falls closest to the ideal type of non-state market-driven transnational governance. It also corresponds very well to the classification of soft law used in this volume, because it relies 'primarily on the participation and resources of nongovernmental actors in the construction, operation, and implementation of [its] governance arrangement' (see page 9). Moreover, participation is voluntary and consensus decision-making is encouraged, although voting procedures are available when consensus fails to develop. Like other soft law institutions, it shares some characteristics of law, and it may possibly form the basis of future binding law. However, its source of authority separates it from more traditional state-centric understandings of international soft law, where authority stems from the stated intention or concern of states.

Forest certification operates according to a market-driven logic, based on the expectation that customers — whether manufacturers, retailers, or end users — prefer sustainably produced forest products and will demand them in the marketplace. Market incentives will then operate to produce forest products sustainably.[6] This ordering of customer preferences is not simply taken for granted. It occurs in the context of increasing environmental awareness and public advocacy on sustainable forestry. Groups involved in such schemes also actively manipulate the preferences of consumers and retailers by promotional activities, including public campaigns and information, as well as the organisation of buyer groups that educate and socialise retailers and also manipulate market incentives. For example, the World Wildlife Fund (WWF) has helped to organise buyers groups in the United Kingdom (1995+ Group), Holland (Hart Voor Hout), Belgium (Club 1997), Austria (Gruppe 98), the United States (the Certified Forest Products Council, renamed Metafore in 2001), and internationally with the creation of the Global Forest and Trade Network in 1997.

Forest certification schemes come in two basic varieties. In one, certification produces environmental rules and restructures influence among business, environmental, and social interests in order to raise standards (Elliott 1999). In the other, certification provides for continual improvement and, in many cases, approval of forestry practices already considered sustainable, as well as the maintenance of business influence in private policy-making processes. The focus on the FSC means this chapter is concerned primarily with the viability of the first type of scheme. Neither conception, however, is immune from a vigorous debate over the ecological and business implications of certification. On one side, some environmental critics fear that nongovernmental eco-labelling schemes such as certification risk sliding into weak forms of self-regulation if governments are unwilling or unable to take a more hands-on regulatory role. Notably, many of the domestic and regional schemes arose as industry-led competitors to the NGO-led FSC. Some of these competitors face criticisms from environmental groups and FSC supporters for being more concerned with industry control than with promoting sustainable forest management. Conversely, many industry and forest company officials argue that their programmes address forest sustainability in a more balanced manner, avoiding unnecessary details and regulations usually associated with public policy regulations.

Prominent FSC competitors include the American Forest and Paper Association's Sustainable Forestry Initiative (SFI) programme, the Canadian Standards Association (CSA) programme initiated with strong support from the Canadian Pulp and Paper Association (now the Forest Products Association of Canada), and the Pan European Forest Certification (PEFC) system, created by landowner associations that felt especially excluded from the FSC process. These programmes, along with newer schemes such as Brazil's Programme for Forest Certification (CERFLOR), created an umbrella organisation, initially under the auspices of the International Forest Industry Roundtable (IFIR) of the World Business Council on Sustainable Development (WBCSD), to establish an international presence to match the FSC's. Efforts by the IFIR to establish mutual recognition among all systems were later transformed by the WBCSD's (2003) CEO Action team into a proposed 'legitimacy threshold model'.

Under this model, the FSC would be accepted as unique, and mutual recognition would depend on establishing process and standard threshold requirements for gaining market-based legitimacy.[7] While the FSC remains hesitant to accept programmes originally designed to offer an alternative vision of governance, the market underpinnings of certification have at once moved the programmes closer together and made the FSC recognise that it, alone, will be unable to achieve global certification. As a result, the FSC is under pressure to find some accommodation with its competitors.

The FSC originated in discussions in 1990 as environmentalists drew the world's attention to threats to the world's tropical rainforests. Increased scrutiny of industry practices and activities that might contribute to deforestation resulted in a demand from timber retailers and distributors for fibre from well-managed forests (Meidinger 1997, 2003). Adding impetus to the need for non-state action, international negotiations to protect and manage the world's forests failed to produce a meaningful international forests treaty at the United Nations Conference on Environment and Development (UNCED, often referred to as the Earth Summit) in Rio de Janeiro in 1992. Environmental groups led by the WWF concluded that rather than expending significant effort to resurrect negotiations for a forest convention, they could instead directly influence the private sector more quickly through the development of private eco-labelling programmes that would recognise those companies practicing sustainable forestry. They created the FSC in Toronto, Canada, in 1993. It became legally registered in Oaxaca, Mexico, in 1995 (Hansen 1998, 19), and recently moved its head office to Bonn, Germany.

The FSC originally created nine principles (later expanded to ten) and more detailed performance-based criteria, including tenure and use rights, community relations, workers' rights, environmental impact, management plans, and monitoring and preservation of old growth forests (Moffat 1998, 44; FSC 2000). To avoid a one-size-fits-all approach to forest management that could foster local or regional resistance, the FSC programme requires the development of regional or national working groups to develop specific indicators and verifiers with which to apply the principles and criteria. Accredited certifiers then audit companies to see if they are in compliance with the regional standards.[8] This institutional dynamic is crucial because it has directed many political struggles over certification to the territories designated as national or regional working groups.

In line with a non-state market-driven identity, the institutionalised decision-making processes of the FSC limit business influence. Its three chambers — environmental, economic, and social — have equal voting weight. Each chamber also gives North and South representation equal voting weight (Domask 2003). The FSC model also forbids government involvement in rule making, and allows wide-ranging policy initiatives. The limiting of business influence as well as the grouping together in the economic chamber of forest owners who must implement forest certification rules with businesses further down the supply chain are key reasons why FSC competitor programmes emerged. Many non-industrial private forest owners and industrial forest companies felt that their interests and concerns were not accommodated in the FSC model (Cashore, Auld, and

Newsom 2004). As a result, most forest owners refused to participate in the FSC consensus process in regional standard-setting working groups in North America, and those who did were often at the losing end of key decisions on standards development (Lawson and Cashore 2001; Cashore and Lawson 2003; Cashore, Auld, and Newsom 2004, chs. 3, 4, 6). By 2003, the FSC's competitor programmes had certified far more companies and forest land in North America and Europe than had the FSC, revealing the indirect influence of the FSC in promoting non-state market-driven governance, even if it was not its own. The FSC has also certified companies in developing countries, including Brazil, Bolivia, and Indonesia, although the attraction to certification remains very weak in the South compared to the North.

Global Legitimacy

Traditional forms of international governance authority are state-based and increasingly multilateral in form, in the sense of being embodied in institutions built by a number of states co-operating to achieve joint gains that recognise a set of reciprocal rules and obligations. As such, these institutions depend on the support of major states for their legitimacy. But what if governance arises precisely in response to the lack of legitimacy and effectiveness of intergovernmental, multilateral processes, as in the case of the FSC? What then is the source of legitimacy?

This section focusses on the requirements for international legitimacy. It argues that transnational authority faces many of the same difficulties in getting and sustaining legitimacy as public, multilateral authority in that both operate within an existing set of international norms that define appropriate behaviour and practices. In the forestry case, those conditions actually work in favour of legitimacy for non-state market-driven governance compared to a multilateral global forest convention, owing to characteristics of the forestry issue area. However, there are also important differences between transnational and international authority, the main one being the relevant audiences where acceptance of a scheme must be achieved.

Scholars of international relations have long noted that international governance and co-operation requires domestic legitimacy in leading states (Ruggie 1982; Keohane and Nye 2001). However, if an institution is based, in the first instance, on markets and their supply chains rather than on territories, and if nongovernmental groups are successful at creating transnational pressures that seem to operate independently of states or in a realm other than the normal political realm of inter-state relations, then looking to democratic processes within states or to the authority of states as a source of legitimacy is too limited, and perhaps even misleading.

The starting point of this chapter, then, is a very broad concept of legitimacy rooted in organisational sociology and constructivist international relations scholarship rather than notions of political legitimacy rooted in democratic theory and practice that tend to dominate the political science literature. The chapter's conclusion considers the limitations of a strictly sociological approach, especially in regard to institutional design. It recognises that procedures and structures that ensure stakeholder access and accountability to those affected by decisions are likely important for achieving legitimate and effective outcomes, because norms of access and accountability reflect increasingly shared understandings of 'right process' transnationally. The design of the FSC clearly takes such concerns into account. Moreover, lack of access or resources to participate in the scheme can seriously undermine it, as the case study of Brazil in Chapter 3 demonstrates. Nonetheless, as a first cut, a more sociological notion of legitimacy moves directly to addressing the empirical and explanatory issue of the requirements for the acceptance of particular forms of governance, while bypassing thorny philosophical issues of what non-state governance should ideally consist of, or whether it should be permitted at all.[9] Put simply, in this view a rule or institution is legitimate if relevant audiences accept it as appropriate (Suchman 1995; Florini 1996). Given the nature and location of the audiences that must grant authority, as described above, legitimacy operates at two levels: global, which corresponds to the transnational marketplace and global normative structure, and national or regional, which corresponds to the immediate marketplace and regulatory regime in which firms compete.[10] Each is addressed here in turn and the interactions among them considered.

The Fit with the Global Normative Environment

At the global level, the legitimacy of non-state governance depends on its fitness or competition with existing international sets of institutionalised norms already accepted as legitimate in the relevant issue areas, whether in the form of specific rules of multilateral hard law, soft law, or uncodified norms nonetheless understood to reflect shared beliefs about appropriate behaviour.[11] This focus stems from the recognition that rules and institutions compete for social fitness, so the focus turns to the conditions for success in that competition. For the normative environment, one must consider broadly accepted norms of global environmental governance and forest use and conservation norms, including emerging norms on sustainable forestry, as embodied in international environmental treaties and declarations, as well as action programmes, leaders' statements, and scholarly opinion. The relevant multilateral rules or hard law examined is primarily trade law, because rules on sustainable forestry remain largely in the realm of soft declaratory law.

In the case of sustainable forest management, Steven Bernstein (2001) has labelled the relevant broader global normative environment 'liberal environmentalism', where international environmental protection is predicated on the promotion and maintenance of liberal economic order. Liberal environmentalism evolved out of a series of North–South compromises, but also an ideational shift in how the international community framed environmental issues and responses over the last 30 years. The main elements of this compromise include state sovereignty over resources within a particular state's borders on the political side, the promotion of global free trade and open markets on the economic side, and the polluter-pays principle (and its implicit support of market instruments over strict regulatory mechanisms) and the precautionary principle on the management side.[12] The polluter-pays principle refers to the idea that the polluting firm ought to shoulder the costs of pollution or environmental damage by including it into the price of a product. Ideally, price signals would reflect the real costs of pollution. This principle thus favours market mechanisms — including labelling and certification — because they operate by institutionalising schemes that incorporate environmental costs into prices. The precautionary principle is a norm concerning risk under uncertainty that essentially says that under conditions of risk of serious environmental harm, a precautionary stance is warranted under conditions of uncertainty.

These norms are embodied most notably in the Rio Declaration on the Environment and Development, the main statement of international consensus on how environment and development concerns ought to be addressed by the international community. States agreed to these norms at the 1992 Rio Earth Summit and subsequently reaffirmed them at the 1997 UN General Assembly Special Session to review the implementation of Agenda 21 (the detailed plan of action adopted in 1992 at Rio) and the World Summit on Sustainable Development in 2002. In addition to being reinforced in Agenda 21 and related action programmes, and in a wide range of declaratory and customary international law, these norms and the principles consistent with them are embodied in major multilateral environmental treaties such as the UN Framework Convention on Climate Change (UNFCCC), the 1997 Kyoto Protocol to the UNFCCC, and the Convention on Biological Diversity, as well as programmes of multilateral institutions, including the World Bank, UN Conference on Trade and Development (UNCTAD), the UN Environment Programme (UNEP), and the Organisation for Economic Co-operation and Development (OECD) (Bernstein 2001, ch. 3).

Arguably, much of the difficulty in reaching an agreement on a global forest convention is precisely because states have been unable to formulate policies that do not conflict with many of these key norms.[13] For example, certification is advantaged over multilateral diplomacy in the case of forestry because it is a market-driven mechanism consistent with environmental cost internalisation and the polluter-pays principle. The logic of certification is that firms internalise the costs of sustainable forest management through the requirements to gain a label, and the market operates to reward sustainable forest practices. Such a mechanism is consistent with liberal environmentalism, and avoids politically charged debates in which developing countries in particular fear that the North will impose standards and unfair trade restrictions on them (Lipschutz 2001; Humphreys 1996, 1999).

Although the debate continues over the effectiveness of economic instruments versus bans or other compliance and control mechanisms in environmental policy (Porter and van der Linde 1995; Esty and Geradin 1997, 1998; Gunningham, Graborsky, and Sinclair 1998; Gunningham and Sinclair 2002; Haufler 2001), a strong policy consensus has nonetheless emerged that favours market mechanisms, reinforced by the work of economists and internationally and domestically sponsored policy research, including in influential international institutions such as the OECD, UNCTAD, the World Bank, and the United Nations Industrial Development Organization (UNIDO), and within the UN system more generally. While market mechanisms have been adopted for issues such as climate change, in the forestry area specific arguments in support of market-friendly or incentive-based mechanisms have also made headway. For example, major environmental NGOs such as the WWF subscribe to the argument put forward in the wider international forestry policy community that logging bans create perverse incentives such as increasing pressure to clear land for other uses or that they do not create incentives to improve practices because they eliminate markets for tropical timber (Hock 2001, 348).

This supportive normative environment is conducive to viewing certification and labelling schemes, and other market mechanisms, as legitimate means to addressing environmental problems, especially as compared to traditional forms of regulation of the command-and-compliance variety, such as bans or quotas. By granting legitimacy, these norms make it more difficult for market players or states opposed to the schemes owing to particularistic or short-term interests to make the case that others should not accept these schemes. Even with a normative context broadly conducive to the acceptance of such schemes, however, they must be designed to avoid running afoul of relevant international hard law, especially highly institutionalised bodies of rules embodied in multilateral institutions.

Consistency with Existing Multilateral Rules and Institutions

The multilateral institution of most relevance to transnational forestry governance is the WTO, specifically the Agreement on Technical Barriers to Trade. It covers labelling schemes, including eco-labelling, and is binding on WTO members. Other potentially relevant institutions include environmental agreements such as the Statement of Forest Principles agreed to at Rio,[14] the 1997 Kyoto Protocol, the Convention on Biological Diversity, and the Convention on International Trade in Endangered Species. However, it should be noted that rules in areas other than trade are still evolving, are often non-binding, and are less institutionalized by most measures. In the case of climate change, for example, the political agreement ironed out in negotiations in Bonn and Marrakesh in July and October-November 2001, which was meant to pave the way for ratification of the Kyoto Protocol, still left open the question of whether the use of carbon sequestration in forests (or other land use, land-use change, or forestry activities permitted) must adhere to principles of sustainable forest management. In the only notable reference in a decision of the parties to the convention, they affirmed a set of principles to govern such activities, including 'that the implementation of land use, land-use change and forestry activities contributes to the conservation of biodiversity and sustainable use of natural resources' (UNFCCC 2001b, 56). Similarly, forest management activities under the protocol are defined as 'those practices for stewardship and use of forest land aimed at fulfilling relevant ecological (including biological diversity), economic and social functions of the forest in a sustainable manner' (UNFCCC 2001a, 4). How these principles will be operationalised remains uncertain. Negotiators were much more focussed on reaching agreement on the conditions and extent to which carbon sequestration in forests (or other land-use changes) could count toward reaching emission targets.

In terms of binding rules, the agreement on TBT stands out as potentially most significant since any labelling of a traded product formally falls within its ambit. States supported the agreement, in its current form, in the Uruguay Round of trade negotiations that created the WTO. Disputes covered by the agreement are subject to the WTO panel process and appellate body rulings. Consistent with the norms that underpin the trade regime, the TBT agreement aims primarily to ensure that technical regulations and standards do not 'create unnecessary obstacles to international trade' (WTO 1994, preamble, art. 2.2). It permits national programmes and standards, including for environmental purposes, as long as they do not discriminate on the basis of national origin, are necessary for the stated objective, and are the least trade restrictive to achieve that objective (art. 2). In addition, the agreement states that recognised international standards (such as an ISO standard), or relevant parts of them shall be used 'as a basis for their technical regulations' except when they would be inappropriate or ineffective for the 'legitimate' objectives covered by the agreement.[15]

The implications of these provisions for non-state market-driven governance are uncertain. The agreement on TBT does cover nongovernmental schemes (WTO 1994, preamble, art. 3). However, whether such a scheme would be recognised as an international standard is less certain, because the agreement only refers specifically to national standards. Nor is it clear what a transnational scheme's status would be if not recognised by states. Halina Ward (1997, 143), for example, argues that 'purely voluntary, non-governmental labelling schemes administered through organizations that do not fall within the scope of the term "standardizing bodies" for the purposes of the TBT Agreement ... do not fall within the scope of the TBT at all'. In addition, ISO standards, by virtue of the WTO's status and procedures, might be recognised automatically, even if they conflict with voluntary standards of a nongovernmental scheme with otherwise widespread support. However, working against their legitimacy, it should also be noted that procedures for determining the ISO 14000 series of standards in particular have been criticised because ISO decision making occurs through working groups of representatives from national standard-setting organisations, which are dominated by government and industry experts. Some non-industry NGOs and developing countries complain this configuration leads to biases and unequal representation (Clapp 1998; Deutsche Gesellschaft für technische Zusammenarbeit GmbH 2000, 18). So far, only two WTO disputes have been launched under the labelling provisions of TBT agreement, and both involved countries that argued for proprietorship of names of traded products — scallops and sardines — not labels that denoted standards regarding health or environment, or that involved nongovernmental labelling schemes. Thus, the interpretation of relevant conditions under the agreement remains untested.

The picture is further complicated by what kind of criteria for a label the TBT agreement deems acceptable. A major bone of contention is whether the criteria cover production and processing methods (PPMs), especially nonproduct-related production and processing methods (as in life-cycle analysis that takes into account values or effects not directly related to production of products for export or import). The Uruguay Round incorporated PPMs into the TBT agreement, so arguably even voluntary schemes that incorporate them are covered, although their status both before and after the Uruguay Round is still disputed (Chang 1997, 140–147). Non-product–related PPMs are not mentioned, and thus do not fall under TBT by almost any interpretation. This creates the ironic situation where certification and labelling schemes that are potentially more expansive in scope, by covering broader inputs and values in the life cycle of a product, are generally considered outside the scope of the WTO as long as governments remain uninvolved. Once governments become involved, such schemes may fall under other rules of the General Agreement on Tariffs and Trade (GATT) or other WTO agreements (Chang 1997).

How trade law treats measures related to PPMs is still evolving. In theory, at least, such standards or measures can operate consistently with free trade norms and the GATT since, if applied fairly, such a standard or measure to protect the environment could simply internalise the cost of pollution. However, from a trade perspective, especially when viewed from the South, such measures face resistance owing to a perception that they potentially violate sovereignty (especially in the case of nonproduct-related PPMs) because they involve guidelines on practices within an exporting state, not just the nature of a product. Moreover, different countries may argue they have different optimal levels of pollution and should not be subject to the same standard.

It should also be noted that outside of the TBT, WTO, and earlier GATT rulings generally have set the bar high in justifying environmental measures, such as under GATT articles XX(b) and (g) on exceptions (for example, for health and safety and including animal and plant life and health, or to protect exhaustible natural resources), especially when it appears that standards are being applied extraterritorially.[16] However, post-1994 Appellate Body rulings, relying on a more literal reading of Article XX exceptions, suggest some easing of the tests for acceptable environmental measures, in effect suggesting that earlier rulings applied requirements that had no basis in the treaty text (Wofford 2000). In addition, rulings in cases such as beef hormones (1998) — a dispute launched by the United States over a European ban on hormone-modified beef — indicate some very limited openness to interpretations of science and risk more favourable to considering environmental concerns, such as those consistent with the precautionary principle, although it is still far too early to judge what such rulings portend in the longer run.[17]

These issues and related concerns remain subjects of controversy within the WTO Committee on Trade and Environment. Committee discussions suggest openness to eco-labelling in principle because it does not deny market access, but developing countries continue to express concern that it could restrict sales. This North-South divide shows little sign of easing.[18]

Where does all this leave forest certification programmes in their non-state market-driven form? As private transnational voluntary schemes operating in the marketplace, they are largely immune from WTO discipline, unless adopted by a government. While some pressure continues to build for a harmonisation of criteria for sustainable forest management and certification schemes, a scheme such as the FSC arguably would not run afoul of basic WTO principles even if the agreement on TBT were extended to include nonproduct-related PPMs and transnational voluntary schemes. In the case of the FSC, its criteria are consistent with equivalency of environmental measures in importing countries (that is, different environmental goals apply to different countries), mutual recognition (that is, if it qualifies in the exporting country, it should qualify in the importing country), as well as general WTO goals such as transparency. For example, Hock (2001, 363–365) argues that the FSC process is transparent, is non-discriminatory (because it applies to all forests and ecosystems as opposed to being aimed mainly or solely at tropical forests), and does not restrict trade because it aims to inform the consumer rather than ban trade.

Under these conditions, the FSC or similar non-state certification schemes seem well placed to gain international legitimacy, or at least not to provoke legitimacy conflicts with existing institutions. Where the FSC and some of its competitor non-state programmes find conditions favourable to gaining legitimacy, a proposed global forest convention did not. A brief discussion of that failure further highlights the comparative advantage of non-state market-driven governance in this case.

Failed Negotiations Toward a Global Forest Convention

International attempts to forge a forests treaty have their roots in the efforts of Northern countries and environmentalists in the 1980s to press for action to protect tropical forests. These efforts originally focussed on securing the inclusion of a mandate to conserve tropical forests and their genetic resources in the International Tropical Timber Agreement (ITTA) of 1983. Even from this early date, forest negotiations were seen largely in the context of international trade and economic development, not primarily environmental protection or conservation. The primary mandate of the International Tropical Timber Organization, the decision-making body of the ITTA, remains the regulation of trade in tropical timber (Humphreys 1996). As it became clear that tropical deforestation was not being addressed adequately within the structure of the International Tropical Timber Organization or other existing initiatives, pressure mounted, especially in the run-up to the 1992 Rio Earth Summit, to develop a more wide-ranging and environmentally focussed treaty on tropical forestry.[19] It quickly became apparent, however, that any international agreement would have to include all forests, not focus only on tropical forestry, and would include the full range of forest values. Moreover, the compatibility of any action on forests with other multilateral institutions was a requirement for success. Following from arguments made earlier, the compatibility requirements can be broadly conceived as consistency with norms of liberal environmentalism that were emerging during this period, and especially with norms of free trade as embodied in the GATT.

It should come as little surprise that a consensus emerged in forest negotiations that trade and environmental policies can be mutually supportive, and this became a central element in the nascent discourse around sustainable forest management. This idea eventually became explicit in Principle 12 of the Rio Declaration and Principle 13 of the Statement of Forest Principles (both agreed to in Rio). The former says that states should 'promote a supportive and open international economic system that would lead to economic growth and sustainable development in all countries' (UNCED 1992), while the latter states that 'open and free international trade in forest products should be facilitated' and that 'removal of tariff barriers and impediments to the provision of better market access … should be encouraged to enable producer countries to better conserve and manage their renewable forest resources' (UN 1992). Also consistent with the norms mentioned earlier, Article 13(c) encourages states to incorporate 'environmental costs and benefits into market forces and mechanisms, in order to achieve forest conservation and sustainable development'.

This consensus, however, masks deep political conflicts that prevented the creation of a binding treaty in 1992. The debate at Rio pitted Northern states such as the United States and Canada, which argued for a 'global responsibility' approach, against Southern states such as Malaysia and India, which argued for 'sovereign discretion'. The latter feared that Northern states would dictate how forests within the jurisdiction of Southern states should be managed (Porter and Brown 1996, 115–129). As a result, negotiations were mired in North-South debates, where the G77/China bloc of developing countries remained suspicious of Northern intentions on several fronts and at the Earth Summit refused even to consider a binding agreement with few economic incentives (Humphreys 1996; Porter and Brown 1996, 115–129; Williams 1993, 7–29). Despite various attempts since 1992 to reinvigorate movement toward a convention, all have failed.[20]

The failure to create a binding treaty at Rio (the Statement of Forest Principles is declaratory) and to make substantial progress toward a forest convention since 1992 is in large part owing to the inability to work out how this compatibility of free trade, development, and protection of forests can work in practice. This poor fit is partly explained by the nature of forest resources themselves, which some perceive as a global commons problem, but which fall under the sovereign jurisdiction of individual states (although many forests cross borders). Thus, proposals to address forestry issues have difficulty gaining legitimacy because they can be easily perceived as violating norms of sovereignty over resources and free trade, especially since developing countries worry that any mandatory rules on sustainable forest management, even if not explicitly trade related, would disproportionately affect developing country exports. Aid or assistance to institute sustainable forest management is the other major stumbling block.

While some have argued (such as Steiner 2001, 98–99) that various soft law principles, including those found in the Statement of Forest Principles, Agenda 21's

Chapter 11 on deforestation, and various government-supported programmes such as criteria and indicator processes, might provide the basis for future action, even in these areas results have been limited at best. Moreover, apart from references to forestry issues in other treaties, interested states have made little progress toward creating binding forestry rules. The Statement of Forest Principles is symptomatic of an issue area where states have tried to include every value in international negotiations and refused to make tradeoffs to give direction on how ambiguities would be addressed in practice. The UN Forum on Forests (UNFF) provides a more secure institutional setting to implement concrete proposals, but the political divides remain as strong as ever.

There is a sense among environmental NGOs that a convention with concrete commitments on sustainable forest management is unlikely in the short or medium term, and could even entrench watered-down rules if a weak regime prevailed, and this sentiment accounts in part for the push toward certification. Such schemes can be kept relatively independent from multilateral processes bogged down in North-South conflict. The inability of forest negotiators to formulate rules that fit with the broader normative context of what would be perceived as acceptable by states, and the ability of nongovernmental schemes to take advantage of opportunities that same normative environment provides, offered an opportunity to promote sustainable forest management where intergovernmental negotiations remained deadlocked.

Implications of Norms and Institutions for the Legitimacy of Certification

The discussion above suggests that the viability and acceptability of certification as a legitimate form of governance largely depend on how certification is viewed in the context of a wider debate on the compatibility of trade and environmental policies. For most states, caution prevails. Fears persist that government-supported sustainable management schemes could lead to unjustifiable restrictions on trade and obstacles to market access. Thus while certification has gained increasing attention in international forest negotiations, so far states can only agree to endorse more research to gain experience. Hence, the UNFF and related intergovernmental work will continue to encourage research with the aim of enhancing comparability of various schemes, transparency, and non-discrimination in operation of programmes, but certification is unlikely to be included in any inter-state agreement on forests in the short or medium term.

Furthermore, developing countries, in particular, would prefer the WTO to take the lead on certification rather than leave it in the hands of forest negotiators who may be less sensitive to these concerns. They especially resist efforts to move certification discussions into environmental forums such as UNEP. But given the slow pace of resolving these conflicts in forums such as the Committee on Trade and Environment, supportive state leaders may simply view schemes such as the FSC or competitor programmes as the most viable way to incorporate abstract principles they have agreed upon, while avoiding the political stumbling blocks that have plagued forest negotiations.

Domestic Legitimacy

The domestic arena constitutes the second level of legitimacy. Its importance arises because land owners and forest companies — the very entities whose behaviour certification programmes seek to modify — make choices at this level over whether to pursue forest certification and, if so, which programme to support. Just why forest companies and forest landowners would agree to a set of rules not required by their own governments is an important question that requires an examination of the evaluations forest companies and landowners make of such programmes.

In order to address this question, Cashore (2002) has drawn on Mark Suchman's seminal work on organisational legitimacy. Suchman (1995) identifies three types of legitimacy that external audiences may grant an organisation: pragmatic legitimacy based on short-term material self interest, moral legitimacy based on a belief that supporting the actions of the organisation in question is 'the right thing to do', and cognitive legitimacy, where external audiences grant legitimacy because to do otherwise is unthinkable or because it is taken for granted. The common element that makes each a type of legitimacy follows from the earlier definition, that legitimacy at its most basic means acceptance as appropriate by relevant audiences. 'Pragmatic', 'moral', and 'cognitive' reflect the reasons for that acceptance; for example 'I view an organisation as appropriate because it benefits me',[21] 'I view it as appropriate because it is the right thing to do', or 'I view it as appropriate because it seems natural to follow its rules because I have internalised them and I would not think to act otherwise'. These distinctions permit a more nuanced understanding of how domestic-level legitimacy works, and the nature of support for the FSC by companies all along the supply chain, as well as by those domestic environmental groups seeking to alter and influence company evaluations.

Comparative survey research and extensive interviews of forest company officials and landowners reveals that support by forest companies and landowners at the domestic level for forest certification operates, for the most part, at the level of pragmatic legitimacy (Cashore, Auld, and Newsom 2003, 2004; Auld 2001; Cashore and Lawson 2003; Auld, Cashore, and Newsom 2002; Lawson and Cashore 2001; Newsom et al. 2002). Firms and companies support FSC-style non-state governance because of short-term material incentives — they either see the benefits of gaining access to a certified market or they support certification in order to avoid costs of boycott campaigns. Similarly, many companies and landowners reluctantly support industry or landowner association non-state certification programmes because they wish to grant these programmes legitimacy over what they view as the more costly prescriptive FSC-style certification programmes. Thus domestic legitimacy of FSC-style non-state governance is currently weak, and often rests on environmental groups maintaining negative economic pressure (boycotts, targeting of companies). Forest companies and landowners grant even FSC competitor programmes a relatively weak form of legitimacy, since support here is often a defensive tactic.

Recognition of the relatively weak, pragmatic form of legitimacy at the domestic level raises important questions about whether and how domestic legitimacy for non-state governance systems might move toward more durable moral or cognitive forms. Suchman (1995) argues that cognitive legitimacy can be attained by routinisation, but just how long companies would have to grant pragmatic legitimacy before it becomes routinised, or whether routinisation is more or less automatic or depends on particular conditions, is uncertain. Depending on these factors, non-state governance systems could easily wither away before cognitive legitimacy was granted.

One promising direction of research in investigating this transition is to focus on 'legitimacy achievement strategies' (Suchman 1995) at the domestic level. FSC and competitor programme supporters are actively engaged in high-stakes efforts to win the legitimacy contest among firms along the supply chain. Here efforts and results diverge dramatically among different cases. In British Columbia (BC) in Canada, the FSC has made significant inroads through active legitimacy achievement strategies, reversing initial forest company rejections. By 2002, seven of the ten largest companies in the province indicated some support for the FSC (Cashore, Auld, and Newsom 2003). But by the summer of 2002, much of this support was dissipating as the FSC working group produced standards in BC that companies felt were too stringent and would hurt them economically if adopted (Bourgeois 2002; Cashore, Auld, and Newsom 2004, ch. 3). In the United States, with few exceptions, large forest companies continue to reject the FSC. Instead, they strongly support the American Forest and Paper Association's SFI programme. In the United Kingdom, state forest owners have reluctantly supported the FSC, while in Germany, most forest landowners support the PEFC system, although the FSC has earned the support of state forest landowners where elected leaders support an environmental agenda (Cashore, Auld, and Newsom 2004, chs. 4, 6; Auld 2001; Newsom 2001). In Sweden, large industrial forest companies support the FSC, while small landowners reject this approach.

A focus on domestic legitimacy is required to answer the question of why support for global non-state governance would vary across countries, or why some firms would opt for the more prescriptive FSC scheme while others would steadfastly support the industry alternatives. Research by Benjamin Cashore, Graeme Auld, Deanna Newsom, and James Lawson reveals the importance of understanding how variations in the features of domestic forest sectors mediate domestic legitimacy dynamics. Since soft law mechanisms such as certification depend to some degree on the success of active legitimacy achievement campaigns, it is important to have a thorough understanding of the process through which domestic legitimacy is achieved, and the ability of actors supporting the FSC and other certification programmes to alter preferences.

The competition between the FSC and its competitors is especially important because even if the FSC loses, it may have succeeded in creating pragmatic and possibly cognitive legitimacy for the general concept of non-state market-driven governance in forestry, even if its preferred rules and procedures do not fully prevail. Indeed, this may very well be the direction of forest certification in the United States, where the entry of the FSC led to little direct support, but to widespread support of its competitor, the SFI. There is emerging evidence that support for the SFI is reaching a cognitive phase among industrial forest owners, becoming an everyday business practice. However, if domestic legitimacy struggles result in the emergence of a non-state market-driven certification system that lacks broad-based support from environmental groups, and faces criticism by them for what they believe are inappropriate standards for promoting sustainable forestry, then the question arises as to whether this form of governance will address the problems for which environmental groups originally created the FSC.

Understanding domestic legitimacy first requires an analysis of the way pressures along the supply chain may reach or affect forest company and forest landowner legitimacy evaluations. Support for various schemes is uneven along the supply chain. Initially, the FSC was more successful than its competitors in gaining support from retailers such as Home Depot, B&Q, Sainsbury, Centex, and German publishing companies, which then urged or required their suppliers to provide them with FSC-certified wood or wood products. Supporters of the FSC have also been successful in creating buyers groups in different countries (see examples above) with the purpose of establishing economic incentives for forest companies by creating a clear demand for certified forest products. One unanticipated consequence, however, is that many forest companies have used this pressure to create their own programmes in an attempt to meet retailers' demands without having to adhere to more onerous FSC rules. Consequently, many of the above-mentioned retailers, as well as others, now have inclusive procurement policies that implicitly or explicitly accept FSC competitors.

While still in its infancy, recent research on the influence of supply chain dynamics in the struggle for domestic legitimacy shows the centrality of different monopoly interests of the FSC and competitor programmes at different parts of the supply chain (Cashore, Auld, and Newsom 2004). At the level of the forest landowner and forest company, FSC competitor programmes have a monopoly interest in gaining support. If a landowner or forest company accepts the rules of the FSC, the competitor programme has lost the legitimacy struggle because it was created expressly to limit the impact of the FSC. However, the FSC does not have a monopoly interest in support at this level — it does not matter to the FSC if a forest landowner or forest company follows another programme's rules as long as it supports and follows FSC rules as well. The reason is that the FSC was created as the most prescriptive and wide-ranging programme, which means it can encompass the rules of competitor schemes.

The situation is reversed, however, further down the supply chain. Take, for example, the case of the retailer Home Depot. The FSC requires that Home Depot only support and grant legitimacy to the FSC in order to create pressure along the supply chain. If Home Depot also supports — that is, grants legitimacy to — a competitor programme, it effectively removes the economic incentive for landowners and forest companies to support the FSC, taking away all of its market leverage required for pragmatic legitimacy. Conversely, based on the same logic, FSC competitor programmes have no monopoly interest at the retailer level at all — all they care about is that their programmes, in addition to the FSC, are deemed to be legitimate. It is for these reasons that at the retailer level FSC competitor programmes formed an umbrella organisation — the IFIR — whose members called for mutual recognition not just of their own programmes, but also of the FSC (Cashore, Auld, and Newsom 2004, ch. 8).

Further complicating this process, the supply chain is often not limited to a single national market. Transnational supply chains can produce different legitimacy achievement dynamics because different players along the chain operate in different national contexts. Acting across borders presents both opportunities and constraints for influencing the support for particular schemes. Cashore and his colleagues found, for example, that commitments from retailers and purchasers of forest products were easier to obtain, and had more direct market impact, when those retailers and purchasers were outside of the domestic unit. For example, German publishers committed fairly quickly and easily, following environmental NGO pressure, to demanding that their BC forest company suppliers commit to the FSC. The same German publishers paid much less attention to requiring that their domestic supply be FSC certified (Newsom 2001). To do so would have meant becoming embroiled in domestic debates about the supremacy of German forestry, and being accused of pandering to foreign demands. As a result, German publishers have more impact on influencing domestic legitimisation struggles in BC than in their own country. A similar dynamic occurred in the case of Scott Paper. Its UK branch quickly demanded that its BC suppliers become FSC certified while Scott Paper Canada took the opposite stance, publicly announcing it would not succumb to unfair environmental group demands that would hurt the Canadian economy (Auld 2001). This should not be surprising given this pattern that the U.S. Home Depot's announcement to favour the FSC had more impact in altering forest company support for the FSC in British Columbia, than it did in the United States. There, a well-organised industry effort has resulted in a stalemate of sorts: as long as the U.S. companies refuse to commit to the FSC, Home Depot will be unable to meet its commitments to the FSC, which puts pressure on Home Depot to eventually support the industry competitor programme as legitimate. Such a move would do much to grant non-state market-driven governance legitimacy in the United States, but at the same time could undermine the FSC conception of forest certification.

Domestic Legitimacy and Durability

While much of the effort to achieve legitimacy is centred on gaining pragmatic legitimacy, early research suggests two important limits to Suchman's arguments. First, contrary to Suchman, Cashore's team found that caution is warranted in assuming that pragmatic legitimacy is always the least durable. They found that for profit-maximising companies, it appears that pragmatic legitimacy is a prerequisite for non-state market-driven governance, making its status more important, at least initially, than that of moral legitimacy. Pragmatic evaluations will trump moral ones, and are thus potentially more durable, since companies need to stay in business. But such support is double edged. While it may mean that companies will support certification when it offers a clear market advantage, the reverse is also true — companies granting only pragmatic legitimacy will walk away if they view certification as a liability. Indeed, understanding the conditions that produce both moral and pragmatic legitimacy appears to be the key to durability.

Moreover, survey findings and interviews of non-industrial landowners in the United States, Germany, the UK, Canada, and Sweden suggest their notions of private property rights and fierce feelings of independence militate against the FSC and industry-created FSC competitor programmes obtaining moral legitimacy since outside rules are deemed largely inappropriate (Auld, Cashore, and Newsom 2002; Newsom et al. 2002). Many landowners even view forest landowner association certification programmes with suspicion (Newsom et al. 2002), unless landowners were specifically involved in the creation of rules and they deemed their participation as purely voluntary. Here, the importance of institutional design, specifically regarding participatory norms and accountability, creeps back in. The issue of non-industrial private forests raises the need to understand how such audiences might grant pragmatic legitimacy, and if and when this support for any programme might eventually be converted to cognitive support.

Second, a focus on domestic legitimacy achievement strategies forces consideration of whether such strategies take conforming or converting approaches (Cashore, Auld, and Newsom 2004).[22] Under converting strategies, the certification programmes and their supporters actively work to change the evaluations of those from whom they seek support by changing the outside audience in some way. The usual strategy is to create economic pressure along the supply chain. But if this economic pressure falls short, certification programmes may undertake legitimacy achievement conforming strategies, where they change their own rules or programmes in order to gain support. For example, the FSC has relaxed its rules governing harvesting ancient forests in order to gain support in British Columbia, and relaxed its rules governing forest plantations in order to get support from companies operating in the tropics and other fast-growing climates. The strategy chosen is important because the extent to which the FSC and its competitor programmes must conform to different audiences in their efforts to achieve legitimacy will affect their ability to remain true to their original goals. In this regard, Cashore and his colleagues found that certification schemes appear to have core audiences who grant moral legitimacy that must be maintained. This finding suggests that the conforming strategies to anchor moral legitimacy in core audiences may be required to gain legitimacy from non-core audiences.

Future Research

Domestic legitimacy contests are far from predetermined. Analysis of how strategic choices interplay with key features of different domestic forest sectors is required to make a clearer case whether or not non-state market-driven governance can gain legitimacy where a forest convention has failed.

One especially important line of inquiry is to investigate the role of traditional domestic state authority in this process, since, as noted above, it still provides the general rules of the game under which these struggles occur. In so far as the state is a landowner or purchaser of wood products, its actions and policies, or the broader regulatory or legal environment it creates, might add to or reduce the ability of non-state market-driven governance to gain legitimacy.

Moreover, a key question is whether non-state market-driven governance institutions might actually replace traditional domestic public policy arenas as the places where forest conflicts are resolved. If an array of actors along the supply chain accepts an FSC — or even a hybrid — conception of governance as legitimate, environmental governance power relations could conceivably be altered in two ways. On the one hand, the FSC grants more decision-making authority to non-business interests than many traditional process for public policy making. On the other hand, it is much more difficult for environmental NGOs to target a firm that abides by FSC rules, and is certified as doing so, as being environmentally unfriendly since it possesses the legitimacy granted to it by NGO-supported governance system.

Conclusion

The domestic-transnational nexus associated with non-state market-driven governance is unique because the very location of authority under it occurs along a supply chain that can transcend borders. Nonetheless, traditional domestic borders are important for two reasons. In the case of the FSC, they are important because the FSC grants national or regional bodies the right to develop appropriate rules domestically through consultation processes. Such domestic-centred processes are important for how rules are developed and how corresponding legitimacy is granted or not granted by different domestic audiences. Furthermore, the FSC has resulted in the emergence of domestically focussed competitor programmes, which means legitimacy contests are largely domestic in nature.

At the same time, schemes that operate transnationally must gain legitimacy at the international level as well, since they come into contact with existing norms, rules, and institutions that define and regulate appropriate behaviour in relevant issue areas. The main goal of this chapter has been simply to show that a scheme must be granted legitimacy at both levels to be successful, and to offer some preliminary explanations as to how and why legitimacy is or is not likely to be granted at those two levels. A sociological approach to the problem of legitimacy is useful in this regard.

One limitation of a sociological approach, however, especially when focussed on the legitimacy of the norms and rules of a scheme, is the tendency to bracket democratic or other procedural aspects of legitimacy that link decision making and outcomes of a regime to the communities that authorise it, and over which it is granted authority. The importance of institutional design and the right process in gaining legitimacy from relevant communities has been noted here, although not explicitly addressed. Indeed, it may be that insights from the democratic and normative literature on legitimacy cannot easily be separated from the focus here on the fitness of norms and rules and legitimacy achievement strategies. The link may come in the form of investigating the shared norms of stakeholders on legitimacy requirements. This poses a particular challenge for non-state market-driven governance since, owing to differences in relevant audiences in terms of identities (producers, consumers, environmentalists), location, or interests, consensus may be lacking on what constitutes either procedural or substantive legitimacy. Thus there may be little agreement on how to design a non-state market-driven institution that could generate authoritative commands. More research is required on these links, particularly if demands for greater democracy and accountability in global governance institutions continue to increase.

The focus on legitimacy should also not obscure the underlying sustainability goals in this case. Thus, this chapter concludes on a cautionary note. Success in gaining legitimacy does not necessarily mean that a scheme is effective at producing sustainable forest management or is desirable from the perspective of all actors. Two examples highlight this danger. First, in the context of North-South debates over global forest protection and use, it is notable that the focus of certification campaigns very quickly shifted from the South to the North (Europe, the United States, and Canada) because of the strength and coherence of environmental groups in the North and because these locations offer the most likelihood of institutionalising non-state market-driven governance (see figures 2-1 and 2-2). Once institutionalised domestically or regionally in the North, such governance will arguably provide leverage, both materially and morally, to force developing countries and emerging markets to accept. While such a dynamic might bypass the thorny political divisions that militate against an international convention, it necessarily risks being perceived as a form of eco-imperialism. It thus appears crucial that the decision-making structure of the FSC (and other programmes) strives to be truly representative, and that its principles and processes remain sensitive to local conditions and needs.

Second, as hinted at above, there is always a risk that a scheme that gains legitimacy among major players, if not driven by a political process that pushes standards higher than some interested actors may desire, and that it may institutionalise lowest common denominator standards, industry preferences, and lax monitoring. This risk is mitigated to the degree that nongovernmental groups manage to entrench themselves as core audiences or continue to apply material pressure. Nonetheless, it may be that a non-state market-driven certification system must first become institutionalised with base-line standards to which most companies eventually adhere, at which point across the board increases could occur without putting those who support forest certification at a competitive disadvantage.

This case also raises broader questions concerning whether non-state governance poses a viable alternative to state or multilateral governance. While the legitimacy dynamics of the FSC and its competitor schemes are instructive, much work remains to be done on how legitimacy dynamics at various levels interact, and whether legitimacy

Figure 2-1 Percentage of Forest Cover under Leading Certification Programmes

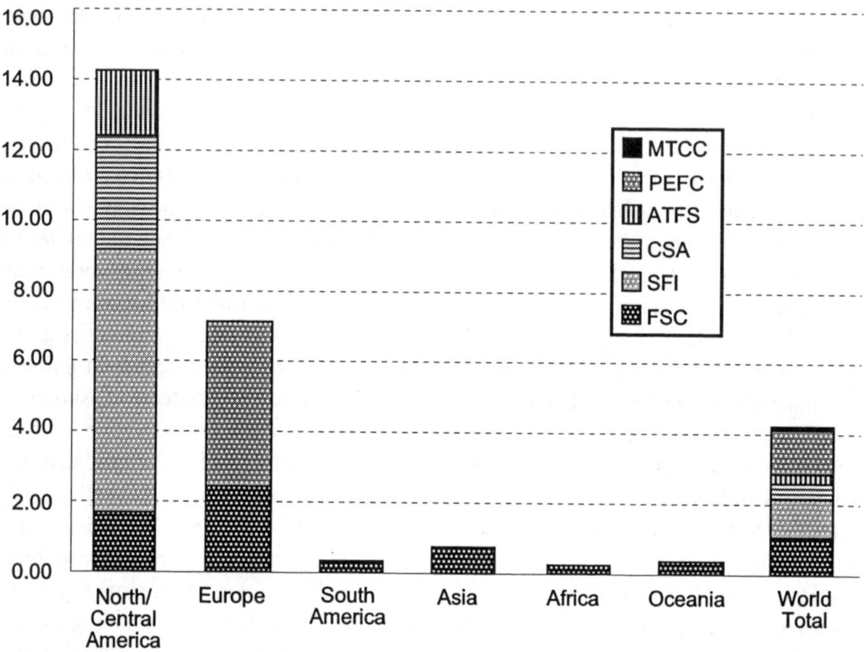

Notes:

1. The percentages of forest cover in this table downplay the influence of forest certification programmes. Data from the Food and Agriculture Organization include all lands with '10 percent forest cover', and include lands generally ineligible for forest certification, such as commercially unviable forest land, forestland used for non-forestry activity, and land with formal protected area status. For this table, certified land by programme is recorded, so the data 'double count' the small amount of forest land certified under more than one programme.
2. FSC: Forest Stewardship Council; SFI: Sustainable Forestry Initiative; CSA: Canadian Standards Association; ATFS: American Tree Farm System; PEFC: Pan European Forest Certification; MTCC: Malaysian Timber Certification Council.

Sources: Data reported as of 1 December 2003. See websites of the Forest Stewardship Council <www.fscoax.org>; Sustainable Forestry Initiative <www.afandpa.org>, Canadian Standards Association <www.sfms.com>, James O'Connor (joconnr@forestfoundation.org) of the American Tree Farm System; Pan European forest Certification <www.pefc.cz>, and the Malaysian Timber Certification Council <www.mtcc.com.my>. Forest cover is reported by the Food and Agriculture Organization, 2000, and identifies land with 10 percent forest canopy.

is enough without the coercive power of the state to back it up. In the face of the limited progress toward sustainability goals by traditional multilateral governance, an investigation of the viability of these new forms is all the more imperative, as is an assessment of how serious a challenge to state authority they pose.

Notes

1. The authors wish to thank Dan Esty, the editors of this volume, and the anonymous reviewers for comments on an earlier draft. The usual disclaimers apply. Steven Bernstein thanks Elizabeth Acorn for research assistance and gratefully acknowledges support from a Social Sciences and Humanities Research Council of Canada grant. Benjamin Cashore thanks Graeme Auld, Deanna Newsom, Emily Noah, Aseem Prakash, Erika Sasser, and James Lawson for their collaboration on related work that has pushed this paper in important directions. He thanks Laura Bozzi for research assistance and is grateful for support from the United States Department of Agriculture's National Research Initiative, the Kohlberg, Merck, and Ford foundations, the Rockefeller Brothers Fund, and the Doris Duke Charitable Trust.
2. Rosenau acknowledges, as do the authors of this chapter, that such new islands of authority are still the exception and exist in oceans of disorder or the more classic order of the sovereign state system.

Figure 2-2 World Area under Major Forest Certification Programmes

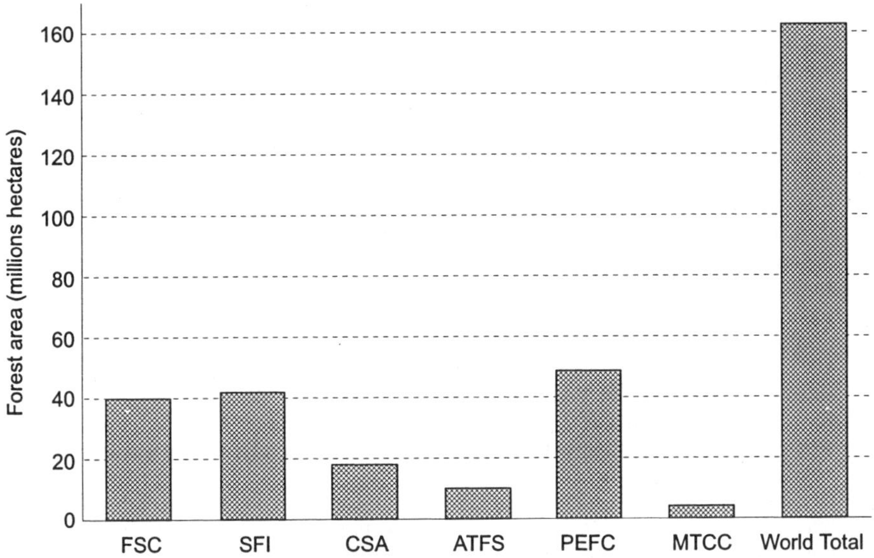

Note: FSC: Forest Stewardship Council; SFI: Sustainable Forestry Initiative; CSA: Canadian Standards Association; ATFS: American Tree Farm System; PEFC: Pan European Forest Certification; MTCC: Malaysian Timber Certification Council.

3. Hard law enforcement mechanisms available to governments under the authority of international law may not be the only means of coercion, but they are the primary source of coercive power and authority in the international system. This source of authority is usually unavailable to non-state schemes unless delegated by states. One of the few alternative sources of direct coercion (as opposed to indirect market pressures or voluntary codes or rules) available in non-state governance resides in the marketplace, where actors' decisions to purchase or not (including a decision to boycott certain products) can put pressure on the targets of governance, whether states or firms, to comply. These marketplace decisions or activities may be legitimated under non-state governance arrangements. Thus non-state governance is not necessarily without coercive elements, an observation that fits with the argument that authority links power and legitimacy over a community. See Bernstein and Cashore (2000) on the use of transnational market pressure to affect policy change.
4. In the strict context of this 'fitness' argument, a traditional understanding of soft international law as largely declaratory or non-binding is applied, but that has wide acceptance or agreement among states and frequently serves as a basis for future hard law (Chinkin 1989; Dupuy 1991). The modifier of 'international' is used here to differentiate it from the broader, non-state focus of soft law in this volume.
5. This table is adapted from Cashore's (2002) categorisation of domestic governance.
6. Owing to competing schemes and the complex science behind sustainable forestry, no consensus definition of sustainable forestry exists. This lack of a universally accepted definition makes an examination of rule development within the FSC and competitor programmes important in order to understand how they may define sustainability.
7. The legitimacy threshold model was developed and promoted by the IFIR following the Forest Dialogue meeting on forest certification in Geneva on 17 October 2002.
8. Because many regional or national working groups are still developing standards, many companies and landowners are audited according to provisional rules developed by certifiers and accepted by Oaxaca.
9. See Bernstein (2003) for a detailed discussion of these broader issues.
10. Although the term 'global' is used here to differentiate this level from a strictly state-centric (international) focus, marketplaces may not be truly global and global norms may not be universally accepted.
11. For a related discussion of specific requirements for forest certification schemes consistent with parts of the argument here concerning the importance of fitness with international rules and norms, see Deutsche Gesellschaft für technische Zusammenarbeit GmbH Forest Certification Project (2000).
12. A detailed discussion of why this set of institutionalised norms is believed to prevail can be found in Bernstein (2001). State sovereignty over resources is widely considered the foundational norm of international environmental law, existing in various forms in legal decisions and documents such as the UN charter, but stated explicitly beginning with the UN General Assembly Resolution 1803/62 (1962) on Permanent Sovereignty over Natural Resources, and later in Principle 21 of the Stockholm Declaration on the Human Environment and Principle 2 of the Rio Declaration. The latter expanded it to include a sovereign right to exploit resources pursuant to a state's own environment and development policies. See, for example, Sands (1994, 7).
13. Given space limitations and the limited focus of this chapter, the argument that this is the reason for failure cannot be defended here. If anything, failure to reach an agreement on a binding treaty is overdetermined since the issue area is characterised by structural conflicts of interest, lack of scientific consensus on the transnational effects of deforestation, conflicts that arise from differences in national forest economies, and so on. See Lipschutz (2001) and Dimitrov (2003) for an exploration of some of these alternative explanations. The argument here is more modest, namely that these norms would enable any agreement that could be

framed to be consistent with them, in effect redefining interests and making it more difficult for those opposed to successfully argue their case.

14. The full title of the Rio agreement is the 'Non-legally Binding Authoritative Statement of Principles for a Global Consensus on the Management, Conservation, and Sustainable Development of All Types of Forest'.
15. 'Legitimate' objectives explicitly mentioned include national security, prevention of deceptive standards, and the protection of human health and safety, animal or plant life and health, and the environment.
16. Trade-environment disputes that touch on the issue include the tuna-dolphin dispute and the shrimp-turtle dispute. See, for example, Wynter (1999) for an argument that the WTO Appellate Body's decision on the shrimp-turtle dispute between the United States and India, Malaysia, Pakistan, and Thailand, over the danger to protected sea turtles by shrimp trawling, left the door open for the use of PPM-based trade measures, if properly applied, under GATT Article XX, even though it ruled against a U.S. restriction on imports of shrimp caught with nets not equipped with 'turtle excluder devices'. The ruling found the U.S. acted in an unjustifiably discriminatory manner, not that PPMs were impermissible.
17. For a succinct summary of the Panel and Appellate Body rulings, see Nordström and Vaughan (1999, Annex 1, paras. 140–147). On possible implications, see Winham (2003).
18. At meetings of the Committee on Trade and Environment during 2000 (WTO 2000), many delegates asked that the committee work on clarifying coverage of eco-labels under the TBT agreement, and some expressed concern they could become an unnecessary barrier to trade. At the same time, delegates supported their use as an effective tool to promote environmental policies. One member state argued that eco-labels that included nonproduct-related PPMs were not consistent with WTO rules. There is general agreement within the committee that eco-labelling schemes (whether mandatory or voluntary) should be developed in a way that is transparent, non-discriminatory (e.g., consistent with rules of national treatment), and least restricts trade to achieve the policy objective. Pressure may build to allow only mandatory schemes, such as those agreed to under multilateral environmental agreements, since voluntary schemes run a greater risk of being considered barriers to trade under WTO rules if there is any government involvement without harmonisation. However, agreement on mandatory schemes is obviously much more difficult and the lack of progress in the committee suggests continued opening for purely private market-driven schemes.
19. After prolonged negotiations on a successor agreement (the ITTA 1994, which came into force in 1997), consumer countries pledged to manage their own forests sustainably by 2000, but outside the context of the International Tropical Timber Organization. They reasoned that a pledge within the organisation would undermine the prospects for a global forest convention, which they wanted to be much broader than the trade-focussed ITTA. Negotiations originally concentrated on tropical forestry (pre-Rio), now on the North and South, producers and consumers, and the full range of forestry issues. A new round of negotiations is underway for a further successor agreement following the ITTA 1994, which will expire in December 2006.
20. A variety of narrower or regional initiatives have taken shape since 1992 in its place. Examples include the Montreal Process on creating criteria and indicators for the conservation and sustainable development of temperate and boreal forests, the Helsinki Process on protecting forests in Europe (which includes a criteria and indicators process), and negotiations toward successors to the ITTA. The only process to focus serious attention on a global convention was the Intergovernmental Panel on Forests in 1995, sponsored by the UN Commission on Sustainable Development. After two years of discussions that failed to reach consensus on a number of key issues, the Intergovernmental Forum on Forests (IFF) succeeded the Intergovernmental Panel on Forests, but obstacles concerning trade, sovereignty, and aid continued to plague those discussions ('Report of the First Session of

the CSD Intergovernmental Panel of Forests' 1995; 'Summary of IPF-2 Geneva' 1998). The IFF process eventually evolved in October 2000 into the permanent United Nations Forum on Forests (UNFF), a subsidiary body of the UN Economic and Social Council with the mandate to carry on work to implement existing agreements and initiatives from the IFF and its predecessor. In addition, it would consider within five years the possibility of a mandate for negotiating a binding convention.

21. Pragmatic legitimacy is therefore not simply the same as material self-interest, because of its emphasis on acceptance of a rule or institution as appropriate. If a company acts simply out of coercion — that is, it does not accept the scheme but denounces it even while it feels compelled to go along to avoid costs — then pragmatic legitimacy has not been achieved. However, given that pragmatic legitimacy still largely rests on short-term cost benefit calculations — the same consequentialist logic as an interest-based argument that does not necessarily imply legitimacy — pragmatic legitimacy is seen as a relatively weak form.
22. The categories are modified from Suchman (1995), with 'converting' replacing his 'manipulation' category.

References

Auld, Graeme (2001). 'Explaining Certification Legitimacy: An Examination of Forest Sector Support for Forest Certification Programs in the United States Pacific Coast, the United Kingdom, and British Columbia, Canada'. M.Sc. thesis. Auburn University, Auburn, AL.

Auld, Graeme, Benjamin Cashore, and Deanna Newsom (2002). 'Perspectives on Forest Certification: A Survey Examining Differences among the U.S. Forest Sectors' Views of Their Forest Certification Alternatives'. In L. Teeter, B. Cashore and D. Zhang, eds., *Forest Policy for Private Forestry: Global and Regional Challenges*. CABI Publishing, Wallingford, UK.

Bernstein, Steven (2001). *The Compromise of Liberal Environmentalism*. Columbia University Press, New York.

Bernstein, Steven (2003). 'The Elusive Basis of Legitimacy in Global Governance: Three Conceptions'. Manuscript under review.

Bernstein, Steven and Benjamin Cashore (2000). 'Globalization, Four Paths of Internationalization, and Domestic Policy Change: The Case of Ecoforestry in British Columbia, Canada'. *Canadian Journal of Political Science* vol. 33, pp. 67–99.

Bourgeois, Bill (2002). 'Letter to FSC Canada and FSC-BC Regional Initiative'. Economic Chamber Member, Forest Stewardship Council, British Columbia Steering Committee, Vancouver.

Cashore, Benjamin (2002). 'Legitimacy and the Privatization of Environmental Governance: How Non-State Market-Driven (NSMD) Governance Systems Gain Rule Making Authority'. *Governance* vol. 15, no. 4, pp. 503–529.

Cashore, Benjamin, Graeme Auld, and Deanna Newsom (2003). 'Forest Certification (Eco-Labelling) Programs and Their Policy-Making Authority: Explaining Divergence among North American and European Case Studies'. *Forest Policy and Economics* vol. 5, no. 3, pp. 225–247.

Cashore, Benjamin, Graeme Auld, and Deanna Newsom, eds. (2004). *Governing through Markets: Forest Certification and the Emergence of Non-state Authority*. Yale University Press, New Haven.

Cashore, Benjamin and James Lawson (2003). 'Private Policy Networks and Sustainable Forestry Policy: Comparing Forest Certification Experiences in the U.S. Northeast and the Canadian Maritimes'. *Canadian American Public Policy* Spring.

Cashore, Benjamin and Emily Noah (2003). 'Developing a Multi-disciplinary Evaluation of an Environmental Policy Innovation: Impacts of Forest Certification within the United States'.

Paper presented at the annual meeting of the Association for Public Policy Analysis and Management, 6 November. Washington DC.

Chang, Seung Wha (1997). 'GATTing a Green Trade Barrier: Eco-Labelling and the WTO Agreement on Technical Barriers to Trade'. *Journal of World Trade* vol. 31, no. 1, pp. 137–159.

Chinkin, C.M. (1989). 'The Challenge of Soft Law: Development and Change in International Law'. *International and Comparative Law Quarterly* vol. 38, pp. 850–866.

Clapp, Jennifer (1998). 'The Privatization of Global Environment Governance: ISO 14000 and the Developing World'. *Global Governance* vol. 4, pp. 295–316.

Deutsche Gesellschaft für technische Zusammenarbeit GmbH Forest Certification Project (2000). 'Institutional Requirements for Forest Certification: A Manual for Stakeholders'. Working Paper 2, June.

Dimitrov, Radoslav S. (2003). 'Knowledge, Power, and Interests in Environmental Regime Formation'. *International Studies Quarterly* vol. 47, no. 1, pp. 123–150.

Domask, Joe (2003). 'From Boycotts to Partnership: NGOs, the Private Sector, and the World's Forests'. In J. P. Doh and H. Teegan, eds., *Globalization and the International Law of the Environment*. Praeger, New York.

Dupuy, Pierre-Marie (1991). 'Soft Law and the International Law of the Environment'. *Michigan Journal of International Law* vol. 12, pp. 420–435.

Elliott, Christopher (1999). 'Forest Certification: Analysis from a Policy Network Perspective'. Ph.D. diss. École Polytechnique Fédérale de Lausanne, Lausanne.

Esty, Daniel C. and Damien Geradin (1997). 'Market Access, Competitiveness, and Harmonization: Environmental Protection in Regional Trade Agreements'. *Harvard Environmental Law Review* vol. 21, no. 2, pp. 265–336.

Esty, Daniel C. and Damien Geradin (1998). 'Environmental Protection and International Competitiveness: A Conceptual Framework'. *International Studies Quarterly* vol. 32, no. 3, pp. 5–46.

Florini, Ann (1996). 'The Evolution of International Norms'. *International Studies Quarterly* vol. 40, no. 3, pp. 363–389.

Forest Stewardship Council (2000). 'FSC Principles and Criteria'. <www.fscoax.org/principal.htm> (April 2004).

Franck, Thomas M. (1990). *The Power of Legitimacy among Nations*. Oxford University Press, New York.

Gunningham, Neil, Peter Graborsky, and Darren Sinclair, eds. (1998). *Smart Regulation: Designing Environmental Policy*. Clarendon Press, Oxford.

Gunningham, Neil and Darren Sinclair (2002). *Leaders and Laggards: Next-Generation Environmental Regulation*. Greenleaf Publishing, Australia.

Hansen, Eric (1998). 'Certified Products Market Place'. In United Nations Timber Committee, ed., *Forest Products Annual Market Review*, pp. 17–28. United Nations Timber Committee, Geneva.

Haufler, Virginia (2001). *A Public Role for the Private Sector: Industry Self-Regulation in a Global Economy*. Carnegie Endowment for International Peace, Washington DC.

Hock, Teresa (2001). 'The Role of Eco-Labels in International Trade: Can Timber Certification Be Implemented as a Means to Slowing Deforestation?' *Colorado Journal of International Environmental Law and Policy* vol. 12, no. 2, pp. 347–365.

Humphreys, David (1996). 'The Global Politics of Forest Conservation since the UNCED'. *Environmental Politics* vol. 5, pp. 231–256.

Humphreys, David (1999). 'The Emerging Forests Regime'. *Global Environmental Change* vol. 9, no. 3, pp. 251–254.

Keohane, Robert and Joseph Nye (2001). 'The Club Model of Multilateral Cooperation and Problems of Democratic Legitimacy'. In R. B. Porter et al., eds., *Efficiency, Equity, and*

Legitimacy: The Multilateral Trading System at the Millennium, pp. 264–294. Brookings Institution Press, Washington DC.

Lawson, James and Benjamin Cashore (2001). 'Firm Choices on Sustainable Forestry Forest Certification: The Case of J.D. Irving, Ltd.' In L. Teeter and D. Zhang, eds., *Forest Policy for Private Forestry*. CABI Publishing, Wallingford, UK.

Lipschutz, Ronnie D. (2001). 'Why Is There No International Forestry Law? An Examination of International Forestry Regulation, Both Public and Private'. *UCLA Journal of Environmental Law and Policy* vol. 19, no. 1, pp. 153–180.

Meidinger, Erroll E. (1997). 'Look Who's Making the Rules: International Environmental Standard Setting by Non-governmental Organizations'. *Human Ecology Review* vol. 4, no. 1, pp. 52–54.

Meidinger, Errol E. (2003). 'Environmental Law for Global Civil Society: The Forest Certification Prototype'. In E. E. Meidinger, C. Elliott and G. Oesten, eds., *Social and Political Dimensions of Forest Certification*. Forstbuch, Remagen-Oberwinter.

Moffat, Andrea C. (1998). 'Forest Certification: An Examination of the Compatibility of the Canadian Standards Association and Forest Stewardship Council Systems in the Maritime Region'. MES, Environmental Studies, Dalhousie University, Halifax.

Newsom, Deanna (2001). 'Achieving Legitimacy? Exploring Competing Forest Certification Programs' Actions to Gain Forest Manager Support in the U.S. Southeast, Germany, and British Columbia, Canada'. M.Sc. thesis. Auburn University, Auburn, AL.

Newsom, Deanna, Benjamin Cashore, Graeme Auld, et al. (2002). 'Forest Certification in the Heart of Dixie: A Survey of Alabama Landowners'. In L. Teeter, B. Cashore and D. Zhang, eds., *Forest Policy for Private Forestry: Global and Regional Challenges*. CABI Publishing, Wallingford, UK.

Nordström, Håkan and Scott Vaughan (1999). *Trade and Environment*. World Trade Organization, Geneva.

Porter, Gareth and Janet Welsh Brown (1996). *Global Environmental Politics*. Westview Press, Boulder, CO.

Porter, Michael E. and Claas van der Linde (1995). 'Green and Competitive: Ending the Stalemate'. *Harvard Business Review* (September-October), pp. 120–138.

Rawls, John (1971). *A Theory of Justice*. Harvard University Press, Cambridge, MA.

'Report of the First Session of the CSD Intergovernmental Panel of Forests'. (1995). *Earth Negotiations Bulletin*, vol. 13, no. 45, 11–15 September.

Rosenau, James (1995). 'Governance in the Twenty-First Century'. *Global Governance* vol. 1, no. 1, pp. 13-43.

Rosenau, James (2003). *Distant Proximities: Dynamics Beyond Globalisation*. Princeton University Press, Princeton.

Ruggie, John G. (1982). 'International Regimes, Transactions, and Change: Embedded Liberalism in the Postwar Economic Order'. *International Organization*, vol. 36, no. 2, pp. 379–415.

Sands, Philippe, ed. (1994). *Greening International Law*. New Press, New York.

Sasser, Erika N. (2003). 'Gaining Leverage: NGO Influence on Certification Institutions in the Forest Products Sector'. In L. Teeter, B. Cashore, and D. Zhang, eds., *Forest Policy for Private Forestry*. CABI International, Wallingford, UK.

Steiner, Melanie (2001). 'After a Decade of Global Forest Negotiations, Where Are We Now?' *RECIEL* vol. 10, no. 1, pp. 98–105.

Suchman, Mark C. (1995). 'Managing Legitimacy: Strategic and Institutional Approaches'. *Academy of Management Review* vol. 20, no. 3, pp. 571–610.

'Summary of IPF-2 Geneva'. (1998). *Earth Negotiations Bulletin*, vol. 13, no. 45, 24 August–4 September.

United Nations (1992). 'Annex III: Non-legally Binding Authoritative Statement of Principles for a Global Consensus on the Management, Conservation, and Sustainable Development

of All Types of Forest'. Report of the United Nations Conference on Environment and Development, 3–14 June, Rio de Janeiro. <www.un.org/documents/ga/conf151/aconf15126-3annex3.htm> (April 2004).
United Nations Conference on Environment and Development (1992). 'Rio Declaration on Environment and Development'. <www.unep.org> (April 2004).
United Nations Framework Convention on Climate Change (2001a). 'Preparations for the First Session of the Conference of the Parties Serving as the Meeting of the Parties to the Kyoto Protocol (Decision 8/CP.4)'. FCCC/CP/2001/L.11/Rev.1. <unfccc.int/resource/docs/cop6secpart/l11r01.pdf> (April 2004).
United Nations Framework Convention on Climate Change (2001b). 'Report of the Conference of the Parties on its Seventh Session, Held at Marrakesh from 29 October to 10 November 2001'. FCCC/CP/2001/13.Add.1. <unfccc.int/resource/docs/cop7/13a01.pdf> (April 2004).
Wang, S. (2001). 'Towards an International Convention on Forests: Building Blocks Versus Stumbling Blocks'. *International Forestry Review* vol. 3, no. 4, pp. 251–264.
Ward, Halina (1997). 'Trade and Environment Issues in Voluntary Eco-labelling and Life Cycle Analysis'. *RECIEL* vol. 6, no. 2, pp. 139–147.
Williams, Marc (1993). 'Re-articulating the Third World Coalition: The Role of the Environmental Agenda'. *Third World Quarterly* vol. 14, pp. 7–29.
Winham, Gilbert R. (2003). 'International Regime Conflict in Trade and Environment: The Biosafety Protocol and the WTO'. *World Trade Review*.
Wofford, Carrie (2000). 'A Greener Future at the WTO: The Refinement of WTO Jurisprudence on Environmental Exceptions to GATT'. *Harvard Environmental Law Review* vol. 24, no. 2, pp. 563–592.
World Business Council for Sustainable Development (2003). 'Forest Certification Systems and the "Legitimacy" Thresholds Model (LTM)'. Discussion Paper. Geneva, 3 July.
World Trade Organization (1994). 'Agreement on Technical Barriers to Trade'. <www.wto.org/english/docs_e/legal_e/17-tbt_e.htm> (April 2004).
World Trade Organization (2000). 'Committee Tackles Market Access Issues'. *Trade and Environment Bulletin* vol. 32 (17 March).
Wynter, Marie (1999). 'The Use of Market Mechanisms in the Shrimp-Turtle Dispute: The WTO's Response'. In K. Bosselmann and B. J. Richardson, eds., *Environmental Justice and Market Mechanisms: Key Challenges for Environmental Law and Policy*, pp. 169–188. Kluwer Law International, The Hague.

Chapter 3

The Forest Stewardship Council: A Developing Country Perspective

Tasso Rezende de Azevedo

In 1993, Toronto witnessed the foundation of the Forest Stewardship Council (FSC), probably the most daring and effective civil society initiative for the promotion of good forest management in the world.

Since the former Woodworkers Alliance for Rainforest Protection (WARP)[1] began organising the first meetings in 1992, the FSC has faced many challenges and difficult times. For those participating in the 1993 Toronto meeting, there were many tense moments during the discussions, and the process was almost aborted three times. The final decision to establish the FSC was only reached at 4 a.m.

By October 2001, almost a decade later, the FSC had achieved an impressive record: almost 25 million hectares of forests certified in 40 countries and more than 1500 companies producing and selling certified products.[2] More than 800 enterprises had joined buyers groups committed to producing, promoting, and marketing products certified according to FSC standards.

An analysis of the FSC-certified areas in 2001 shows that 86 percent were located in more developed regions (Europe, Canada, and the United States) and only 14 percent in developing countries (Africa, Latin America, the Pacific, and Southeast Asia).[3] However, the great majority of certified areas are managed by corporations, with less than 10 percent managed by communities.

In Brazil, the corresponding figures also show a marked imbalance: 70 percent percent of all certified forests are made up by plantations that are managed by vertically integrated industrial corporations and only 40 percent are natural forests.[4]

In the absence of any deeper analysis of context, these figures have led many authors to label the FSC certification system as discriminatory and ineffective in promoting forest management and market access to small forest producers and enterprises in developing countries. However, a more detailed examination of the history of certification and its impact reveals that certification has been the driving force behind many changes that are occurring in forest management throughout the world, helping the global forest sector to advance toward a new reality that is much closer to sustainability.

This chapter will demonstrate that certification has been a powerful catalyst of good forest management practices in Brazil by way of an analysis of the history of FSC forest certification in that country, as well its impacts, influence, and limitations. Limitations nonetheless remain, and must be eliminated through complementary mechanisms.

The Brazilian Context

Brazil has about 169.7 million inhabitants distributed over a geographic area of 8.5 million square kilometres, which is larger than the continental U.S. More than 4.5 million square kilometres are forest, including 3.5 million square kilometres of Amazon forest covering nine Brazilian states, from Acre to Maranhão, the largest tropical forest on the planet. In terms of forest management, the two main forest formations in Brazil are the Amazon Tropical Rainforest and the Atlantic Tropical and Subtropical Forests and the dry forests in Cerrado and Caatinga.

The Amazon Tropical Rainforest represents more than 90 percent of the country's production in natural forests. In addition, it supplies more than 200 non-timber forest products, such as latex obtained from *Hevea brasiliensis*, Brazil nuts, special oils (such as copaíba, amapá, and andiroba), frozen fruit pulp, palm heart, and many plants used in the preparation of medicine and cosmetics. More than 4000 primary processing industries operate in the region. Three states — Pará, Mato Grosso, and Rondônia — produce more than 78 percent of all Amazon lumber and veneer (Smeraldi and de Oliveira Veríssimo 1999). There is a much demand for deforestation in the Amazon region to make land available for agriculture, especially in southern Pará, northern Mato Grosso, Tocantins, and Rondônia. During the dry season, from June to October, the owners of small and large farms use fire to clear the land, destroying much forest. In 2000 alone, more than 19 000 square kilometres were deforested in the region. Current legislation, which has been in effect for many years, requires that at least 80 percent of the natural forest cover of individual rural properties be maintained. However, there is constant pressure to reduce this limit to 50 percent, even as low as 20 percent.

The Atlantic Tropical and Subtropical Forests initially covered 1.3 million square kilometres, along the Brazilian coast from Rio Grande Norte to Rio Grande do Sul. Today only 7 percent of this forest remains (SOS Mata Atlântica 2001). A large part of the 6 million hectares of forest plantations in Brazil today are located in this part of the country. Until the 1980s, this region supplied most of the logs to Brazilian forest industries: Imbúia (Brazilian walnut), Jacarandá da Bahia (rosewood), Ipê (Pau Lope), and Paraná pine were the main commercial species. Today, except for a few small operations in Santa Catarina and in southern Bahia, industrial processing of wood from the Atlantic forest is prohibited. Since this region is heavily populated, with 70 percent of the country's population, it suffers great pressure from predatory extractivism (such as palm heart and tree ferns) and conversion of land for agricultural purposes and urban development. Economic alternatives for managing the forest, such as non-timber forest products and eco-tourism, have been promoted by different organisations that work in the region.

Cerrado and Caatinga are two forest ecosystems present in the central and northeast of Brazil. They represent 23 percent of the forests, characterised by drier and lower forest formation, which are important for the production of biomass for energy and non-timber forest products. In the northeast region, the biomass coming from Caatinga is responsible for more than 30 percent of the energy supply.

Other forest formations such as Pantanal — the lowlands of the Paraguay River system in the west-central region — and the Pampas — the savannahs in the extreme southern region — are less important in terms of forest production, but are nonetheless essential for the conservation of water resources and biodiversity.

About 19 percent of Brazilian forests is on land set aside for indigenous populations and less than 10 percent is in conservation areas, including parks and forest reserves.

In 1998, the category of Reservas Particulares do Patrimônio Natural (RPPN; Privately Owned Natural Conservation Areas) was approved for forest conservation by legislation. This designation allows private owners to maintain private conservation areas that become an integral part of the national system of conservation units. The distinction is important because the structure of land ownership in Brazil is based on private property. A consequence of the Portuguese colonization, this system led to the concentration of large forest areas held by private companies and individuals who later promoted their subdivision and sale as part of colonization plans. In contrast to most of the other countries in the Amazon region, a large portion of the area available for forest management in Brazil is in the hands of private interests.

However, the process of illegal occupation of government land (*grilagem*) led the Instituto Nacional de Colonização e Reforma Agrária (INCRA; National Institute for Colonization and Agrarian Reform) in 1999 to start revising all land titles of properties larger than 10 000 hectares. Consequently, in 2001 the Ministry for Agrarian Development and INCRA jointly announced that more than 20 million hectares in the Amazon region had been reclaimed by the government due to improper title documentation. Since 2002, a good portion of these lands is becoming public production forest.

Approximately 1.15 million square kilometres of areas could be placed under forest management thanks to an number of adequate attributes: these areas are not in protected zones, they present commercial forest cover, they are in sparsely populated areas, and they are located near wood-processing industries (Schneider et al. 2000). Approximately 700 000 square kilometres of managed forests — 14 percent of the legal Amazon total surface area — could permanently supply the region's log demands.

In the Amazon, public production forests amount to approximately 83 000 square kilometres (8.3 million hectares), which represent 1.6 percent of the total area of the region. However, there are neither forest concessions nor commercial logging in this area. The first forest concession, in the form of a pilot project involving 5000 hectares of the Flona Tapajós (Tapajós National Forest), occurred in 2000. Brazil's National Forest Programme, launched in 2000, included a plan to start establishing concessions of national forests in 2002, with more still to be created; forest certification was specified as one criterion to be considered in the analysis and selection of concessionaires.

Legislation and Governance

The Brazilian legislation that governs forest management is based on four general fundamentals:

- the forest code that defines permanent preservation areas, the limits of conversion of forest areas into other land uses, and the requirement of an approved management plan before logging operations begin;
- the instructions and regulations set forth by the Instituto Brasileiro do Meio Ambiente e dos Recursos Naturais Renováveis (IBAMA; Brazilian Institute of the Environment) for management plans and forest operations;
- environmental licensing carried out by the state environmental authorities; and
- the environmental law that classifies the various crimes against the environment and sets the respective penalties.

Traditionally, the Brazilian forest legislation has discouraged forest activities and encouraged the conversion of forests into other land uses. One of the main reasons for this is that it recognises some actions carried out by the farmer, such as building roads and clearing trails, construction, and making clearings for cattle or agricultural crops, as proof of the farmer's right to the land. However, the requirements and red tape involved in getting authorisation for implementing a forest management plan are much more complicated than those for clearing the forest.

Put simply, the official structure governing forest activities consists of the following:

- The Comisión Nacional del Medio Ambiente (CONAMA; Brazilian National Environment Council) defines global environmental policies for the country. Its members are representatives of the public and private sectors and of society at large, including representatives from many nongovernmental organisations (NGOs).
- IBAMA evaluates and approves management plans and also monitoring them. It is responsible for all activities related to conservation areas, coastal and inland waters, licensing of new construction, and many other issues related to environmental matters.
- State environmental authorities are in charge of environmental licensing and, in some states, carry out activities normally the responsibility of IBAMA, such as the approval and monitoring of forest management plans.

Historically, control measures applied by IBAMA and by environmental authorities of individual states on forest activities have not been very efficient or effective. For example, IBAMA ordered an independent assessment of more that 3000 management plans approved for the Amazon region, which found that less than 15 percent of these plans were sufficient to allow forest activities to continue as planned. These plans had been formally approved by IBAMA and, in theory, were completely valid. Based on this outcome, IBAMA is undergoing reforms to improve efficiency. The creation of an independent forest agency, focussed exclusively on managing the country's forest resources, is also being considered.

Geography of Wood Production for Brazilian Industries

Forest production in Brazil consists basically of two types: plantations in the south and south-east, and logging of natural forests, especially in the Amazon region.

Although there is no accurate information, it is estimated that forest plantations cover about 5 million hectares, mainly pine and eucalyptus, with some other species such as teak, pine, and acacia planted on a smaller scale. These forests are located mainly in the southern part of the country, although there are also plantations in the states of Bahia, Mato Grosso, Pará, Amapá, and Maranhão. They supply almost all the raw material used by the pulp and paper and reconstituted panel industries, totalling about 25 million cubic metres per year, 70 percent of the wood used as charcoal by the steel industry, and 40 percent of all industrial round wood consumed in Brazil, mainly for the production of lumber and plywood.[5] The use of plantation wood for lumber production has been increasing steadily since the early 1990s.

The pulp and paper sector, as well as the reconstituted panel industry, requires large capital investments in equipment. These industries are vertically integrated and depend mostly on their own plantations for raw material. Plantations generally cover tracts of land as large as 10 000 hectares, but tree-planting programmes, in partnership with small landowners, are often conducted in industrially controlled areas.

The lumber industry, furniture manufacturers, and other small-scale wood-processing industries obtain their raw material from their own plantations, which are much smaller — 500 to 5000 hectares — and also buy wood from independent producers.

In the Amazon natural forests, approximately 28 million cubic metres of logs are processed each year, mainly for the production of hardwood lumber, veneer, and plywood (Smeraldi and de Oliveira Veríssimo 1999).

The states of Pará and Mato Grosso are the main producers, with 11.9 and 9.8 million cubic metres per year respectively. Approximately 86 percent of all wood produced in the Amazon region is used by the domestic market, and the state of São Paulo is the largest consumer. The industries now operating in the Amazon have migrated from the southern and south-eastern regions where they had large-scale operations in the period of 1950 to 1980, before exhausting their raw material sources.

Forest Certification in Brazil

From Boycotting Tropical Products to Establishing the Forest Stewardship Council

In the late 1980s, products from tropical forests were being targeted by boycotts in the United States and Europe. Many governments and corporations stopped buying tropical forest products because they were alleged to cause deforestation in their place of origin. In 1992, more than 30 towns in the United Kingdom, 200 towns in Germany, and at least half of the counties in Holland had banned the use of tropical timber, using the logic that, with no demand for its products, tropical forests would be automatically protected.

Brazilian products were a particular target. The impact of these boycotts on the rate of deforestation in tropical countries was negligible. In Brazil, for example, deforestation rates continued to increase throughout the 1990s, for several reasons. First, more than 85 percent of all wood production in the tropics is absorbed by the domestic market. Second, a forest owner deprived of the income from selling wood looks to other ways to generate cash — resulting almost always in clearing forest to create land for pasture and agricultural crops. Third, logging was not itself the direct cause of deforestation, although it has some effect insofar as forest roads and trails built on unpopulated areas can provide access to settlers.

Nonetheless, boycotting affected the supply of tropical wood to consumers in countries such as the United States and the United Kingdom. In New England, U.S. artisans who produced furniture and small wooden objects, and who specifically needed tropical species for their work, began to fear for the survival of their craft. They established the Woodworkers Alliance for Rainforest Protection and, in partnership with the Rainforest Alliance, an NGO based in New York, worked toward identifying forest projects around the world where logging did not destroy forests. Starting with these efforts, WARP began publishing a list of wood producers that became known as the 'Good Wood List'. This list grew rapidly and expanded to include sources outside the tropics. Soon, questions were raised about the criteria to be followed by producers who wanted to be included in the list and how these criteria could be verified.

Between 1991 and 1992, a series of meetings involving NGOs, wood producers, and consumers were held in Washington, Honduras, and San Francisco, among other places, with the objective of preparing a proposal for establishing standards for good forest management and for the creation of an international organisation to accredit producers. Those were the first steps toward the foundation of the Forest Stewardship Council.

Brazil Enters the Discussions

An international consultation process was conducted between 1991 and 1993 in order to establish forest certification standards. As part of this process, specific surveys were carried out in ten countries, including Brazil. Early in 1993, four sectoral meetings were organised involving forest industries based on plantations, those based on natural forests, and environmental and social NGOs. In addition, a broad postal survey was carried out in order to evaluate the possibility of establishing an international system of forest certification and to solicit contributions to the first draft of international standards for good forest management. As a result of this consultation process, two informal working groups were established, one involving the social and environmental sectors and the other the economic sector — that is, forest enterprises.

When the meeting to found the FSC was held in Toronto in 1993, more than ten Brazilian delegates attended — more than the delegations from any country other than the host, Canada. Two Brazilian delegates were elected to the first board of directors: Amantino Ramos de Freitas, representing the Economic Chamber/South, and Juan Carlos Rueda, representing the Social Chamber/South. In 1998, Sandra Faillace was elected to the FSC board.

Between 1994 and 1996, there were some attempts to create an informal working group to implement a national initiative to adapt FSC standards for Brazilian conditions. But it was only in 1997 that this initiative became a reality, when the FSC Working Group was established in Brazil. This group had 18 participants from the three chambers representing the environment, social, and economic sectors and an executive secretariat located in Brasília.

The working group decided that national standards would be prepared according to the main types of ecosystem and forest management, starting with forest management of plantations and Amazon dryland forests. From 1997 to 2000, three series of consultations and workshops were organised to draft these standards. After public meetings and field tests had been carried out, the standards were approved by the working group in January 2001 and then, in October 2001, by the FSC's board of directors in Oaxaca.

The Brazilian Forest Management Council — an NGO that transformed the previous informal working group into an organisation officially established according to the Brazilian laws — was founded on 18 September 2001. More than 60 participants, representing NGOs, forest corporations, social movements, federal and state governments, universities and research institutions, and trade associations were the original members of this new organisation.

The work continues with the development of standards for certifying Atlantic coast forests (Mata Atlântica) and Brazil nuts produced in the Amazon. Recent activities have included setting standards for the cerrado, which are the dry, open forests of the Brazilian central plateau, and for the flood plain forests of the Amazon region.

The Evolution of Forest Certification in Brazil

The first forest certification initiatives in Brazil started in 1993 with assessments of the forest operations of Amacol, in the state of Pará, and Duratex, in the state of São Paulo. The first FSC certifications occurred in 1995; since then, the area of certified forests in Brazil has grown significantly.

In October 2001, there were 875 000 hectares of certified forests in the country, with 15 certified areas in the seven states of Amazonas, Mato Grosso, Minas Gerais, Paraná, Rio Grande do Sul, Santa Catarina, and São Paulo.[6] Plantations represented the largest share of certified forests, with 590 000 hectares, about 10 percent of all plantations in Brazil, having been certified and another 330 000 hectares undergoing certification.

There are significant advances in the certification of natural forests as well, but these take place at a much slower pace. It takes drastic changes in attitude for the forest industry to adapt to the standards that ensure an environmentally sound and socially fair forest management. Until 2000, only one area was certified according to FSC standards — Mil Madeireira Ltda., a subsidiary of the Precious Wood Swiss Group that had been certified in 1997 and was located in Itacoatiara. In 2001, four additional areas were certified, three in the state of Pará and one in the state of Amazon.

Two certified companies in Brazil involve non-timber forest products, among the first ones in the world. Muaná Alimentos, located in Muaná in the state of Pará, manages 4000 hectares of forest to produce palm heart and fruit pulp from the açaí palm. Klabin Produtos Florestais, located in Telêmaco Borba in the state of Paraná — home of the largest pulp and paper industrial complex in Latin America — produces cosmetics and medicine from approximately 40 plants that grow in the certified areas of the company. This type of product line was the first in the world to receive FSC certification.

In addition to the 15 certified enterprises listed in Table 3-1, 77 companies in Brazil have received chain of custody certification (CoC)[7] and can therefore label their products as certified. From only one in 1997, by the end of 2001 there were 80 manufacturers with CoC certification, an increment of 65 percent in the previous 12 months (see

Table 3-1 Certified Forests in Brazil (October 2001)

Company	State	Area (ha)
Planted Forests		
Floresteca Agroflorestal	Mato Grosso	8519
Duratex	São Paulo	47 904
Eucatex	São Paulo	48 962
Klabin do Paraná	Paraná	221 000
Plantar S/A	Minas Gerais	9 420
Mannesman Florestal	Minas Gerais	235 886
Faber Castell	Minas Gerais	8987
Flosul de Madeiras	Rio Grande do Sul	6764
Juliana Florestal	Santa Catarina	4143
Rohden Artefatos de Madeira	Santa Catarina	1481
Natural Forests		
Mil Madeireira	Amazonas	80 571
Gethal Amazonas	Amazonas	40 862
Muaná Alimentos	Pará	4012
Cikel Brasil Verde	Pará	140 658
Juruá Florestal Ltda.	Pará	12 000

Source: Forest Stewardship Council (2001).

Figure 3-1). These companies process between 200 000 and 300 000 cubic metres of lumber per year to produce more than 100 different products, including furniture, flooring, hammocks, kitchen woodenware, musical instruments, and even jewellery.

Developing the Market for Certified Products in Brazil

Except for the pulp and paper sector, which exports more than 40 percent of its production[8] — Brazil is the world's largest exporter of short-fibre pulp — most of the country's forest production goes to the domestic market.

Brazil is the world's largest producer and consumer of tropical forest products. About 86 percent of all wood produced in the Amazon region is consumed by the domestic market (Smeraldi and de Oliveira Veríssimo 1999). The state of São Paulo alone uses more tropical wood than all the countries that make up the European Community. Similarly, with regard to most non-timber forest products, 80 percent of palm heart production stays in the country.[9] The figures are about the same for essential oils and products obtained from medicinal plants.

Therefore, because it is a marketing tool that encourages good forest management, certification can only be effective in Brazil when Brazilian consumers can be convinced to buy certified products.

Figure 3-1 Evolution of Forest Management and Chain of Custody Certification in Brazil

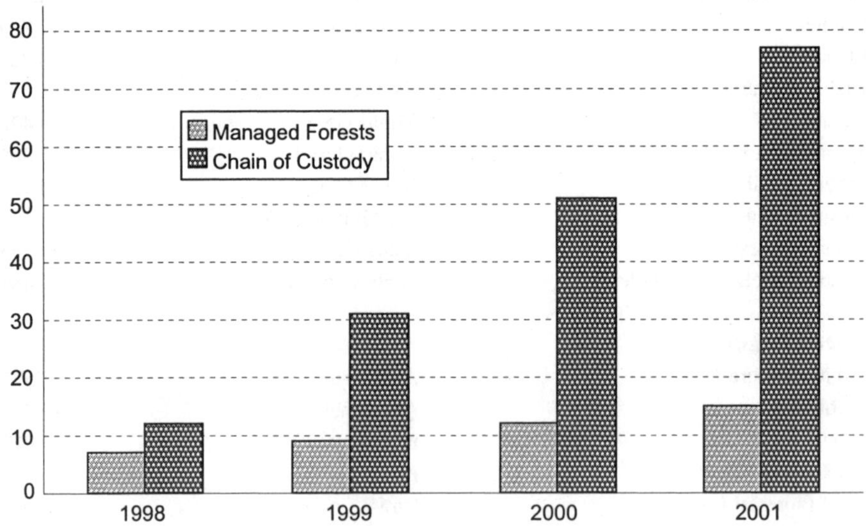

Source: Imaflora data based on FSC list of certified operations.

Interest in Certified Products

A 1998 market survey by the Brazilian Institute for Public Opinion and Statistics showed that 68 percent of the population considered deforestation to be the biggest environmental problem in the country. This is particularly remarkable in light of the fact that 70 percent of the population lives in urban areas, where there are problems with the basic infrastructure such as water supply and garbage and sewage treatment and disposal.

In 2001, Instituto de Manejo e Certificação Florestal e Agrícola (Imaflora; Brazilian Institute for Agricultural and Forest Management and Certification), the Instituto do Homem e Meio Ambiente da Amazônia (Imazon; Amazon Institute for Man and the Environment), and Friends of the Earth (FOE) conducted a joint study of 1100 companies that use or sell wood in the state of São Paulo, including lumber yards, processing industries, and construction companies. Among other findings, this study estimated the size of the market that could respond to or be affected by certification.

As happens in many other countries, in Brazil the promotion of certification is limited by the availability of certified wood, which, in turn, is limited by the existence of certified forests. However, the lack of a clear option of the market for certified products is also a limiting factor for operations that choose to certify their forests. This 'chicken and the egg' situation hinders the organisation of more elaborate campaigns to educate consumers. There is no sense in promoting a product that is not readily available. To resolve this situation, institutions concerned with promoting certification work closely with potential agents of change of consumer opinion, such as architects, designers, and secondary processing industries.

Buyers Groups

In 1996, Imaflora interviewed 20 renowned architects and designers in the state of São Paulo about certification. The results showed that 100 percent were ready to change their wood suppliers if they could get wood from guaranteed sustainable sources, and 90 percent agreed to pay between 5 percent and 20 percent extra for this guarantee. Although market intention studies are uncertain and controversial, given that they deal with decisions to be taken in the future, these results are comparable to similar studies carried out in European countries.

Based on these findings, in 1997–99 Imaflora, Imazon, and FOE's Brazilian Amazon Programme developed a series of bilateral agreements with wood industry and trade companies, with the objective of ensuring their commitment to certified forests. The first of these commitments was signed by Tok & Stok, the largest retailer of furniture and household items in Brazil.

In November 1999, wood sellers and buyers met in Manaus to discuss the demand and availability of certified products. Buyers were able to confirm the existence of good forest management projects with certification potential, and producers recognised that there was real demand for certified products. One of the results of this meeting was the decision to create a certified products buyers group, which was established in 2000 and co-ordinated by FOE. This group, which promotes the preferential use of certified products, includes 64 companies of various sectors (furniture, flooring, and prefabricated housing manufacturers, publishers, retail stores, charcoal producers), two townships (Guarujá and Niteroi), and two state governments (Amapá and Acre).

Depending on the type of product or raw material used, different types of commitments were signed to facilitate the transition to certified products. In the case of charcoal and plantation wood, for example, participants agreed to aim for 100 percent certified supplies by 2005. In the case of wood from natural forests, such as those in the Amazon region, the goal was 50 percent by 2005. In all cases, participating companies announced that they would gradually replace suppliers that showed no interest in certification. The companies that took part in the buyers group until end of 2001 had a combined demand for 1 million cubic metres of wood per year, a volume not yet reached by certified forests in Brazil.

Although the amount of wood exported is much lower than the amount used domestically, exports represent an important source of income to forest industries, especially when the dollar is strong compared to the local currency. Consequently, because the export market requires certified wood, many companies are encouraged to seek certification. Since for every cubic metre of export-quality wood about one or two cubic metres go to the domestic market, in the long run the demands of the international market will affect the domestic availability of certified raw material positively.

Non-Timber Forest Products and the Cosmetics Industry

Because the Brazilian market for non-timber forest products is large, many companies — especially in the area of cosmetics and phyto-therapeutic products — use such raw materials in their production. By and large, however, these companies have little or no knowledge about the origin of the raw material. Usually they receive it from suppliers specialised in producing fragrances, refined essential oils, and concentrates at specified strengths. The production chain between the cosmetics industry and its source of raw materials is long, often passing through six to eight intermediaries.

On the other hand, non-timber forest products are often produced by people in local communities, such as riverside dwellers, indigenous peoples, coastal fishers, and *quilombolas*, who use traditional knowledge that adds social value to the raw material.

A few significant companies in this sector are carefully considering their sources of raw materials in order to determine whether they are environmentally sound and socially fair, perhaps to promote social responsibility or just to take advantage of a market opportunity. Two examples are worth mentioning: Natura and Klabin Forest Products.

Natura is Brazil's largest manufacturer of cosmetics, with a sales force of more than 260 000. In 2000, it launched a new product line based on the biodiversity of the Brazilian flora. In order to ensure the sustainability of its raw material sources and the well-being of the communities involved in its production, Natura made a commitment to seek and encourage certification according to FSC standards. This involvement has made suppliers more aware of the components in their production chain, and they are encouraging their own sources to seek certification.

Klabin Forest Products is a subsidiary of Indústrias Klabin de Papel e Celulose (IKPC), the largest producer of pulp and paper in the southern hemisphere. It is the result of a health programme developed by IKPC to provide medicine based on local flora to its employees and neighbouring communities of Telêmaco Borba in the southern state of Paraná. The programme has been expanded to serve a population of 80 000 people, with a success rate of 95 percent. A large portion of plant medicine production of Klabin Forest Products is based on raw forest materials obtained from the IKPC's natural forests, which cover approximately 80 000 hectares, and their plantations, covering 130 000 hectares, which were certified in 1998. In 2001, when Klabin Forest Products decided to introduce its products in the domestic market and abroad, it obtained CoC certification and thus became the first enterprise in the world to hold FSC certification for plant medicine and cosmetic products.

Why Do Forest Enterprises Seek Certification?

There are many motivations for forest-based industries to seek FSC certification. One reason is that it opens up access to new or current markets without losing traditional markets. Certified products appeal to a special kind of consumers who use only certified products. In the international arena, this group is primarily represented by buyers groups, in the United Kingdom, Belgium, Germany, and the Netherlands. Within a pre-determined time frame, these groups, which include industries, trading companies, and retailers, commit themselves to buying 100 percent of their forest products from certified sources.

Certified products also draw a higher price, which is probably a circumstantial benefit caused by a strong demand not adequately met by the market. Nonetheless, it has been a significant factor for companies that were pioneers in obtaining certification. Some have benefited from premiums of up to 40 percent for their certified products.

Furthermore, it is expensive to introduce new species and products in the market, but buyers are more willing to try them when they come from certified forests. There will be greater demand for certified products, and sellers do not want to invest time and money on products that may be short lived. In other words, certified products minimise risks.

Another motivation for community forest operations or those with limited capital and resources for marketing is the common identity carried by the FSC label. Through the label, these products are aligned with those of other certified enterprises, or other producers already well established in the marketplace. As a result, producers previously excluded from the market can sell their certified production in more favourable conditions.

Certification also provides access to capital investment from green funds, a significant amount of financial resources channelled to forest operations by investors who base their decisions on the social and environmental performance of the project. Certification becomes an indicator of such performance. In at least three certified enterprises in Brazil, certification had a direct bearing on the decision to make new investment capital available to consolidate and expand forest operations.

Certification has been proven to improve the corporate image of an enterprise, particularly within local communities, and to attract a great number of visitors. Although this may result in added costs, it also brings valuable gains to the company's public image.

Because certification requires long-term commitments to forest management, it brings the possibility of long-term business and a reduced role for intermediaries. This commitment encourages the buyer to establish an extended relationship with the supplier, which in turn increases the direct contact between the two parties, thus eliminating many of the intermediaries that usually take part in the commercial transactions and, consequently, lowering costs. The CoC certification reveals the origin of the wood, so it is easy for that direct contact to be made. Prefabricating housing manufacturers and contractors, in particular, have found this direct contact to be a primary benefit, since they often suffer from unstable sources of wood.

At the policy level, government environmental authorities have begun to use certification in analysing requests for changes in management plans and technical proposals for silvicultural treatments, species selection, and diameters, and even for export.

Limitations to Certification Growth

Despite these many motivations for companies to obtain certification and the fact that market demand for certified products is strong, the number of certified forests in Brazil remains small and highly concentrated in large plantations. Although in some specific cases, the cost of certification is a limiting factor, the primary impediment is the social, environmental, and economic performance of forest management itself.

Large reforestation projects of 10 000 hectares or more are usually managed by vertically integrated enterprises that have a high degree of technological expertise and require enormous volumes of wood, pulp and paper, reconstituted panels, and steel. Between 1967 and 1987, these industries benefited from a generous fiscal incentive programme, but because they must now comply with the requirements of the Programa Integrado Floresta-Indústria (Integrated Industry-Forest Plan for Raw Material Supply), they have developed a solid technology for forest production. This sector is the best organised of all forest-based enterprises in the country. It sees forest certification as a mix of opportunity and challenge. It functions as a bloc and can adapt relatively quickly to any change in the marketplace, often using formal structures and procedures implemented to comply with ISO 9000 and 14000. This sector also has the financial resources required to cover the direct and indirect costs of certification.

Although the number of certified companies is small and the certification process can take as long as one year, at least half of the large pulp and paper and reconstituted panel industries have already obtained FSC certification or are in the process of doing so. However, these companies face several obstacles in obtaining certification, including the use of agrochemicals, differing attitudes toward subcontracted workers and their own workers, conflicts over land tenure rights, and the use of genetically modified organisms.

With regard to small and medium-sized forest plantations, the limiting factor is usually the low social and environmental performance of the companies. Normally, forest activities lack an adequate degree of planning. In addition, unsafe and unhealthy working conditions are common, as are the degradation of soil resources and the deterioration of permanent preservation areas.

When such plantations are part of vertically integrated industrial complexes, they are usually related to industries that are not self-sufficient in raw material. For this reason, they are considered to be storage of standing timber and not fundamental to the sustainability of the industry. Otherwise, they sell their wood at low prices, with small profit margins not conducive to effecting change in management standards.

However, small and medium-sized forest plantations support a good portion of southern Brazil's furniture industry, which is especially geared toward international markets — exactly those markets that highly value certification. Although this industry gets some of its wood directly from large certified plantations, in order to keep its markets on the long run, it must certify its own forests and those of its smaller suppliers. Consequently, a series of good forest management initiatives have been launched in southern Brazil. Indeed, certification in such cases constitutes a valuable reward to significant changes in the focus and attitude on the part of many enterprises.

In 2001, only five areas of natural forest were under certified management in Brazil.[10] Only four out of the 4000 saw mills operating in the Amazon region used raw material from certified sources. Certification processes in this region have been quite long and currently there are more than 20 management projects getting ready to be certified.

In the case of natural forests, many factors contribute to the low quality of management, including insufficient knowledge of forest management techniques, complex changes required in high-impact operations, training, uncertainties about land tenure rights, the nature and impact of the relationship between the forest enterprise and local communities, bureaucracy and corruption, and access to lines of credit. In fact, all these factors are related to the lack of an adequate forest culture in Brazil.

Community forest management began in the mid 1990s. Today there are more than 20 initiatives dealing with wood products, with fewer than half operating commercially, although almost all are undergoing certification. Some projects have serious problems related to their social organisation, market access, and training and capacity building. In many such cases, forest certification is not the best way to promote good forest management. Other options, such as fair trade, may prove to be important tools in market promotion, paving the way for future certification.

The Impact and Influence of Certification

Certification has had significant impacts on how forests are managed in Brazil. These effects may be classified as direct and indirect. Direct impacts are those related specifically to the scope of certified forest operations; indirect impacts are those affecting the forest sector as a whole, including public policies and credit concession.

Direct Impacts of Certification

Certification has a direct impact on working conditions in the forestry sector. Logging activities are among the most dangerous and unhealthy of many industrial sectors. Normally, people working in this sector do not have the professional qualifications to find better jobs. However, the requirements for improving working conditions in order to obtain certification have changed the way companies treat their workers.

One example of how conditions have improved can be seen in logging camps. Usually, workers sleep onsite in flimsy tents made of black polyethylene, without adequate sanitation or a proper place to eat their meals. In certified operations, logging camps are equipped with proper sleeping quarters, toilets, eating areas, an office, a first-aid room, and a leisure area, perhaps also a television or sports facilities. In Manicoré, Gethal Amazonas provided workers with portable camp structures that could be pulled by a motor vehicle from one area to another, thus reducing the time needed to move crews through the forest and get them ready to work.

In terms of safety, until 1995 the use of personal protective equipment in forest activities was limited in Brazil to a few plantation-based industries, as a result of organised pressure from labour unions and the Ministry of Labour. In the Amazon region, it was almost impossible to find a single worker who wore protective gear. Forest entrepreneurs argued that there was no use in providing it to workers because safety trousers and boots and hard hats were too uncomfortable in the tropical climate, and that ear protectors hindered harvesting operations. So, with the complicity of the workers themselves, things never changed.

However, certification requires safe working conditions. All workers must wear protective equipment supplied by the company in good condition and adequate to the specific activities. Workers must be properly trained to carry out forest activities safely and efficiently. Certified companies usually organise internal safety committees to promote a safe work environment. In addition, they offer medical and hospital assistance to their workers and provide regular medical check-ups.

Until 1996, the municipality of Itacoatiara, in the state of Amazon, had the highest rate of fatal, serious accidents related to forest activities in Latin America. Every year, two or three people died working for Madeireira Itacoatiara Ltda., the principal local forest company. Since its certification in 1997, there has been just one fatal accident, caused by the fall of a dead tree in 2000. This event was so unusual that employees mourned for two days, and on the accident's first anniversary all activities were halted for half a day to pay tribute to the deceased worker and to discuss safety-related issues.

Another area of improvement under certification is job stability and formal work contracts. Forest activities, especially in natural forests, are seasonal: work must be interrupted during the rainy season. In the Amazon region, logging takes place from June to December, a period locally known as summer, because there is less rainfall than in the rest of the year, known as winter. Consequently, employees are overworked during summer and lose their jobs during the winter. Because forest companies could never count on recruiting the same workers every year, there was no incentive for investing on training and capacity-building programmes. In order to avoid paying overtime, employers paid according to production and never formalised work papers, leaving workers without many benefits, such as unemployment compensation, paid vacation, and the '13th salary' (one month's pay paid annually or proportional number of months worked, and mandatory in Brazil).

However, because certification requires good working conditions, training programmes, and formal work contracts, the companies worked with labour unions and the Ministry of Labour to establish a compensation mechanism called the 'bank of hours', a concept similar to flex-time. During the dry season, workers work one additional hour every day. This extra time accrues during the summer and allows for the continuity of the work contract during the months when forest activities stop. All workers are legally hired, with work contracts properly specified in their work papers. In fact, this situation benefits everyone: workers have job stability and guaranteed income throughout the year, and employers have lower staffing costs and can invest in improving the technical capacity of their employees. Since there is less turnover, workers become more skilled as they stay for longer periods in the company. As a result, there are significant gains in efficiency over the years. Two good examples of this positive impact of certification are Mil Madeireira and Gethal, the first two organisations certified in the Amazon region; the productivity level of the workers at both companies increased by at least 20 percent in the first two years after certification. The concept of the bank of hours is now being adopted by several companies operating in the Amazon, and even by agricultural companies that deal with seasonal activities.

Another benefit is the stability extended to the families of workers who join the bank of hours, as exemplified by the forest workers union of Itacoatiara. One indicator of this fact is many workers decide to build a house for their families, an initiative encouraged by the forest enterprises.

FSC considers that certification principles and criteria apply equally to all workers carrying out activities in the forest management unit, be they hired directly or by another company. This way, differences in treatment between permanent workers and the subcontractor's workers must be minimized. Subcontracting is a recent trend in plantations in southern Brazil for harvesting and maintenance, and in general has made working conditions worse in uncertified enterprises: the worker ends up losing benefits, especially long-term job stability, because the company considers itself responsible only for its own workers.

For example, in 1997 Klabin do Paraná had excellent working conditions for all its workers, and those for the subcontractors' workers were not as good. To improve this situation, a series of indicators were established, such as the use of protective equipment, the quality of food and transportation, work contracts, quality of logging camps, and the availability of medical assistance. These indicators were used to bring subcontractors in full compliance with the requirements for certification. The results obtained through these indicators were presented to the labour union and to the certifier. By 2000, the difference in treatment between the two types of workers had decreased considerably.

The Impact on Forest Operations

It is well recognised that forest activities have a significant impact on the forest, especially the activities involved in logging. As a rule, insufficient planning is the cause of such effects. Some certification requirements have led to a number of interesting improvements in forest operations in Brazil.

For example, in the Amazon region, forest management plans are often works of pure science fiction. Many are prepared in comfortable air-conditioned offices far away from where the action is, sometimes without even a single visit to the forest. Often, they contain gross miscalculations that raise cutting volumes to absurd levels. Several consulting firms have specialised in producing management plans and getting them approved by the proper authorities. Some essential preparatory activities, such as a full inventory of all the commercial trees, were considered unfeasible and never carried out. Management plans were filed away in the offices, while logging activities carried out without any operational plan. Not surprisingly, this lack of planning had an enormous impact on harvesting, which takes longer for the forest to recover.

To get certified, however, companies must present a complete management plan. The plan must include at least four components: management objectives; a description of the forest resources to be managed, the environmental and land use limitations, and the socioeconomic conditions of the management unit and its vicinity; a description of the management system based on the ecological characteristics of the forest and on data collected in forest inventories; and annual operational plans specifying all activities to be conducted in the management unit, including annual cut volume, species and diameter selection, procedures for monitoring forest growth, and forest dynamics. This information is used to justify the cutting cycle chosen, measures to attenuate identified environmental impacts, and the places of nesting and reproduction of fauna, as well as the identification of rare, threatened, or endangered species. The management plan must also include maps or charts indicating the location of the forest resources, protected areas, main forest types, topography, hydrology, pre-existing infrastructure, current use of soils from company areas and neighbouring areas, and the justification of equipment and harvesting techniques chosen in order to reduce their impact.

The management plan of certified enterprises must also state the long-term strategy for supplying existing industrial processing units, and the mechanisms for ensuring economic sustainability during the whole estimated cycle of recuperation of the forest, usually between 20 and 30 years.

One successful example is Cikel, a forest enterprise located in Paragominas in the state of Pará and was certified in 2001. Its management plan led to a complete revision of its strategy for using its 120 000 hectares of highland forests, including its strategy for supplying the industrial processing unit over an estimated period of 25 years necessary for the forest to recover before it can be logged again.

When there was no worry about the impacts caused by logging on the forest, harvesting activities were implemented without any planning. Power saw operators cut trees without any attention to where they would fall, causing serious damage to the remaining trees. Skidder operators searched for trees within the forest without reference to precise locations, resulting in very low productivity and damage to the forest's ability to regenerate. Many hollow trees were cut unnecessarily and others were left behind, because of inadequate information about their location. The effect of these activities was that the forests would take twice as long recover as they would with careful planning (Amaral et al. 1998).

To reduce the impact of logging, in 1995 Mil Madeireira implemented the first commercial-scale project in the Amazon region with full operational planning carried out before harvesting operations started. All potentially commercial trees were identified and plotted on a map. Trees were selected for harvest based on biological and economic qualifications. At least 20 percent of all commercial-sized trees of each species are left untouched; on average, only five to seven trees per hectare are logged in a cycle of 20 to 25 years. All fragile areas and areas along watercourses are preserved. Power saw operators are equipped with detailed maps that show the exact location of each tree to be harvested. Before cutting each tree, the power saw operator carefully considers the best felling direction, in order to minimise damage to the remaining trees and to facilitate the extraction of the log from the stump site. Skidding crews, also equipped with adequate maps, travel on established forest roads and trails and use cables to bring the logs to the skidding trails. The result of such careful planning of logging operations is that the total canopy opening is only about 4 to 7 percent.

Based on the results of Mil Madeireira's experience and pilot projects carried out by the Tropical Forest Foundation in the Brasil Verde area owned by Cikel, since 1999 a complete map of all commercial trees and planned trails is now mandatory before logging can commence in whole Amazon region.

An interesting aspect of the planning system for reduced-impact logging is that forest roads become part of the permanent infrastructure in the managed areas. Because they involve long-term planning, the roads are designed to last much longer and are therefore considered long-term investments, similar to logging equipment and other machinery, rather than as operational costs.

Along watercourses and other sensitive areas, such as swamps and steep-grade terrain, conservation is unusual, even though it is necessary and required by Brazlian law. In the case of forest plantations, many enterprises plant and harvest trees right up to the river bank, ignoring the protected zoning. In certified operations, however, the areas must be protected and restored with a permanent forest cover. In the municipality of Caçador in the state of Santa Catarina, for example, Juliana Florestal left at least 10 percent of two forest areas untouched in order to allow the restoration of permanent vegetation.

For natural forests, some of the main problems involve getting machinery and equipment across waterways; another problem is the felling of trees in permanent preservation areas. Thus, in certified operations, all watercourses must be identified and plotted on maps, so that logging activities can be organised so as to exclude trees within the preservation areas. The road and trail system can be designed to avoid crossing watercourses, or can allow for adequate crossing structures, such as bridges and culverts, to be constructed.

Another important requirement in certified operations but seen rarely in other areas is the stipulation that the enterprise also must set aside at least 5 percent of the total area under management to be preserved, in addition to any sensitive areas or watercourses. This area must reflect the ecosystems being managed and thus serves as a benchmark for future impact assessments. When such absolute conservation areas and permanent preservation areas of certified enterprises in the Amazon region are combined, it is easy to demonstrate that, on the average, from 25 to 40 percent of the total area remains free from logging activities.

The conversion of high biodiversity forests into simplified or monoculture systems is one of the most controversial issues related to forest activities. In Brazil, there are many examples where natural forests have been replaced by plantations, mostly eucalyptus and pine. According to FSC standards, however, any such plantations established after 1994 cannot be certified.

Yet in the management of natural forests in some situations stands can be managed in such way that they become very simplified. An example is the management system used on the island of Marajó in the state of Pará for the production of the fruit and palm heart of the açaí (*Euterpe oleraceae*). The traditional management system for this crop is based on the continuous elimination of the regeneration of competitive species. With the passing of the years, these areas become açaí monocultures. The certification process led to a radical revision of this practice.

In order to comply with the requirements of good forest management, Muaná Alimentos, the first enterprise to obtain FSC certification for the production of palm heart, planned its activities and defined limits for the number of palms per hectare according to the original conditions of the forest. Strata with different management treatments were defined in each plot of 2.5 hectares, with the objective of ultimately creating a mosaic of areas with varying concentrations of palms per hectare.

Impact on Community Relations

By and large, forest operations in Brazil are the object of conflicts between enterprises and local communities. Because FSC certification procedures focus on aspects related to land tenure rights and community relations, these two subjects are fundamental to the assessments carried out for certification purposes. The resolution of these conflicts points to new directions in forest management: the establishment of relations between forest enterprises and the community. The following section describes some of the problems found during certification assessments and how certification brought about adequate solutions.

In theory, 19 percent of the surface area of Brazil consists of indigenous lands, but in practice very few areas have been demarcated; examples of native populations having sovereignty over their territories are rare. Lumber companies frequently invade indigenous areas to cut mahogany and other valuable species; the same often happens in extractivism reservation areas.

Forest operations seeking certification must be completely separate from indigenous areas, even if they have a legitimate claim to land tenure rights and demarcation is in its initial phases. During the evaluation process to certify Gethal, it was noted that a 4000-hectare parcel of the forest management area of the company was located in an area declared eligible for establishing an indigenous reservation. For many years, the company had considered putting that area under management but, as part of the certification process, the land had to be completely eliminated from its forest management unit. The legal status of this area has now been established by the National Foundation for Indigenous People.

Another example is the Xikrin do Catete Indigenous Nation, located in Marabá in the state of Pará. The solution adopted in this case was to prepare a forest management plan to be carried out by the indigenous people themselves, the first such plan in Brazil, as a means to guarantee control over their forest resources and prevent logging companies from stealing the wood.

According to the Brazilian standards approved by the FSC board of directors in Oaxaca, Principle 3 and its criteria apply not only to indigenous peoples but also to other traditional communities such as *quilombolas* and extractivism communities.

With regard to traditional rights, many forest operations have adverse effects on the subsistence of local communities. For an enterprise to be certified, such impacts must be avoided. Any that occur must be mitigated and adequate compensation offered to the population affected.

Manicoré is a municipality located in the state of Amazon, about two days by river from Manaus. The people in this town have always based their income on extractivism activities, including rubber tapping, Pau Rosa oil, and Brazil nuts. In 1998, Gethal Amazonas bought 45 000 hectares of land there to establish a forest management plan to supply its Itacoatiara plywood plant, 30 hours down the river. The areas where families lived along the river were left untouched and their land tenure rights were guaranteed. The company authorised the families to gather Brazil nuts in this area, which is their traditional principal source of income, and it does not harvest nut-producing trees; nonetheless, logging activities affected the activities of the nut gatherers. Tree felling and skidder trails hindered their access to the nut-producing areas and their extractivism activities. The end result was that the productivity and income of the nut gatherers decreased.

A loosely organised community would have difficulty requesting compensation for the losses caused by the presence of such a company in an area that has provided that community with income over the years. If it relied on public authorities, the case could take a long time to be solved. As a certified enterprise, however, Gethal had to behave proactively, identifying the impacts and proposing solutions. It is now carrying out a survey of all families involved with nut gathering and their respective income, and identifying the effects of its logging activities in the field. Among the alternatives under study, Gethal intends to supply maps to facilitate the location of nut-producing trees, plan logging activities for after the nut-collecting season has passed, and offer transportation facilities to nut gatherers to inaccessible areas. Gethal is also studying the possibility of assisting the community to shell the nuts locally and promote their sale.

Land use rights constitute another extremely complex subject in Brazil. There are at least 17 different types of land titles and they very often overlap each other. The problem of land ownership and land tenure rights causes many conflicts that have resulted in armed confrontations and, in many cases, fatalities. Furthermore, it threatens the quality and the survival of forest resources as a result of forest fires and predatory logging.

One of the most remarkable aspects of FSC certification is the requirement that the situation of the forest area be made absolutely clear with respect to land tenure rights, its use by traditional populations, and the absence of conflicts that may threaten the physical integrity of individuals and of the forest resources.

The case of Mil Madeireira is an example of the resolution, through the process of certification, of a dispute over land tenure rights that remained pending. When it requested certification of 82 000 hectares, the company was aware that there were families living in company areas, but it did not know them nor had it maintained any relations with them. During the certification assessment it became clear that, although there was no apparent conflict, the potential for such situation existed. These families used fire to clear their agricultural plots, a fact that could threaten the forest resource under management. Moreover, they did not recognise Mil Madeireira as the legitimate owner of the area. As one of the certification conditions, no forest management activity could take place in the forest compartments bordering the occupied areas. (A forest compartment is an area to be harvested once in every 20 to 30 years, during which time it remains untouched in order to allow the forest to recover.) The company was required to carry out a complete survey of all families living on its land and submit a proposal for formally recognising their land tenure rights. After two years of work with the state government to formalise the proper legal situation of the families with regard to their land tenure rights, Mil Madeireira decided to work with the communities to demarcate the areas they occupied, including some forest areas. As a final step to close the issue, the company gave each family a letter officially recognising its land tenure rights over the area.

Indirect Impacts of Certification

Brazil has little forest culture and tradition. Its forest-based industries derived from the predatory exploitation of the forest resource. The rich reserves of Araucaria and the mixed broadleaved forests of southern Brazil and the lush forests of the coastal range, known as Mata Atlântica, were almost completely destroyed during the last century without any consideration to their sustainability. The first experiments with low-impact management only started in the 1980s, but they were unable to develop into full commercial operations.

As a rule, most enterprises, government agencies, and even NGOs considered sustainable forest management a good idea but difficult to be implemented because of cultural, technical, and financial problems. The first cases of certification in 1997 demonstrated that good forest management is possible, and this fact influenced the behaviour of the various actors.

Certification has had many other indirect impacts.

NGOs that had previously campaigned against forest activities became more selective. On the one hand, they had to present more objective demands against companies trying to carry out forest management activities, and, on the other, they could identify in certified companies examples to be used as a reference for introducing the changes in the forest sector that they wanted to see. A good example is the partnership between Greenpeace and Mil Madeireira. In turn, forest companies feel obliged to answer the questions presented by society at large with objectivity, because they can no longer state that forest management is not feasible. This situation has led a number of companies, such as Amaplac/WTK, Edai, and Eldorado, to sign written commitments to good forest management.

As forest companies started to apply forest management concepts on a commercial scale, new questions came up in the field during the certification process. Consequently, research institutions recognised the need for an urgent and more objective research agenda. Issues such as the assessment of the impacts of forest management on the soil, water, fauna, and flora are today the subject of dozens of theses and dissertations on certified enterprises. Research on species ecology now focusses on the main species considered in forest management plans. Furthermore, researchers received a welcome bonus by being able to have access to data generated by the inventory of thousands of hectares — information that would normally not be accessible to a research institution.

In the 1980s, official banks in Brazil blocked credit to logging companies because of the high risk of potential environmental claims. Certification allowed these banks to identify good examples of responsible behaviour in the forest sector, so they could provide investment capital at lower risks. In 2001, BNDES, the largest investment bank in the country, resumed its investments in the Amazon forest sector, but made them conditional to certification.

Similarly, private funds seek not only lower environmental risks but also long-term guarantees for their investments. At least two environmental funds have been established in Brazil to invest in sustainable forest management projects, also under the condition of FSC certification. Axial Renewable Resources, in partnership with the American group GMO, has invested US$15 million in Gethal, which was certified in 2000. Other commercial banks are planning to make investment in forest management projects under the same conditions.

As certified enterprises incorporate technical advances, they improve the level of forest management standards. As a consequence, forest legislation is keeping up with these developments and becoming more strict. In addition to the 100 percent inventory that was incorporated as a legal requirement in 1999, annual operational plans and maps before and after harvesting are now mandatory. Pioneer work carried out by certified enterprises, such as the annual operation plans used by Cikel and post-harvesting maps used by Mil Madeireira, has served as a model to forest regulation authorities.

In fact, legislation has been simplified in order to encourage forest management. In other words, as people start believing that good forest management is feasible and that companies can spontaneously decide to adopt it, legislation tends to adapt itself and become less bureaucratic. A good example is the proposal for eliminating inspections before management authorisations have been issued; it is more efficient to check the management plan when field operations begin, when the proposed actions can be verified on the spot.

Environmental control organisations such as IBAMA have realised that forest certification can be an important ally in monitoring forest enterprises. Such organisations are more likely to accept innovative proposals coming from certified enterprises, as happened when logs of durable species were exported to Germany and the Netherlands for marine construction.

As a rule, the forest departments of wood-based industries have played a secondary role. All efforts were concentrated in improving the quality and lowering the cost of production in order to guarantee and increase market share. Because certification establishes performance standards for forest production, forest departments have acquired higher status within forest corporations. It is not uncommon for foresters working in certified enterprises to achieve executive positions within the company's administration, with decision-making power equal to their colleagues' from the industrial or marketing areas. Furthermore, shareholders can see that investments in the forest-producing divisions are becoming more important, since the quality of their management may affect market conditions.

Until 1997, there were no forest management training programmes at the operational level in the Amazon region. Thanks to certification, however, forest management activities have become highly valued and new professionals have had to be hired in order to implement good forest management. As a result of this demand, two forest technician courses have been created; moreover, about a dozen courses on low-impact logging have been offered under the co-ordination of the Tropical Forest Foundation.

In terms of civil society, organised groups have been concerned with debating forest management and its impacts for some time. These groups, either through their participation in public consultations related to certification processes or through claims against certified operations or operations undergoing certification, can now monitor how forest operations are working.

Companies that have CoC certification and commercialise certified products are completely barred from having any relation with forest products coming from illegal or controversial sources. This fact has encouraged a large number of manufacturing and retail companies to demand that their suppliers legalise their forest operations, thus reducing competition with wood from illegal sources.

Limitations and Challenges

Certification may lead to many solutions to forest management problems but it has also presented some limitations that must be overcome. Many of these limitations are related to problems that certification is not able to solve on its own, nor do these limitations exist only in Brazil. Certification must meet the challenge of making access more democratic, especially with regard to forest management carried out by local populations or small forest enterprises. Some of these challenges are described below.

Access to Information and Certifiers

Certification is only available to those who have information about certification schemes and access to certifiers. Information on certification is not always available in easily understandable language. This problem seems to be more serious in countries with no national initiatives or those where such initiatives are just beginning, such as the Guyanas, Venezuela, Argentina, Paraguay, Ecuador, and Panama. With regard to certifiers, only a few local organisations operate as certifiers. Only four countries have local organisations active in carrying out assessment for certification purposes: Brazil, Bolivia, Costa Rica, and Mexico. When local certifiers are not available, contacts must be established with international companies, with implications of language and communication barriers, and higher costs.

Compliance with Standards

Compliance with FSC requirements usually represents three barriers:

- the incomprehensibility of the standards. With their 10 principles and 56 criteria, FSC standards are very lengthy. Wherever they exist, national standards tend to be even lengthier and more difficult to understand. Even so, the way the FSC's principles and criteria are presented makes them very hard to understand for the majority of traditional populations.

- the inadequacy and invalidity of the standards. Generally speaking, standards are developed and written in a style more appropriate for medium-sized and large enterprises, with safeguards for cases of higher social and environmental impacts involving large private companies and concessions.
- insufficient information on forest management. In many situations, forest activities that involve logging are not traditional but are carried out by local populations as an alternative source of income. Whatever these populations know about forest management is usually empirical and transmitted from one generation to the next. When a new activity is introduced, especially when it might generate serious social, environmental, and economic impacts, access to information that may reduce the level of empirical decisions — or promote a faster process of knowledge transfer and absorption — becomes fundamentally important.

Costs

Certification costs can be direct — related to the assessment and certification processes as developed by the certifier — or indirect — associated with changes in the current forest management in order for the company to attain certification.

Indirect costs are mostly associated with activities necessary to bring current management practices to the level of those considered good forest management. Existing distortions usually refer to the requirements set by the standards (see the discussion of access above).

Direct certification costs represent a long-term sustainability problem. A community forest management (CFM) initiative is unlikely to have considered the direct costs of certification an impediment to achieving certification. Bilateral aid agencies and NGOs provide substantial support toward meeting these costs. However, most CFM projects in Latin America are unable to finance maintaining certification. Many alternatives have been used to cover such costs but long-term sustainability depends on new mechanisms that make the income of CFM compatible with the direct costs of certification.

In countries with no local certifiers, these costs are even higher. Ruth Nussbaum and her colleagues (2001) propose setting a minimum direct cost of US$1000 for certification. In Latin America, direct costs rarely remain below US$4000; usually they are more than US$8000. These are very high costs to be absorbed by CFM projects.

Markets and Benefits

Most CFM projects face serious difficulties in selling their certified products in a way that earns a profit from the benefits of good forest management and certification. In local communities (*ejidos*) in Mexico, three out of four expectations of certification relate to the market: finding a market niche, finding markets for lesser known species, and fetching higher prices on international markets (Robinson 2000). None of these three expectations has been fully met by certification, and, in the case of obtaining higher prices, results have been frustrating. The market's lack of response to certified products causes an automatic disincentive to certification and even to good forest management.

In contrast, buyers are always claiming there is not enough certified wood and products on the market to meet existing demand. In fact, this reveals a lack of communication among market, buyers, and producers. In more demanding markets such as the United States and Europe, except for a few honourable cases, buyers declare they are commited to buy certified products but demand the same quality, the same species, the same delivery times, the same quantities, and, if possible, lower prices. Although this type of problem affects all certified producers, it has greater impact on CFMs because they have poorer infrastructure, smaller production scale, and not much experience in tapping these markets.

Because certified products do not identify their origin, FSC certification has not offered an adequate mechanism to help market competition become more equitable. A fair trade mechanism could be provide a possible solution.

Conclusion

Instead of demonstrating the lack of success of certification, the fact that the number of certified enterprises is fairly modest reveals the hard reality of forest management today. Forest management is still new; in general, it performs at a level not sufficient to meet the minimum requirements defined by the international community through the FSC principles and criteria.

Despite the fact that in 2001 in Brazil there were only 15 forest areas with a little more than 85 000 hectares certified, FSC certification has played an important role in many of the changes in the forest sector. Certified enterprises were able to improve the level of minimum requirements regarding working conditions, including safety and health aspects, reduced negative impacts of logging activities, and community relations. Certification has also helped to demonstrate that forest management is feasible when the perception of various actors of the Brazilian society with regard to forest activities can be influenced in a positive way.

Nonetheless, there are many challenges and situations in forest management where certification is not the best solution. Even where forest certification is a good mechanism, the limitations may require a more global and complete approach, such as the case of community forest management and small-scale forest operations. However, it is a long road to travel before all the possibilities of certification, as an instrument of change in Brazil's current situation, can be fully realised.

Notes

1. In 1995, WARP became the Good Wood Alliance, which in turn merged with the Certified Forest Products Buyers Group in 1997 to form the Certified Forest Products Council.
2. By October 2003, the certified area totalled 39.7 million hectares in 58 countries with more than 3000 companies having the chain of custody certification.
3. In 2003, approximately 81 percent of the certified forests were in North America and Europe and 19 percent were in the other regions.
4. By October 2003, there were 33 forest operations and 1 376 943 hectares certified in Brazil. Plantations made up 65 percent of the area certified (23 operations) and the remaining 35 percent consisted of natural forest (ten operations). The first three community operations were certified in 2002.
5. See the website of the Sociedade Brasileira de Silvicultura at <www.sbs.org.br>.
6. By the end of 2003, the total area of forests certified in Brazil was 1 376 943 hectares in 33 forest operations. The area of forest plantations certified was 806 863 hectares (Forest Stewardship Council 2003).
7. The FSC has two types of certification: forest management, which certifies that the forest has been managed in a way that is environmentally adequate, socially fair, and economically viable according to FSC principles and criteria, and custody certification, which certifies that the raw material, or the material used in the manufacture of a product, comes from a certified forest.
8. See BRACELPA — Associação Brasileira de Celulose e Papel [Brazilian Association for Pulp and Paper] (2001), on the website of the Sociedade Brasileira de Silvicultura at <www.sbs.org.br>.
9. National Association of Palm Heart Manufacturers (2000). Personal communication.
10. By the end of 2003, there were ten certified natural forests.

References

Amaral, Paulo, Adalberto Veríssimo, Paulo Barreto, et al. (1998). *Floresta para Semre: um Manual para Prodção de Madeira na Amazônia*. Imazon, Belém.

Forest Stewardship Council (2003). 'FSC in Brazil: The Brazilian FSC Initiative'. <www.fsc.org.br/ingles.htm> (April 2004).

Nussbaum, Ruth, Michael Garforth, Hannah Scrasse, et al. (2001). 'Getting Small Forest Enterprises into Certification: An Analysis of Current FSC Accreditation, Certification, and Standard Setting Procedures Identifying Elements Which Create Constraints for Small Forest Owners'. ProForest, Oxford.

Robinson, Dawn (2000). 'Certification in Communally Managed Forests: Perspectives from Mexico'. *Forest, Trees, and People*, November, pp. 28–31.

Schneider, Robert R., Eugene Y. Arima, Adalberto Veríssimo, et al. (2000). 'Amazônia Sustentável: limitantes e oportunidades para o desenvolvimento rural'. World Bank and IMAZON <www.imazon.org.br/pdf/bm.zip> (April 2004).

Smeraldi, Roberto and Adalberto Veríssimo (1999). 'Hitting the Target: Timber Consumption in the Brazilian Domestic Market and Promotion of Forest Certification'. Imaflora, Imazon, and Friends of the Earth, São Paulo. <www.gtz.de/forest_certification/download/d32e.pdf> (April 2004).

SOS Mata Atlântica (2001). 'Atlas da Mata Atlântica'. SOS Mata Atlântica, São Paulo.

Appendix 3-1
Forest Stewardship Council

The Forest Stewardship Council (FSC) is an international nonprofit organisation founded in 1993 to support environmentally appropriate, socially beneficial, and economically viable management of the world's forests. It supports the development of national and regional standards to be used to evaluate whether a forest is being managed well.

It is an association whose members represent a diverse group of representatives from environmental and social organisations, forest and product industries, indigenous people's organisations, community forestry groups, and certification bodies from around the world. Membership is open to all who are involved in forestry or forest products and share its aims and objectives.

FSC's head office is in Oaxaca, Mexico. It is governed by an elected board, made up of people from industry, environmental, social, and labour groups, indigenous people's groups, and others.

The Forest Stewardship Council's Principles and Criteria

After three years of extensive global consultation, at the FSC's founding meeting in 1993 ten principles and complementary criteria were agreed upon. The principles and criteria are universal in nature and apply to tropical, temperate, and boreal forests. They serve as a guiding framework for developing standards appropriate to social, ecological, and economic conditions at national and regional levels.

Principle No. 1: Compliance with Laws and FSC Principles. Forest management shall respect all applicable laws of the country in which they occur, and international treaties and agreements to which the country is a signatory, and comply with all FSC principles and criteria.

Principle No. 2: Tenure and Use Rights and Responsibilities. Long-term tenure and use rights to the land and forest resources shall be clearly defined, documented, and legally established.

Principle No. 3: Indigenous Peoples' Rights. The legal and customary rights of indigenous peoples to own, use, and manage their lands, territories, and resources shall be recognised and respected.

Principle No. 4: Community Relations and Workers' Rights. Forest management operations shall maintain or enhance the long-term social and economic well-being of forest workers and local communities.

Principle No. 5: Benefits from the Forest. Forest management operations shall encourage the efficient use of the forests' multiple products and services to ensure economic viability and a wide range of environmental and social benefits.

Principle No. 6: Environmental Impact. Forest management shall conserve biological diversity and its associated values, water resources, soils, and unique and fragile ecosystems and landscapes, and, by so doing, maintain the ecological functions and the integrity of the forest.

Principle No. 7: Management Plan. A management plan — appropriate to the scale and intensity of the operations — shall be written, implemented, and kept up to date. The long-term objectives of management, and the means of achieving them, shall be clearly stated.

Principle No. 8: Monitoring and Assessment. Monitoring shall be conducted appropriate to the scale and intensity of forest management — to assess the condition of the forest, yields of forest products, chain of custody, and management activities and their social and environmental impacts.

Principle No. 9: Maintenance of High-Conservation Value Forests. Management activities in high-conservation value forests shall maintain or enhance the attributes that define such forests. Decisions regarding high conservation value forests shall always be considered in the context of a precautionary approach.

Principle No. 10: Plantations. Plantations shall be planned and managed in accordance with Principles and Criteria 1 to 9, and Principle 10 and its Criteria. While plantations can provide an array of social and economic benefits, and can contribute to satisfying the world's needs for forest products, they should complement the management of, reduce pressures on, and promote the restoration and conservation of natural forests.

Chapter 4

Indigenous Rights and Forest Certification in British Columbia

Chris Tollefson

The challenge of protecting the rights of indigenous peoples while promoting more sustainable use of the world's forest resources is receiving growing international recognition. It is a nexus that forms a key component of the Forest Stewardship Council (FSC) certification system, and one that is accorded special recognition in FSC Principle 3 and its corollary criteria. Principle 3 obliges forest managers to recognise and respect 'the legal and customary rights of indigenous peoples to own, use, and manage their lands, territories and resources'. However, as with other international initiatives aimed at protecting indigenous rights, the challenge of translating this laudable aspiration into measures that will be effective on the ground looms large. This is particularly so in jurisdictions where legal recognition of indigenous rights is inchoate, and that lack a strong domestic judicial and political tradition of recognising and respecting such rights.

A key theme of this volume concerns the relationship between hard and soft law, and the normative and practical implications of this often-murky distinction. It is a distinction that has emerged and, to this point, has been primarily debated within the realm of international law. Despite this, it is also a distinction with relevance and resonance in the domestic legal domain.

Laws are said to be hard when they impose generally applicable, binding obligations that are articulated with a relatively high degree of precision and are enforceable by delegated authorities (whether courts or specialised tribunals). Consequently, it has been suggested that the realm of soft law 'begins once legal arrangements are weakened along one or more of the dimensions of obligation, precision, and delegation' (Abbott and Snidal 2000, 422).

Within international law circles, soft law is often derided as being ineffective, allowing states to preserve the *status quo* while appearing to take action on pressing global problems. Of late, however, various scholars have been coming to soft law's defence. Some of these defenders contend that soft law should be seen as an indispensable way station on the road to harder and more durable forms of legalisation. On the international stage, whether in the realm of environment, human rights, or business standards, illustrations of soft law performing this role abound (Morais 2002). Similar arguments have been made about the function of soft law in the domestic realm. It has been contended, for example, that federal environmental assessment law in Canada reflects an evolution from soft to hard(er) law (Kennett 1993).

Others have argued that it is important not to fall into the trap of regarding soft law as inferior to hard law, or as merely serving as a platform for it. Indeed, in some circumstances, it is claimed that soft law 'is preferable on its own terms [in that it] … offers many of the advantages of hard law, avoids some of the costs of hard law, and has certain independent advantages of its own' (Abbott and Snidal 2000, 423). These advantages could include lending support for social learning and adaptive management, engaging a broad array of non-state actors in the law-making process, and facilitating compromise (particularly between actors with differing interests, values, and degrees of power).

Does FSC certification offer a viable model of forestry development that is sustainable and protects indigenous rights? Moreover, can model serve as a way station *en route* to an enduring hard law resolution of hongstanding indigenous claims to forest resources on the domestic front? Answers to these questions may emerge in British Columbia (BC), a jurisdiction in which conflicts among government, industry, and indigenous peoples over rights to the forest have been a discouragingly prevalent feature of the political landscape for decades.

For BC First Nations one of the most attractive features of the FSC model is its potential to protect their rights and interests in the forest resource, and to secure benefits from the use of that resource, without the necessity of judicial or political intervention. For the forest industry, the FSC model also has strong appeal given its potential to reduce the costs and uncertainties associated with the unresolved status of Native land claims throughout the province. BC's forest companies rely heavily on timber harvested on Crown lands that are subject to pending aboriginal rights claims. These companies increasingly face market pressures from buyers concerned about the province's failure to recognise and respect aboriginal rights. By securing FSC certification, they hope to reduce or avoid the costs and market impacts associated with First Nations–sponsored blockades, protests, and litigation that have, of late, been an inevitable feature of carrying on business.

But although the market appetite for FSC-certified BC forest products may be strong, and forest companies and First Nations within the province have shown considerable interest in FSC certification, to date only a very small percentage of BC's forestlands have secured certification. Indeed, FSC certifications of Canadian forest operations, as a proportion of the total forestlands now certified under FSC worldwide, are less than 0.5 percent, lagging far behind other major forest-product exporting countries such as Sweden, Brazil, and the United States.[1] For the FSC to succeed in BC, the supply of the volume of FSC-certified BC forest products must increase dramatically.

Much hinges on the fate of the Regional Standards set by the Forest Stewardship Council of British Columbia (FSC-BC) that received preliminary approval in July

2003 (FSC-BC 2002a). This regional standard, which prescribes in detail how FSC's international principles and criteria will be implemented within BC, emerged out of a seven-year-long process of consultation and negotiation under the direction of FSC-BC (the FSC's regional affiliate).

This chapter reflects upon Principle 3 (and the BC Regional Standards adopted thereunder) as a soft law vehicle for responding to the failure of the Canadian political and legal system to resolve aboriginal rights — particularly, rights related to forest resources — in British Columbia.

To set the stage for this analysis, the first part of this chapter examines the complex intersection of indigenous rights and forest certification in the global context. The second part then reflects on the troubled relationship between BC First Nations and Canadian governments, with particular attention to the manner in which this relationship has been defined and mediated through the legal system. The third part examines the legal and practical implications and challenges associated with the negotiating and implementing the newly enacted FSC-BC's Principle 3 standard. The final part offers an analysis of that standard as a form of soft law in terms of the nature of the obligations it creates, the precision with which these obligations are articulated, and manner in which interpretive and enforcement authority is delegated.

Indigenous Rights and Forest Certification: The Global and British Columbia Contexts

The Global Context

Driven by concerns about tropical deforestation, for much of the 1980s and early 1990s international forestry initiatives focussed on forest management and conservation. In more recent years, however, there has been a growing appreciation of the need to develop a broader strategy that takes account of the interests of the world's poorest 1.2 billion inhabitants, 90 percent of whom depend on forest resources for their livelihood.[2] The imperative of including 'people as well as trees' in sustainable forest planning is a central theme of the World Bank's ongoing forest policy review: 'adopting major targets for forest conservation and management is not enough: the overall challenge is [to create a] linkage to the Bank's mission for poverty alleviation and sustainable growth and development' (quoted in Colchester 2000; see also World Bank 2001).

While the term 'indigenous peoples' is by no means uncontroversial, it is generally used to refer to politically marginalized, territory-based, ethnic groups that are culturally distinct from majority populations and that recognise themselves as indigenous (Inter-Commission Task Force on Indigenous Peoples 1997, 27).[3] Other defining characteristics include a close attachment to ancestral territories and natural resources, language or languages distinct from the national tongue, the presence of customary social and political institutions, and high levels of dependence on subsistence-based economic production (Colchester 2000).

According to these criteria, there are an estimated 300 million indigenous people worldwide. This number includes First Nations, Native and aboriginal peoples in the Americas and the Pacific, tribal and minority nationalities in Asia, and ethnic minorities in Africa. The traditional lands and territories inhabited by indigenous peoples include large and ecologically critical areas of the world's tropical, temperate, and boreal forests. Virtually without exception, these are peoples who have suffered extreme hardship due to colonialism and its legacy of large-scale concession-based industrial forestry (Colchester 2000). In many instances, they have been brought to the brink of extinction by forcible resettlement, the systematic undermining of their social, political, and cultural practices, and the destruction of their traditional economies. Frequently, these impacts have been a direct consequence of losing control of the forest resources to foreign-owned companies.

But while indigenous peoples have made significant progress in the international arena in securing recognition of their rights in the soft and hard law realms,[4] they have been much less successful in securing domestic recognition for these rights in practice. Indeed, in most instances, domestic governments have strongly resisted the expanded recognition of indigenous rights, fearing detrimental impacts on revenues and investor confidence as well as non-indigenous backlash.

Because of the obstacles and inertia associated with achieving meaningful change through domestic political and legal channels, indigenous organisations are increasingly interested in exploring the opportunities associated with forest certification. Although various competing certification systems are available (Gale and Burda 1998, 282; see also Certified Forest Products Council 2003), the undisputed favourite among indigenous groups is the Forest Stewardship Council.[5] In large measure, this is due to the breadth of the protections that are accorded indigenous rights under the FSC model. Principle 3 requires forest managers to respect the legal and customary rights of indigenous peoples over their lands, territories, and resources, regardless of whether these rights are accorded domestic protection. Moreover, under the Principle's corollary criteria, forest development on indigenous lands and territories must not occur without the free and informed consent of affected indigenous peoples; sites of special cultural, ecological, economic, and religious significance to them must be protected and, where traditional indigenous knowledge is to be used in forest operations, consent must first be secured and appropriate compensation paid. The FSC model has also gained favour, particularly among Canadian First Nations, by demonstrating a commitment to including indigenous voices in FSC governance. For example, in FSC Canada, a separate governance chamber dedicated to representing the interests of First Nations has been created. Moreover, First Nations representatives play a growing leadership role in developing and implementing policy at the regional, national, and international levels of FSC.

The British Columbia Context

To understand the development of the BC's Principle 3 standard and reflect on its future prospects and implications, an understanding of the context is critical. This section is intended to convey the richness of this context by touching on some of the key anthropological, political, and legal features that combine to make the challenge of developing the standard so unique to the province.

The First Nations of British Columbia, as with First Nations throughout the Americas, descend from peoples who are believed to have migrated from Asia by way of a land bridge over what is now the Bering Strait between 12 000 and 20 000 years ago. Physical evidence of a human presence in what is now British Columbia dates back approximately 10 000 years, to the period immediately following glacial retreat (McMillan 1995, 187). For the next 5000 years, the province came to be inhabited by nomadic bands, dependent on foraging the abundant marine and terrestrial resources that were available.

Anthropologists estimate that at the time of European contact, the First Nations population in BC was between 200 000 and 300 000 (Muckle 1998, 37). Initial European contact occurred in 1774 with an exploratory voyage of a Spanish ship up the coast. Soon after, British and U.S. ships frequently began to ply the same route seeking trading opportunities with coastal nations. A vibrant fur trade soon developed on the coast; by the early 1800s, a land-based fur trade with interior nations had also been established (Miller 1991, 141).

Within a few short decades of contact, First Nations populations went into rapid decline, in large part due to their susceptibility to unfamiliar European diseases (McMillan 1995, 221; Boyd 1994). By 1835, the indigenous population had fallen to 100 000; less than a century later, it had plummeted to 23 000, a reduction of more than 90 percent from pre-contact (Muckle 1998, 60). Not until the 1930s did Native populations began to recover. Today there are approximately 110 000 status and 75 000 non-status Indians living in BC.[6] About half the status Indian population currently resides on reserves. These reserves, of which there are more than 1600 in the province, account for less than 0.5 percent of the provincial land base. Many reserves are uninhabited; those that are inhabited usually consist of small communities with populations of fewer than 200 (Muckle 1998, 5).

Canada's *Indian Act* remains the principal legislative vehicle through which the country's Native population is governed. It allocates funds and responsibilities to band councils elected by status Indians. This model of governance has been criticised for reflecting a paternalistic approach that imposes serious constraints on the ability of First Nations to achieve self-sufficiency and autonomy (Monture-Angus 1999). It has also been criticised as imposing a system of governance inconsistent with First Nations social and political organisation (Mathias and Yabsley 1991, 38). As such, in many Native communities there are two distinct sets of political leadership: an elected band leadership in accordance with the Act and a hereditary leadership vested with authority based in custom and practice.

Currently, about 200 BC bands are recognised under the *Indian Act* (Muckle 1998, Appendix 1). In contrast, the number of BC First Nations is the subject of continuing anthropological and political debate. Historically, the province's Native population was often grouped into one of ten First Nations. However, this approach has often been accused of ignoring important ethnic differences. BC First Nations are now more commonly seen as falling into one of 30 or 40 ethnic groups based on a variety of anthropological criteria including shared territory, language, and culture (Muckle 1998, 7). The number and defining characteristics of these groupings remains undetermined for a variety of reasons, including conflicting historical records and disagreements among First Nations as well as within the anthropological community.

There is also considerable uncertainty as to which specific nations belong to which ethnic groups. These uncertainties are the product of changes within aboriginal communities that include shifts in the boundaries of traditional territories and intermarriage. As such, while in some cases the ethnic groupings correspond to what indigenous people consider a single 'nation', in many other cases these groupings comprise several such nations, and in some cases a nation is linked to more than one ethnic grouping.

Legal and Political Relations with the First Nations of British Columbia

Prior to the 1850s, relations between the Native population and Europeans in what is now BC were based on trade and generally amicable. During these early colonial years, relations between the British Crown and the First Nations took place on a nation-to-nation basis consistent with the spirit of the Royal Proclamation of 1763 issued by King George III of England.[7] Between 1850 and 1854, the British government entered into 14 purchase treaties, primarily with respect to lands on lower Vancouver Island (Miller 1991, 141, 145).

Soon, however, the imperial government instituted a much more adversarial treaty policy that denied the existence of aboriginal title and characterised reserves as 'gifts from the Crown to Aboriginal peoples' (McKee 2000, 19). Relations between the Crown and Native peoples deteriorated even further after BC entered Confederation with Canada in 1871. Soon after, the province prohibited aboriginal peoples from voting and imposed restrictions on the allocation of reserve lands. Meanwhile, the federal government embarked on a policy of assimilation inaugurated with the passage of the *Indian Act* in 1876, establishing rules on reserve use, Native governance, and education that remain on the books today. It was also under this legislation that the federal government outlawed many First Nations ceremonies and traditions, including the Potlatch (Muckle 1998, 72). In 1927, as First Nations began to organise to seek redress through the courts, the Act was amended to prohibit fundraising for aboriginal rights claims, which lasted until 1951 (Mathias and Yabsley 1991, 35).

Until relatively recently, Canada's governments stubbornly maintained the notion that aboriginal rights and title are contingent rights that derive exclusively from the state in the form of legislative or executive action. This notion was given credence by the 1888 decision of the Judicial Committee of the Privy Council (then Canada's highest court of appeal) in the *St. Catherine's Milling* case.[8] In this decision, the court held that aboriginal rights and title originated with the Royal Proclamation of 1763 and were 'dependent on the good will of the Sovereign'. According to this analysis, aboriginal rights and title may exist, but only within the areas covered by the Proclamation and only because they had granted by the British Monarch (Howlett 2001).[9]

For close to a hundred years, this interpretation of the *St. Catherine's Milling* case provided the legal justification for the continuing refusal of federal and provincial governments to negotiate treaties with First Nations in large areas of Canada to which the Proclamation was thought not to apply, including throughout all of British Columbia — where initial European contact had yet to occur when the Proclamation was first enacted.

In 1973, the Supreme Court of Canada rendered a decision that abruptly dispensed with the prevailing contingent rights approach to aboriginal rights and title. The *Calder* case was brought by the Nisga'a people of BC's northwest coast, who claimed aboriginal title to their traditional lands on the basis of their historical, uninterrupted use and occupation of those lands.[10] The court held that aboriginal title could exist in common law independently of the Royal Proclamation or other legislative or executive action, based on proof of continuing use and occupation.[11] *Calder* left many questions unanswered, including under what circumstances aboriginal title could be infringed upon or extinguished, questions that the Court would not revisit until more than two decades later. The lasting importance of this decision, however, was that it introduced into Canadian law the notion that aboriginal rights and title could be inherent in nature and, as such, pre-exist European assertions of sovereignty in Canada.

The *Calder* decision ushered in a new era in First Nations activism. In 1982, First Nations secured another milestone victory. This time it was in the political realm, with the incorporation of a constitutionally guaranteed protection for 'existing aboriginal rights and title' in section 35 of Canada's newly repatriated Constitution. The years that have followed can be seen as an ongoing struggle to clarify and, in many ways, redefine the relationship between First Nations and the Canadian governments. To date, this process has unfolded on two fronts: at the treaty table and in the courts.

The Contemporary Treaty Process

In the period following *Calder,* treaty negotiations intensified and began to produce significant results. The federal government established a claims commission tasked with the responsibility of negotiating comprehensive land claim treaties with respect to lands that had until then had remained un-treatied. From the late 1970s to mid 1990s, the process led to the settlement of seven outstanding land claims in Quebec, Labrador, and in Canada's North (Howlett 2001, 123). In contrast to the progress achieved elsewhere in Canada, the Indian Claims Commission made little headway in

BC, due in large part to the continuing refusal of the provincial government to refuse to recognise aboriginal title.

The BC government finally reversed this policy in the late 1980s, establishing its own provincial treaty commission in 1992 (McKee 2000, 29, 33). To date, however, this process has yielded little in terms of concrete results. While almost 50 BC First Nations have participated in 40 distinct sets of treaty negotiations, none of these negotiations has concluded in final agreements.[12] As a result, several First Nations have withdrawn, or have threatened to withdraw, from negotiations, often citing concerns about the time and expense associated with the process. First Nations have also criticised the negotiation process for failing to incorporate adequate 'interim measures' that would protect the lands and resources at stake during the process. Moreover, the provincial government has come under fire for requiring, as a condition of participating in the negotiation process, that First Nations refrain from seeking to enforce their rights through litigation.

In 2002, the future of the BC treaty process was further clouded by a province-wide referendum held by the newly elected Liberal government that was ostensibly designed to provide a mandate for future negotiations. The referendum was uniformly decried by First Nations as an attempt to rewrite the rules of the process mid-stream, opening the door to constitutionally guaranteed minority rights being determined at the ballot box. When their legal challenges to the referendum plan failed, many First Nations and their supporters boycotted the vote. Ultimately, about 30 percent of eligible BC voters cast ballots. The polling results indicated strong support for the government to take a harder, more bottom-line approach in negotiations. Consequently, the future of the BC treaty process — which has on its docket pending claims with respect to a land base larger than the combined total of similar claims in the rest of the country — is highly uncertain.[13]

Judicial Developments

In the 1980s and 1990s, there was also a dramatic upsurge in litigation relating to aboriginal rights and title issues. Much of the resulting jurisprudence arose in cases where First Nations invoked their rights under section 35 defensively, as a means of resisting governmental actions or decisions. Falling into this category are dozens of cases in which aboriginal peoples have relied on their rights under section 35 to defend against regulatory prosecutions (particularly under the federal *Fisheries Act*), and to enjoin government licensing decisions with respect to the development of natural resources on their traditional lands. In only a very few instances have First Nations initiated litigation with a view to securing legal recognition of their aboriginal rights or title, in large measure due to the costs and complexities of presenting the evidence necessary to establish such claims legally.

Nonetheless, particularly in decisions rendered by the Supreme Court of Canada, First Nations have made significant progress toward building on the legal foundation of aboriginal rights and title provided by *Calder* and section 35 of the Constitution.

Important judicial milestones in this process include the Supreme Court's judgements in *Sparrow* (1990),[14] *Van der Peet* (1996),[15] and *Delgamuukw* (1997).[16] Based on these decisions, it is now clear that neither the province nor the federal government can unilaterally extinguish (through regulation or other means) aboriginal rights already in existence when section 35 came into force, that governmental infringements of such rights must meet strict justificatory criteria that include a duty to consult, accommodate, and usually compensate affected First Nations, and that aboriginal title exists as an distinct 'inalienable, communally held' species of aboriginal rights.

Despite the court's pronouncements in these and other cases, however, significant and vexing questions about the precise nature of the legal relationship between First Nations and Canadian governments remain; indeed, in many respects, these pronouncements have engendered new uncertainties. With respect to the rights of First Nations to use and protect the forest resources, these questions fall into three broad categories.

One category of questions relates to the implications of these decisions for the constitutional division of powers with respect to the regulation of forestry on lands subject to pending or established claims by First Nations. *Delgamuukw* established the principle that a province cannot extinguish aboriginal rights and title, but it is less clear whether and to what extent a province is entitled to enact laws that in purpose or effect entrench upon federal jurisdiction over Indians and Indian lands.[17]

The second category relates to the procedural rules that govern the manner in which First Nations may raise aboriginal rights and title in legal proceedings. Pursuing aboriginal rights and title is extremely expensive and time consuming. The *Delgamuukw* case was one of the longest and most costly cases in BC history, but even trials much narrower in scope can easily cost upward of a million dollars. As a result, First Nations have sought to raise aboriginal rights arguments in ways that would allow courts to recognise and protect these rights without recourse to a full trial. However, these efforts have proven largely unsuccessful.[18] First Nations have nonetheless enjoyed some success in securing judicially ordered financial support from governments to assist in defraying the cost of litigating aboriginal title cases.[19]

The third category of questions concerns the nature and extent of the duty of the provincial government to consult First Nations prior to or in connection with the authorisation of forest development, including when the duty to consult arises and what this duty entails. A key issue is when this duty arises. First Nations contend that the existence of the duty to consult should not depend on whether the aboriginal title or right in question is the subject matter of a treaty or has previously been given formal judicial or governmental recognition. They assert that to exempt governments from the duty to consult with respect to proposed forest developments that will affect lands that are the subject of potential aboriginal rights or title claims (and which in some cases are the current subject of treaty negotiations) is a clear violation of the Crown's fiduciary obligations to First Nations, an obligation that is given constitutional expression under section 35.

Until very recently, the position of the BC government has been that the constitutional duty to consult, together with accompanying duties to compensate and accommodate, is not triggered until aboriginal rights or title are judicially affirmed or formally recognised by government. This position has led to considerable litigation in recent years, particularly with respect to provincial decisions purporting to authorise forestry development on lands claimed by First Nations.

Early trial-level decisions tended to support the posture adopted by the BC government, holding that a breach of the duty to consult cannot be invoked to quash or enjoin such decisions where aboriginal rights or title have been asserted but not proven through a trial process, even where the lands in issue were the subject of ongoing treaty negotiations.[20] Two recent BC Court of Appeal decisions, however, have repudiated this approach: *Haida Nation* v. *British Columbia and Weyerhaeuser* (the *Haida* decision) and *Taku River Tlingit First Nation* v. *Ringstad* (the *Taku* decision).[21] These decisions hold that the duty to consult exists where there is evidence that aboriginal rights or title may exist, notwithstanding whether such rights received formal legal recognition.

The *Haida* case arose out of a challenge by the Haida Nation to various forest licensing decisions made by the provincial government with respect to Haida Gwaii, also known as the Queen Charlotte Islands.[22] For more than a hundred years, the Haida Nation has asserted aboriginal title to Haida Gwaii, a title it claimed had never been surrendered by treaty or extinguished by federal legislation. The Haida's lawsuit alleged that the province had authorised logging on Haida Gwaii in breach of its duty to consult. In upholding this claim, the Court of Appeal concluded that there was compelling evidence of aboriginal rights and title to the lands in question that vested the provincial government with a duty to consult. Another significant feature of the *Haida* decision is that it represents the first time that a court has extended the duty to consult to private third parties. In this regard, the decision held that the forest company to which the logging rights had been granted was, like the provincial government, legally obliged to consult and seek accommodation in good faith with the Haida Nation.

Recent pronouncements from the BC Court of Appeal have also addressed the content of the duty to consult. Once again, these decisions represent a broadening of the approach previously endorsed by lower courts. Lower court decisions had held that the province could meet its consultative obligations by providing affected First Nations with relevant information pertaining to the proposed development activity and an opportunity to comment before final approvals are made. In the *Haida* case, the Court of Appeal offers a contextual approach to defining the scope of the duty to consult that is 'proportional to the potential soundness of the claim for aboriginal title or aboriginal rights'.[23] Where the evidence of aboriginal rights is compelling, this requirement may entail securing the consent of the affected First Nation. At a minimum, however, the consultation must be undertaken in good faith, and intended to substantially address 'the concerns of the aboriginal peoples whose lands are at issue'.[24]

Lessons from the Treaty Table and Litigation

The experience of BC First Nations in seeking vindication for their rights, whether through the treaty process or in the courts, has been a frustrating one. In both arenas, the costs of participation — in terms of time, money, and other resources — have been enormous and the benefits difficult to quantify.[25] While, for last decade, the BC treaty process has finally provided the First Nations with a venue in which they are able to articulate and pursue their agenda, governments have to date been unwilling to incorporate into these processes measures that provide meaningful interim protection for their rights or means through which they can begin to derive enhanced benefits from forest and other resources while negotiations continue. Progress at the treaty table has also been hampered by federal-provincial disputes over who is to fund and implement ensuring settlements.

While First Nations have secured important victories in the courts, litigation has not proved any more fruitful in terms of securing interim protection for First Nations resources or immediate benefits from the development of these resources. By its nature, litigation has tended to yield outcomes that are either relatively narrow in their application or alternatively so conceptually broad that the actual impact of such decisions on government policy has been relatively modest relative to the costs that have been incurred. Moreover, even where hard law victories have been achieved, recent experience has shown that these are vulnerable to judicial rollback.[26]

For these reasons, First Nations across Canada, and particularly in BC, have increasingly begun to look to new strategies and approaches to secure recognition and respect for their rights. Can the FSC-BC standard succeed where negotiation and litigation have failed? Does it offer a vehicle for First Nations to exercise new control over forest developments on their traditional lands in a way that will provide immediate protection to those resources, as well as benefits from them? Does it do so in a way that will not compromise rights they are continuing to pursue at the treaty table and through the courts? In short, can soft law provide a means of securing these goals where durable hard law solutions have proven elusive?

First Nations and the FSC-BC Regional Standard

Introduction

The Forest Stewardship Council (2000) has established a set of ten broad principles of forest stewardship and a number of criteria for each. Principle 3 and its four subsidiary criteria state as follows:

> *Principle No. 3: Indigenous Peoples' Rights*
> The legal and customary rights of indigenous peoples to own, use, and manage their lands, territories, and resources shall be recognised and respected.

3.1 Indigenous peoples shall control forest management on their lands and territories unless they delegate control with free and informed consent to other agencies.
3.2 Forest management shall not threaten or diminish, either directly or indirectly, the resources or tenure rights of indigenous peoples.
3.3 Sites of special cultural, ecological, economic or religious significance to indigenous peoples shall be clearly identified in cooperation with such peoples, and recognised and protected by forest managers.
3.4 Indigenous peoples shall be compensated for the application of their traditional knowledge regarding the use of forest species or management systems in forest operations. This compensation shall be formally agreed upon with their free and informed consent before forest operations commence.

Regional affiliates, such as the FSC-BC, are tasked with developing more specific indicators and verifiers for each principle and criterion, and together these constitute the standards against which operators seeking certification in that region will be measured. Following seven years of public consultations and development work, the FSC-BC Regional Standard received preliminary approval in July 2003.

For First Nations, the threshold challenge confronted in this regional process was to design a Principle 3 standard for British Columbia that ensured proactive protection for aboriginal rights and title — an aspiration reflected in the commentary by the BC Standards development team:

> before anything else, Principle 3, its four Criteria, and all its Indicators and Verifiers must unequivocally ensure protection of aboriginal rights if BC's First Nations' participation in FSC processes is to become a reality. The Standards that deal with aboriginal rights must confirm those rights. And...the Standards must give First Nations the power to permit or deny certification based on the quality, the depth and the sincerity of that protection (FSC-BC 2001, 128).

By securing a standard that met this threshold challenge, First Nations hoped to avoid many of the uncertainties and expenses associated with seeking recognition and respect for aboriginal rights and title through legal and political channels. Moreover, many were optimistic that such a standard would ensure that First Nations communities derived significant new benefits in terms of jobs and revenues from the forest resource. To achieve these goals, there was a broad recognition of the need for provisions that would allow First Nations to develop new, mutually beneficial relationships with the province's forest companies.

While First Nations were uniformly of the view that a prerequisite to relationship building was the development of a standard that dramatically enhanced their ability to negotiate with forest operators and participate in forest management, not all agreed on how to pursue this end. Some argued that the legacy of mistrust between First Nations and the forest industry militated in favour of a standard that defined the nature and parameters of this relationship in detail. Others argued for flexibility, contending that a standard would only work if it eschewed legalisms and mandatory requirements, creating the opportunity for new relationships to evolve and adapt to local conditions, interests, and needs.

First Nations were also divided over whether and to what extent the standard should specify requirements governing how proposed forest management plans secured approval within and beyond the First Nation community directly involved. For some, a paramount concern was to avoid the certifying operations where the operation faced substantial internal opposition within a First Nations community or where the operation did not have the support of a First Nation that shared jurisdiction over the territory to be certified (a 'shared territory' situation). Some First Nations participants therefore argued for a standard that specifically addressed and sought to minimise these risks. Others, however, contended that such an approach would undermine aboriginal self-government by requiring FSC certifiers to adjudicate the merits of First Nations decision-making processes. It was also argued that to incorporate such requirements would, in many instances, hamper — if not entirely thwart — the certification process by conferring a veto power on disaffected groups or individuals.

Forest managers involved in negotiating the BC standard also had significant interests at stake. Heavily reliant on Crown timber that is the subject of pending aboriginal claims, BC forest managers are acutely aware of the costs associated with the continuing uncertainty created by the impasse with respect to the resolution of such claims. For many, particularly those with markets in Europe and United States, developing a standard that would reduce or eliminate the aboriginal-sponsored blockades, protests, and litigation now common to carrying on business in the province was seen as a key priority. To achieve this goal, forest operators were cognizant of the need for a standard that fostered the development of constructive, new relationships with First Nations communities. They were also generally supportive of a standard that enhanced the ability to First Nations to participate more meaningfully in the forest planning and management.

Like some First Nations, however, forest managers favoured a standard that established general parameters and goals as opposed to one that imposed specific requirements. This preference flowed from two closely related concerns: viability and cost. Of particular concern on the former front was the spectre of a standard that imposed requirements with respect to internal community support and shared territory situations. The viability of a highly detailed standard was also an issue for some small operators concerned about their capacity to meet the requirements. In addition, a frequently voiced concern was whether, on a cost-benefit analysis, a detailed standard could create adequate incentives for forest operators to seek certification.

British Columbia's Principle 3 Standard

Given the competing perspectives and approaches just described, the standard that ultimately emerged out of the BC Regional process is one born of compromise. As such, in several key respects, both in terms of content and style, it departs from its predecessor draft standards. In crafting these compromises, the goal of the drafters was 'to provide security for those First Nations who require it, flexibility for those First Nations who desire it, and opportunity for improvement of relations between First Nations and forest managers' (FSC-BC 2002b, 1). Confident that they achieved these goals, the drafters described the Principle 3 standard as being one of 'the strongest and clearest' of its kind anywhere in the FSC realm.

The section considers first the nature of the rights recognised and protected by the standard and then analyses the process by which it contemplates protecting these rights.[27]

Nature of the Rights Protected Principle 3 confers upon First Nations a variety of protections associated with each of its four criteria. These oblige forest managers to:

- recognise and respect First Nations legal and customary rights to their lands, territories, and resources (Criterion 3.1);
- maintain the resource and tenure rights of First Nations (Criterion 3.2);
- identify, recognise, and protect sites of special cultural, ecological, economic, or religious significance (Criterion 3.3); and
- compensate First Nations for the application of their traditional knowledge (Criterion 3.4).

The drafters of BC's Principle 3 standard faced several daunting challenges. One was to develop language that defined the nature of these rights in a manner that provided sufficient guidance that it could readily be applied on the ground. Another was that judicial and political definition of these rights remains very much in its infancy and is still the subject of considerable controversy. As such, the drafters sought to define the relevant rights in a manner that referred to current 'minimum legal requirements' but, at the same time, made it clear that managers were expected to commit themselves to exceeding these same requirements. Furthermore, there was the question of who would ultimately control how these rights were defined. As already explained, until recently the dominant approach taken by BC courts has been to vest First Nations with the burden of proving their rights — an approach that effectively presumes that aboriginal rights do not exist until they have been confirmed judicially or recognised by the state. The Principle 3 standard reverses this presumption, giving First Nations affected by forest operations the ability to define the scope and nature of their relevant rights. Yet another challenge was the issue of designing and implementing a process by which these rights and interests are recognised and respected.

In the next section, the lattermost question will be explored. First, however, it is important to consider how the BC Principle 3 standard defines the rights contemplated by various criteria and who controls that process of definition.

Legal and Customary Rights to Lands, Territories, and Resources (Criterion 3.1) The FSC has approved a generally applicable international definition for the subject of this criterion. In the BC standard, the term 'lands, territories, and resources' bears an internationally approved meaning, namely the 'total environment of the lands, air, water, coastal sea, sea-ice, flora and fauna and other resources which [indigenous peoples] have traditionally owned or otherwise occupied or used'.[28] It is the 'legal and customary rights' of First Nations with respect to this broad subject that Principle 3 requires forest managers to respect.

In the BC Regional Standard and First Nations, legal and customary rights are synonymous with 'aboriginal rights and title'. The standard's glossary defines aboriginal rights in a two-sentence paragraph that summarises the ruling of the Supreme Court of Canada in *Van der Peet.*[29] Similarly, it defines aboriginal title in a paragraph that summarises the court's reasons in *Delgamuukw*.[30] The glossary also notes that aboriginal rights must be construed as including the right of consultation, which is defined by reference to recent caselaw on the subject, in particular the BC Court of Appeal decisions in the *Taku* and *Haida* cases.[31]

The BC Principle 3 standard does not specifically address how these broad-brush glossary definitions are to be applied on the ground. The absence of guidance in this regard is of less consequence where the scope and nature of aboriginal rights are defined by treaty. In this situation, the standard contemplates that the terms of the treaty will define the applicable aboriginal rights or title.[32] However, where treaties do not exist, as is the case throughout most of BC, the standard is silent on how these definitions will be implemented. This silence appears to have been deliberate on the part of the FSC-BC and motivated by a least two key considerations. First is the impossibility of developing a more prescriptive standard given, among other things, the fluidity of current caselaw and the diversity of contexts in which the standard will be applied. Second is the need to provide latitude to managers and First Nations to negotiate over these issues, at the same time recognising that, at the end of the process, First Nations must lend their consent to the proposed forest development by affirming that their rights have been respected. How this consent requirement would operate is spelled out in detail in a following section.

Maintenance of Resources and Tenure Rights and Protection of Special Sites (Criteria 3.2 and 3.3) Criteria 3.2 and 3.3 impose particular requirements on managers with respect to aboriginal resources and tenure rights (RTRs), and sites of special cultural, ecological, economic, or religious significance to First Nations (referred to as special sites). Here again the Principle 3 standard incorporates a made-in-BC definition of the subject matter of the right. The glossary definition of RTRs is extremely broad, framed in the form of non-exhaustive, illustrative list (FSC-BC 2002a, 37). This list includes references to various natural resources (water, fisheries, timber and forests, non-timber forest products, subsurface mineral and oil deposits), site or area-based amenities (access routes to resources, hunting and gathering areas, Tribal Heritage Parks, traplines), and activities (guiding operations and cultural tourism). Special sites are also defined in the form of a lengthy catalogue of illustrative sites.

There has been no attempt to tie these definitions to rights currently recognised in Canadian law. Instead, it would appear that BC's Principle 3 standard vests in affected First Nations the responsibility for identifying what RTRs and special sites they wish to have protected.[33] Once identified, the manager is obliged to devise means to maintain and protect RTRs and special sites from being impaired by forestry activities.[34] Where such activities lead to the loss or diminishment of RTRs or special sites, the First Nation must be satisfied with mitigative or compensatory measures that a manager proposes (FSC-BC 2001, 19).

Compensation of the Application of Traditional Knowledge (Criterion 3.4) As with RTRs and special sites, the Principle 3 standard provides a glossary definition of traditional knowledge that takes the form of a non-exhaustive, illustrative list (FSC-BC 2002a, 41). This list includes references to a variety of forms of indigenous knowledge potentially relevant to forest management. It covers knowledge of local fish, wildlife, and plant life and cycles, climatic changes and cycles, local ecosystem information including responses to natural or human disturbances, as well as qualitative information about the human uses of medicinal or edible plants.

Once again, the standard contemplates that First Nations will control the process of defining how this definition is to be applied. Where a First Nation and a forest manager agree, the standard provides that traditional knowledge will be incorporated into the management plan. Moreover, it stipulates that First Nations retain control over their traditional knowledge and be fairly compensated, to their satisfaction, where that knowledge is used by a manager.

The Principle 3 Process as Implemented Implementation of the Principle 3 standard is expected to occur in four stages: negotiation of a protocol agreement, sign-off on the protocol agreement, development of a management plan either through a process of consultation or joint management, and securing the consent of First Nations as a final prerequisite to certification. The protocol agreement defines, in broad terms, the 'nature of the relationship between the Parties'; it is the duty of the manager to negotiate and conclude the agreement (or agreements) with 'all relevant First Nation(s)'. Mandatory components include provisions that address how the parties will establish and conduct their relationship, describe their respective roles, responsibilities, and interests, identify the respective 'appropriate decision-making authorities for all parties', and provide a 'framework for subsequent agreements necessary to give effect to the protocol' (FSC-BC 2002a, 48). Failure to meet these requirements is deemed to be a major failure, a designation reflecting the critical status of the protocol agreement within the standard (49). (A major failure ordinarily precludes certification.)

Where a First Nation so elects, the protocol agreement must contain protective provisions that specify the agreement is without prejudice to 'treaty, land claims settlements, or agreements' that a First Nation may reach with government, and that the agreement shall not be construed to constitute acceptance of provincial Crown title, extinguish aboriginal title, or otherwise derogate from their aboriginal rights

(FSC-BC 2002a, 49). (The standard also enables a First Nation to insist that similar provisions be contained in the management plan that is ultimately developed.)

Once relevant First Nations have signed off on the protocol agreement, the forest manager may proceed to develop a management plan. Development of such a plan can occur either by means of a process of consultation with affected First Nations or through a joint management process at the discretion of the forest manager. The standard's appendix observes that generally speaking 'consultation is better suited to small operators, while joint management are better suited to large operators' (FSC-BC 2002b). It cautions, however, that the decision should also take into account the capacity and interests of the First Nation involved.

The expectations placed on a forest manager who pursues the consultation route are spelled out in the glossary under 'Consulting with First Nations' (FSC-BC 2002a, 18–19). This section commences with a short summary of the legal requirements relating to the duty to consult emerging from recent decisions by the BC Court of Appeal in the *Haida* and *Taku* cases. It then provides a shopping list of what are described as 'characteristics of a good consultation process'. Many would be anticipated to form part of any effective consultative process (that is, that the process is designed by and agreed upon by the parties, that it provides for an adequate schedule for consultation, and that concerns that are raised are appropriately recorded and reflected in the plan that emerges from the consultation). Other characteristics relate to the development of strategies to protect First Nation resources, tenure rights, special sites, and traditional knowledge that are recognised in Indicators 3.2, 3.3, and 3.4. The definition concludes by suggesting that 'financial, technical or logistical capacity-building support, in proportion to the scale and intensity of operations' be made available to affected First Nations to allow them to participate in the consultation process (19).

In contrast, a forest manager who decides to develop a management plan jointly with a First Nation must first negotiate a joint management agreement. Under such an arrangement, the parties embark on an enhanced form of consultation that engages the parties in a process of 'into jointly setting goals, objectives, strategies, implementation, restoration and monitoring of the forest within the management unit' (FSC-BC 2002a, 28–29). It is contemplated that joint management agreements will take a variety of forms, ranging from agreements that deal with 'a relatively few areas of common interest' to those that entail a 'thorough integration of industry and First Nation ideas throughout the whole management plan' (27).

Mandatory components of a joint management agreement include measures to protect First Nation resources, tenure rights, special sites, and traditional knowledge described in Criteria 3.2, 3.3, and 3.4, processes to develop objectives and strategies collaboratively on matters of importance to the First Nation and to develop 'all or a part of the management plan', and efforts to ensure appropriate consultation with respect to any part of the management plan that is not collaboratively developed. Recommended characteristics of an appropriate joint management agreement include, among other things, that it be approved by the decision-making bodies identified in the relevant protocol agreement and that 'financial, technical or logistical capacity-building support, in proportion to the scale and intensity of operations' be made available to First Nations to assist them in its development.

Once a management plan has been developed, either by way of consultation or under the auspices of a joint management agreement, the final stage of the process involves securing the necessary consent from the First Nations involved. Under the BC Principle 3 standard, a manager must secure two types of consent. The first is the 'free and informed consent' of the First Nations to the management plan (FSC-BC 2002a, 49). According to the glossary, such consent must be given both freely (without manipulation, undue influence, or coercion) and knowledgeably (based on all relevant and necessary information) (23). To determine whether such consent has been validly secured, the Principle 3 standard contains a verifier requiring the certifier to ascertain that the First Nation possessed the financial, technical, and logistical capacity adequate to participate on an informed basis in planning and decision making (49).

A forest manager must also convince the certifier that a second form of consent is present. This consent arises out of Verifier 3.1.1(i), which requires that First Nations are satisfied that their 'legal and customary rights over their lands, territories and resources have been recognized and respected' by the manager (FSC-BC 2002a, 48). To meet this test, the certifier must have evidence that the First Nations involved have 'formally [indicated], clearly, unambiguously and normally in writing' that such consent has been given (48). The certifier must also be satisfied that, objectively speaking, First Nations interests and concerns are incorporated in the management plan. Failure of a manager to establish the existence of this second form of consent has been designated as the second (and only other) major failure in the Principle 3 standard.

Two additional consent-related issues merit attention: the role of consent post-certification and the necessity of consent in shared territory situations. As the appendix notes, the concept of consent is not to be construed as a one-time, static event (FSC-BC 2002b, 5). This is evidenced by the requirement that a management plan record 'the conditions under which consent has been given and under which it may be withdrawn' (49). As for situations where more than one First Nation is affected by proposed forestry operations, the standard stipulates that the consent of each affected First Nation be 'ordinarily required'. The appendix, however, emphasises the discretionary nature of this requirement, noting that a certifier must use judgement in assessing the 'directness' of the interest being claimed by referring to factors such as the affected group's proximity to the proposed operation (5).

The FSC-BC Principle 3 Standard as Soft Law

At the outset of this chapter, soft law was defined in terms of the extent to which it departs from the three hallmark features of hard law. Hard law, it was contended, is characterised by clearly enunciated legal obligations, articulated with a relatively high level of precision, in circumstances in which interpretive and enforcement responsibilities are delegated to an independent, legally constituted adjudicative body. Soft law arrangements, therefore, typically entail a softening or weakening of the legal framework along one or more of these three dimensions.

In order to evaluate FSC-BC's Principle 3 standard as a form of soft law, it is instructive to consider its structure and likely operation along these three dimensions. Considered in this light, although the standard is doubtlessly soft law, insofar as the obligations it imposes are self-selected, in many other respects it tilts toward the hard law end of the legalisation spectrum. This observation is bolstered by an assessment that traverses the indicia of obligation, precision, and delegation.

First, one must consider the nature of the obligations it requires managers to undertake. Several features of the standard are noteworthy in this regard. It describes a series of steps that managers must follow before embarking on forest development on lands claimed by First Nations. In undertaking this process, managers are required to respond to and satisfy concerns identified by First Nations. In other words, while managers have formal conduct of the process, they are obliged to meet a satisfaction test defined from the perspective of the First Nations engaged in the process. Moreover, if a manager fails to navigate the process successfully, substantive consequences can arise. This occurs due to the fact that the standard's requirements relating to the protocol agreement and informed consent are designated as major failures that would preclude certification. No other Principle 3 standard currently in force in the FSC world contains analogous substantive bars to certification (Tollefson, Gale, and Haley 2004).

The Principle 3 standard is also intriguing when considered in terms of the legal precision with which it articulates not only the obligations it imposes but also how those obligations are to be met. As already noted, in substantive terms a manager's obligations are to be assessed by the certifier against measurable outcomes (that is, the negotiation of a protocol agreement and the existence of informed consent). Managers must also accomplish other procedural outcomes (that is, the negotiation of a management plan) that are not given the formal designation of major failure.

Nonetheless, in drafting the final Principle 3 standard, FSC-BC deliberately set out to balance the need for precision (and testable outcomes) against the competing need to avoid over-prescriptiveness and legalisms. To this end, and with a view to promoting flexibility in implementation, the final text of the standard is less than half the length of its predecessor draft. As a result, a variety of requirements that appeared in earlier iterations have been eliminated. These include obligations that were seen to be controversial or difficult to implement, such as a proposed obligation on managers to demonstrate that the proposed management plan has broad support within affected First Nation communities.

The decision of FSC-BC to reduce the prescriptiveness of the standard is likely to have two main consequences. On the one hand, insofar as the parties to a certification bid operate within the broader framework of the standard, they will enjoy greater latitude to develop situation-specific solutions. This is, of course, considered to be one of the strengths of soft law that not only promotes private autonomy but also has the potential (through the piloting of new approaches) to support social learning and consensus building. On the other hand, to the extent that the standard fails to enunciate clear measures against which compliance is to be assessed, it effectively devolves to certifiers a broad discretion in terms of their ultimate task: the decision whether to recommend that a company should be given FSC certification.

Finally, it is instructive to reflect on the Principle 3 standard in terms of the delegation of authority. In the international domain, more than other single factor, the reluctance of states to delegate adjudicative authority to independent bodies accounts for the relative paucity of hard law as opposed to soft. This reluctance almost invariably stems from concerns about state sovereignty. Although the literature in this regard in the domestic context is much sparser, there is every reason to expect that governments eschew hard law in favour of soft law for similar reasons (even though this phenomenon is more often framed in the desire to preserve executive discretion as opposed to state sovereignty *per se*).

Given the foregoing, what observations can be made about the FSC regime as it is likely to operate in British Columbia? Companies that elect FSC certification reserve the right to withdraw from FSC jurisdiction at any time without notice. As such, their decision to delegate the power to interpret and enforce all applicable FSC-imposed obligations to FSC-accredited certifiers (during the application process) and latterly to the FSC (once certification is secured) is conditional. Yet while the so-called exit option is ever-present, it is worth recognising that the same is true for states that undertake obligations under international treaties (admittedly to a much lesser degree).

It is also worth considering the delegation issue from the perspective of First Nations that participate in FSC certification of their traditional lands. As with the companies certified to operate on their lands, the Principle 3 standard preserves the exit option for First Nations. This option is reflected, among other things, in the concept that the existence of informed consent is an ongoing requirement for companies wishing to maintain FSC certification.

The incipient terminability of the legal relationship crystallised in the Principle 3 standard renders the legal relationship between the parties tenuous, yet at the same time arguably creates powerful incentives for a certified company and relevant First Nations to negotiate and resolve their differences. In other words, both parties stand to lose if FSC certification is withdrawn. For the company, losing its certification is likely to have adverse market and reputational effects. Equally, from a First Nations perspective, if a company is forced or voluntarily decides to leave the FSC, the unique bargaining power accorded the company under the FSC model with respect to how and when forestry operations are run is lost.

In these respects, the FSC model would appear to capture benefits, and create incentives for negotiation and compromise, not found in hard law models (founded on more inflexible authority-delegation principles). This said, the foregoing analysis is premised on the FSC having the capacity to respond and apply the BC Principle 3 standard appropriately and decisively, especially in relation to allegations that certified companies have failed to abide to obligations they have undertaken to First Nations. At this juncture, and in the context of this chapter, all that can be said in this regard is that the future of the FSC as a self-governing entity remains uncertain and remains an question that merits close attention (Tollefson, Gale, and Haley 2004).

Conclusion: The Future of British Columbia's Principle 3 Standard

That FSC-BC was able to develop and unanimously endorse the Principle 3 standard is a considerable accomplishment. It is an accomplishment that speaks to the commitment of the parties involved to tackle the challenge of finding a workable, new approach to reconciling the legitimate aspirations of First Nations with the interests of those who now and in the future rely on the province's forest resources.

Although the standard is less prescriptive and legalistic than it was in earlier drafts, by virtue of its strong consent requirements it meets the threshold concern of most First Nations: the need to ensure unequivocal protection of aboriginal rights. At the same time, the standard provides a viable foundation for the development of new relationships between First Nations and the province's forest companies. It accomplishes this by carefully balancing the need to create a framework for these relationships to develop without hampering the process by which this occurs through unnecessary prescription.

This said, there is little doubt that implementation of the Principle 3 will present significant challenges in British Columbia. Many of these challenges are likely to flow from the complex politics of First Nations. Unlike some other FSC regional standards, the BC standard does not attempt to define who or what is a First Nation (such as the U.S. Rocky Mountain Standard). This is perhaps not surprising given the complex anthropological and political realities of BC First Nations, particularly the distinction between traditional and elected First Nations leadership. Similarly, the drafters have chosen, at the apparent urging of First Nations themselves, not to address the question of what type of community approval must exist within First Nations community as a prerequisite to certification. Furthermore, while the standard does address the issue of shared territories, it does so in a brief and relatively cryptic fashion that provides little guidance to how the standard should be implemented where more than one First Nation has legitimate legal interests in the area proposed for certification (a situation that will likely occur relatively frequently).

Another looming challenge concerns the financial, technical, and logistical capacity of First Nations to participate effectively in the certification process. Throughout the standards development process, this concern was raised repeatedly. In earlier drafts of the standard, it was addressed by a formal requirement that First Nations be provided with such support without specifying from what source this support was to come; in the final version, the presence of such support has evolved into a verifier that certifiers must consider in deciding whether the manager has secured the informed consent of the affected First Nation.

Thwarted by legal and political forces and institutions in their quest to regain control over forestry operations on their traditional lands, and to secure meaningful economic benefits from them, BC First Nations have increasingly come to regard the FSC-BC standard-setting process as a promising strategic opportunity. Engagement of First Nations leadership in the process has yielded a standard that has the potential to significantly alter how forestry within the province is managed.

Rather than presuming that aboriginal rights and title do not exist, the standard presumes they do. Rather than forcing First Nations to resort to costly and uncertain litigation to protect their rights, tenures, and sites of special significance, the standard obliges forest managers to negotiate with First Nations over how and to what extent these interests will be protected, and to reach agreement with them over compensation and mitigation measures. And rather than forcing First Nations into a defensive, adversarial posture when dealing with forest companies, the standard allows them to develop working relationships with forest managers based on traditional Native protocols of respect and mutual recognition.

Now, more than ever, the imperative of achieving a lasting hard law resolution to the longstanding aboriginal rights and title claims of BC First Nations cannot be denied. The collective failure to achieve this resolution has set the stage — and raised the stakes — for a bold experiment in soft law as crystallised in the FSC-BC Principle 3 standard. Given the complexity of the social, political, economic, and cultural conditions under which it will take place, the outcome of this experiment is fraught with uncertainty. It is hoped, however, the existence of this overarching imperative will create strong incentives to make the experiment work.

Notes

1. The Forest Stewardship Council publishes many statistics on its website at <www.fscoax.org>.
2. For example, in *Our Forests, Our Future*, the report of the World Commission on Forests and Sustainable Development (1999, 155), Resolution 2 states 'that conserving and sustainably managing the world's forests go hand in hand with the priority of reducing poverty and sustaining livelihoods of millions of poor people and numerous communities who depend heavily on forests'.
3. In this chapter, the terms 'indigenous peoples', 'First Nations', and 'aboriginal peoples' are used interchangeably; however, in reference to a particular indigenous 'nation', the term 'First Nation' is used.
4. In the mid 1990s, the United Nations declared 1995–2004 to be the 'Decade of Indigenous Peoples'. A variety of indigenous rights are gaining recognition in international law, including

the right to own and control traditional territories and resources, the right to exercise customary law, the right of self-government, and the right to control and be compensated for the use of traditional knowledge. In some instances, these agreements have crystallised into hard law that is enforceable by indigenous peoples. An illustration is a decision of the ruling by the Inter-American Court of Human Rights that the Government of Nicaragua violated the American Convention on Human Rights by authorising a Korean forest company to log on traditional lands of an indigenous community without its consent.

5. See, for example, section 4.3 of the 'Assessment of the Need for Aboriginal Compliance with Sustainable Forest Management and Forest Product Certification Systems' published by the National Aboriginal Forestry Association (1996).
6. The terms 'status' or 'registered' derive from the *Indian Act*, enacted in 1876 by the federal government, which has exclusive jurisdiction with respect to 'Indians and Indian lands' under section 91(24) of the Constitution. Non-status Indians are of Indian ancestry but, through intermarriage or by abandoning their status rights, have lost their legal status while retaining their Indian identity. Canada's provincial governments are vested with a broad jurisdiction over property and civil rights within the province (section 92[14]), with exclusive powers to make laws with respect to the development, conservation, and management of natural resources within a province, including forest resources (section 92A[1]), and with title to Crown lands within the province (section 109). The statistics quoted here are from Basic Departmental Data reports published by Indian and Northern Affairs Canada (2002); see also Muckle (1998, 4).
7. The Proclamation was intended to serve 'multifarious concerns' including curbing the territorial expansion of the United States, consolidating political control over the newly acquired territory of Quebec, and preventing conflict between white settlers and Natives as European settlement moved west (Rotman 1996, 28; Miller 1991, 73–74). Under its terms, an immense tract of land bordered by Rupert's Land on the north, by the Thirteen Colonies on the east, and by the Mississippi River to the west was designated as Indian hunting grounds. British settlers were forbidden from settling on or purchasing these lands from Indians without prior Crown permission.
8. *St. Catherine's Milling & Lumber Co.* v. *R.* [1888], 14 App. Cas. 46 (P.C.).
9. This interpretation of the Royal Proclamation has been vigorously disputed by First Nations. For example, John Borrows (1994) has argued persuasively that if the historical context and aboriginal perspective are taken into account, it becomes clear that the Proclamation is really part of a treaty of mutual non-interference between First Nations and the Crown. Such a 'perspective illustrates that First Nation people never contemplated relinquishing their sovereignty or the underlying title to their land' (44).
10. *Calder* v. *Attorney-General of British Columbia* [1973] S.C.R. 313.
11. For a discussion of the anthropological evidence in such court cases, see Kew (1993–94, 95).
12. See page 7 of the 'BC Treaty Commission Annual Report 2001' (British Columbia Treaty Commission 2001a). The single treaty to have been concluded is the Nisga'a land claim agreement of 2000. Negotiations leading up to this treaty, however, commenced before initiation of the BC treaty process and were concluded outside of the process.
13. See page 5 of 'Looking Back, Looking Forward: A Review of the BC Treaty Process' (British Columbia Treaty Commission 2001b). Although the referendum process was broadly regarded as undermining relations between the provincial government and First Nations, negotiations have continued and several new agreements in principle have been announced. These are awaiting ratification by the First Nation communities involved (see <www.bctreaty.net>).
14. *R.* v. *Sparrow* [1990] 1 S.C.R. 1075.
15. *R.* v. *Van der Peet* [1996] 2 S.C.R. 507.
16. *Delgamuukw* v. *British Columbia* [1997] 3 S.C.R. 1010.

17. This has been an issue in two recent Supreme Court of Canada decisions including *Paul* v. *British Columbia* 2003 SCC 55 (a constitutional challenge to the jurisdiction of provincial forestry tribunal) and *Kitkatla* v. *British Columbia C* 2002 SCC 31 (a constitutional challenge to a provincial heritage conservation law).
18. The BC Supreme Court rejected an attempt by a First Nation to seek a judicial declaration of its site-specific aboriginal right to fish on the basis that it was not prepared to make such a determination unless and until the First Nation was prepared to embark on a full trial of the issue as framed within the context of specific allegations of government infringements of its asserted right; see *Cheslatta Carrier Nation* v. *British Columbia* [1999] B.C.J. No. 2639 (B.C.S.C.), affirmed in *Cheslatta Carrier Nation* v. *British Columbia* [2000] B.C.J. No. 2030 (B.C.C.A.).
19. See *British Columbia (Minister of Forests)* v. *Okanagan Indian Band* [2001] B.C.J. No. 2297 (C.A.) and *Tsilhqotin Nation* v. *Canada (A.G.)* [2002] B.C.J. No. 1652 (B.C.C.A.).
20. On the trial process, see *Haida Nation* v. *British Columbia (Minister of Forests)* [2000] B.C.J. No. 2427 (B.C.S.C.), *British Columbia (Minister of Forests)* v. *Westbank First Nation*, [2000] B.C.J. No. 1613 (B.C.S.C.), and *Chief Stanley Thomas* v. *British Columbia*, Oral Reasons by Bauman J., 11 July 2000 (B.C.S.C.). On ongoing treaty negotiations, see *Haida Nation* v. *British Columbia (Minister of Forests)* and *Chief Stanley Thomas* v. *British Columbia.*
21. *Taku River Tlingit First Nation* v. *Ringstad et al.* [2002] BCCA 59; *Haida Nation* v. *British Columbia and Weyerhaeuser* [2002] BCCA 147.
22. See *Haida*, note 224.
23. See *Haida (CA)* at paragraph 51 *ibid.*
24. Haida citing from *Delgamuukw* at paragraph 38.
25. Many of the costs incurred by First Nations in treaty negotiations are funded as an 'advance' by the federal government against the proceeds of any settlement that is ultimately achieved.
26. A good illustration is provided by a recent decision of BC's provincial court that struck down as contrary to the equality provisions of the *Charter of Rights and Freedoms* a federal pilot programme designed to assist aboriginal fishers by setting aside aboriginal-only fishing 'openings' (see *R.* v. *Kapp* [2003] B.C.J. No. 1772 [July 2003]). The decision is on appeal. In the meantime, however, the federal government has closed down the programme.
27. For a very useful guide to this and related issues, see Collier, Parfitt, and Woolard (2002).
28. This definition is borrowed from Article 26 of the Draft United Nations Declaration of the Rights of Indigenous Peoples (United Nations Economic and Social Council 1994).
29. The glossary definition states: 'Aboriginal Rights: A practice, custom or tradition integral to the distinctive culture of the aboriginal group claiming the right. Often aboriginal rights, including site specific rights, can be made out even if title cannot: based on *R.* v. *Van der Peet* [1996] 2 S.C.R. 507; *R.* v. *Adams* [1996] 3 S.C.R. 101' (FSC-BC 2002a, 15).
30. The glossary definition states: 'Aboriginal Title: The unique title to the First Nation's lands, territories and resources which arises from occupation before the assertion of British sovereignty, or which arises from and reflects the pattern of land holdings under aboriginal law. Aboriginal title confers more than the right to engage in site-specific activities. What aboriginal title confers is the right to the land itself. If a Nation has aboriginal title, the land may be used for a variety of activities that need not be elements of a practice, custom or tradition integral to the distinctive culture of the aboriginal group claiming the right: based on *Delgamuukw* v. *British Columbia* [1997] 3 S.C.R. 1010' (FSC-BC 2002a, 15).
31. See the glossary definition of 'Consulting with First Nations' (FSC-BC 2002a, 18–19). For citations to cases mentioned in text, see note 20 above.
32. This is explicit in Verifier 3.1.1(i)(b) contained in Draft 2 of the BC Standard (FSC-BC 2001, 16).

33. The obligation on First Nations to identify such sites was explicit in Draft 2; see Indicators 3.2.1(i) and 3.3.1(i) (FSC-BC 2001, 19). This obligation would appear to be implicit in Draft 3.
34. The obligation on a manager to develop such a strategy was explicit in Draft 2; see Indicators 3.2.1(iii) and 3.3.1(iii) (FSC-BC 2001, 19). This obligation would appear to be implicit in Draft 3.

References

Abbott, Kenneth W. and Duncan Snidal (2000). 'Hard and Soft Law in International Governance'. *International Organization* vol. 54, no. 3, pp. 421–456.

Borrows, John (1994). 'Constitutional Law from a First Nation Perspective: Self-Government and the Royal Proclamation'. *University of British Columbia Law Review* vol. 28, no. 1, pp. 1–47.

Boyd, Robert (1994). 'Smallpox in the Pacific Northwest'. *BC Studies* no. 101 (Spring).

British Columbia Treaty Commission (2001a). 'Annual Report 2001: The Year in Review'. <www.ainc-inac.gc.ca/bc/ftno/bctcar1_e.html> (April 2004).

British Columbia Treaty Commission (2001b). 'Looking Back, Looking Forward: A Review of the BC Treaty Process'. <www.bctreaty.net/files_2/pdf_documents/review_bc_treaty_process.pdf> (April 2004).

Certified Forest Products Council (2003). 'Comparison of Forest Certification Systems'. <certifiedwood.org/search-modules/CompareCertSystems.asp> (April 2004).

Colchester, Marcus (2000). 'Indigenous Peoples and the New "Global Vision" on Forests'. <greatrestoration.rockefeller.edu/21Jan2000/Colchester.PDF> (April 2004).

Collier, Russell, Ben Parfitt, and Donovan Woolard (2002). *A Voice on the Land: An Indigenous Peoples' Guide to Forest Certification in Canada*. Ecotrust Canada and the National Aboriginal Forestry Association, Ottawa. <www.ecotrustcan.org/pdf/VOICE_fullbook.pdf> (April 2004).

Forest Stewardship Council (2000). 'FSC Principles and Criteria'. <www.fscoax.org/principal.htm> (April 2004).

Forest Stewardship Council of British Columbia (2001). 'Regional Certification Standards for British Columbia'. Draft 2. <www.fsc-bc.org> (April 2004).

Forest Stewardship Council of British Columbia (2002a). 'Regional Certification Standards for British Columbia'. <www.fsc-bc.org> (April 2004).

Forest Stewardship Council of British Columbia (2002b). 'Regional Certification Standards for British Columbia'. Draft 3.

Gale, Fred and Cheri Burda (1998). 'The Pitfalls and Potential of Eco-Certification as a Market Incentive for Sustainable Forest Management'. In C. Tollefson, ed., *The Wealth of Forests: Markets, Regulation, and Sustainable Forestry*. University of British Columbia Press, Vancouver.

Howlett, Michael (2001). 'Policy Venues, Policy Spillovers, and Policy Change: The Courts, Aboriginal Rights, and British Columbia Forest Policy'. In B. Cashore, ed., *In Search of Sustainability: British Columbia Forest Policy in the 1990s*. UBC Press, Vancouver.

Indian and Northern Affairs Canada (2002). 'Basic Departmental Data'. <www.ainc-inac.gc.ca/pr/sts/bdd02/bdd02_e.html> (April 2004).

Inter-Commission Task Force on Indigenous Peoples (1997). *Indigenous Peoples and Sustainability: Cases and Actions*. International Books, Utrecht.

Kennett, Stephen A. (1993). 'Hard Law, Soft Law, and Diplomacy: The Emerging Paradigm for Inter-Governmental Cooperation in Environmental Assessment'. *Alberta Law Review* vol. 31, no. 4, pp. 644–661.

Kew, Michael (1993–94). 'Anthropology and First Nations in British Columbia'. *BC Studies* no. 100, pp. 78–105.

Mathias, Chief Joe and Gary R. Yabsley (1991). 'Conspiracy of Legislation: The Suppression of Indian Rights in Canada'. *BC Studies* no. 89 (Spring).

McKee, Christopher (2000). *Treaty Talks in British Columbia: Negotiating a Mutually Beneficial Future*. 2nd ed. UBC Press, Vancouver.

McMillan, Alan D. (1995). *Native Peoples and Cultures of Canada: An Anthropological Overview*. 2nd ed. Douglas & McIntyre, Vancouver.

Miller, J. R. (1991). *Skyscrapers Hide the Heavens: A History of Indian White Relations in Canada*. Rev. ed. University of Toronto Press, Toronto.

Monture-Angus, Patricia (1999). *Journeying Forward: Dreaming First Nations' Independence*. Fernwood, Halifax.

Morais, Herbert V. (2002). 'The Quest for International Standards: Global Governance vs. Sovereignty'. *University of Kansas Law Review* vol. 50, no. 779.

Muckle, Robert J. (1998). *The First Nations of British Columbia: An Anthropological Survey*. UBC Press, Vancouver.

National Aboriginal Forestry Association (1996). 'Discussion Paper: Assessment of the Need for Aboriginal Compliance with Sustainable Forest Management and Forest Product Certification Systems'. <www.nafaforestry.org/cert> (April 2004).

Rotman, Leonard I. (1996). *Parallel Paths: Fiduciary Doctrine and the Crown-Native Relationship in Canada*. University of Toronto Press, Toronto.

Tollefson, Chris, Fred Gale, and David Haley, eds. (2004). *The Certified Forest: The Politics, Implications and Prospects for Forest Stewardship Council Standards in British Columbia*.

United Nations Economic and Social Council (1994). 'Report of the Sub-Commission on Prevention of Discrimination and Protection of Minorities on Its Forty-Sixth Session'. <www.unhchr.ch> (April 2004).

World Bank (2001). 'A Revised Forest Strategy for the World Bank Group'. <www.iucn.org/themes/fcp/activities/publications/worldbank_forestpolicynov02.pdf> (April 2004).

World Commission on Forests and Sustainable Development (1999). *Our Forests, Our Future*. Cambridge University Press, Cambridge.

PART II
SETTING STANDARDS FOR LABOUR

Chapter 5

Codes of Corporate Conduct and the Labour Regulatory State in Developing Countries

Adelle Blackett

Contemporary initiatives for setting soft law standards for labour must take into account the specificity of labour regulation, at both the national and international levels. Despite their apparent pragmatic potential to improve often deplorable conditions, such initiatives on labour are deeply called into question. Labour is a field once largely filled by a dense combination of industrial relations regulatory machinery at the domestic level, particularly in states that espoused embedded liberalism, and a plethora of international labour standards, meant largely to guide the creation and strengthening of state labour regulatory power. It is also a field in which governmental, business, and union initiatives of both the hard and soft law varieties have coexisted, each buttressing the other to provide robust, workplace-sensitive regulatory results. Yet the challenges to state regulatory capacity, as well as the capacity of traditional domestic workplace actors in labour deeply familiar to developing countries, are increasingly commonplace in industrialised countries facing the competitive pressures of a particular model of economic globalisation. The understandable resistance of the International Labour Organization (ILO) to move beyond its sophisticated incentives-based monitoring to a more robust sanctions-based regime leaves it incapable of acting alone as an effective counterbalance to the decreased state capacity.[1] In the face of highly publicised portrayals of labour rights violations committed by major multinational corporations (MNCs), new forms of soft law — notably codes of corporate conduct — have proliferated.

Codes of corporate conduct vary significantly in nature and in scope (Vallée et al. 2003); but whether they are individual company codes or co-ordinated efforts of groups of nongovernmental organisations (NGOs) and MNCs, they constitute emerging forms of labour regulation (for a thorough discussion, see Blackett 2001). They all embrace one basic paradox: they seek to offer varying levels of non-state regulation via the territorially unconstrained product to ensure that labour, wherever the workplace may be, is not commodified. As new regulatory mechanisms, codes of corporate conduct should be justified on the basis of the underlying goals and principles of labour regulation.

This chapter draws on three critiques of labour regulatory initiatives such as codes of corporate conduct (see Blackett 2001). First, the initiatives fail to consider the impact of self-regulatory approaches on the particular places in which they are meant to apply. Most notable are export processing zones (EPZs), where actual or *de facto* deregulation of labour policy has taken place. Corporate self-regulatory initiatives lead to the development of a new form of legality — corporate self-regulation by way of the product. Second, the new forms of labour regulation fail to live up to labour law's basic goals: worker protection and democratic participation. Indeed, they may further hamper the ability of traditional labour law approaches to be adapted to the circumstances of increasingly integrated societies. Third, creative approaches to the transnational regulation of labour may usefully de-centre the traditional approaches to the all-powerful unitary state; however, the regulation of labour across state borders calls for careful consideration of the importance of representation through democratic governance.

This chapter applies these critiques to a widely circulated policy discussion entitled 'Ratcheting Labor Standards: Regulations for Continuous Improvement in the Global Workplace', prepared by Charles Sabel, Dara O'Rourke, and Archon Fung (2000; see also Fung, O'Rourke, and Sabel 2001a, 2001a). Sabel, O'Rourke, and Fung agree that codes of corporate conduct addressing labour standards offer a potentially powerful but currently limited approach to labour regulation. They propose a competition-based alternative they call 'ratcheting labour standards' (RLS). The decidedly low-tech metaphor of a ratchet wrench in the context of economic globalisation seems intended to focus attention on the historical parallels with nineteenth-century industrialisation: at both moments, a shift in the paradigm of economic growth has occurred and, at both moments, shocking labour conditions in newly constructed spaces have led to a need to rethink existing regulatory approaches. The authors contend that their third way can improve the social performance of firms, by calling upon four principles discussed below — transparency, competitive comparison, continuous improvement, and sanctions. Although their quest for alternative regulatory models is compelling, this chapter responds that re-regulatory initiatives in relation to labour must be held up to the light of central labour regulatory goals. When held up to these goals, the measures are revealed to be deeply limited.

The Ratcheting Labour Standards Approach to Codes of Corporate Conduct

Sabel, O'Rourke, and Fung's proposed approach is not meant to displace the plethora of individual or co-ordinated self-regulatory initiatives. Rather, it seeks to add other organisational layers to rationalise and integrate national and intergovernmental organisations into the flurry of private sector activity. First, it would require corporations to disclose a range of information to monitors. The monitors, in turn, would rank those firms according to a set of indicators for labour standards. The monitors would be trained and themselves monitored by intergovernmental organisations acting in the role of super monitors or umpires alongside NGOs and international confederations of trade unions; Sabel, O'Rourke, and Fung (2000, 33) refer in detail to the World Bank and in passing to the ILO, suggesting that they should 'coax national governments to adopt [RLS]-compatible labour law by developing model legislation and international covenants'.[2]

The four principles mentioned above would be the cornerstones of the RLS framework. First, transparency would be required for accurately providing information to consumers. High-profile firms would compete to build their reputation as socially responsible actors, on the basis of information comparisons — hence competitive comparison. Furthermore, 'social and market pressures would push firms, perhaps with public agencies, NGOs or unions' to improve their labour performance continuously, leading to improvement in the very nature of the labour standards that are respected (Fung, O'Rourke, and Sabel 2001b, 21). The authors concede that forceful regulatory institutions that have the power to sanction are necessary, in addition to more informal approaches such as forms of ethical consumerism and boycotts. Under RLS, this translates into using 'public power to dislodge information that would in turn trigger more nuanced and ultimately powerful incentives to improve social performance'(21).[3]

The authors' competition-based model focusses on the image and profit-oriented motives of leading firms to ratchet best practices upward, on a continuous basis. The proposal is meant to provide a building-block or rolling-rule approach to improving labour standards compliance in developing countries. The ability to obtain, disseminate, and ensure that the information provided is credible, reliable, and acted upon by potentially information-overloaded consumers is, of course, central to their scheme.[4] But, in short, Sabel, O'Rourke, and Fung argue for an approach that would put in place the conditions to achieve upward competition.

The RLS proposal is both compelling and problematic. It is compelling because Sabel, O'Rourke, and Fung offer a sophisticated understanding of multinational firms and seek to apply contemporary research on firm theory to improve working conditions across borders. They also refuse a deterministic approach to game theory's prisoner's dilemma, by recognising that the outcome in the labour context is not an automatic race to the bottom; instead, they aim to put into place the conditions that would promote an ongoing race to the top through the ratcheting upward of labour standards. Their work also shows the real merit of grappling with the special regulatory challenges associated with the vexing problem of the so-called informal sector in developing countries.

The proposal is problematic, however, from the perspective of labour regulation. It fails to grapple with the specificity of labour, that is, on why regulatory action in the field of labour relations needs to be addressed differently from many other fields of inquiry, notably on the environment. The RLS framework overlooks the underlying justifications for — and principles of — labour regulation. This leads the authors to overstate the ways in which firms may compete to promote certain kinds of protective standards, but to underestimate the importance of broader, locally supported labour relations systems. And, in their attention to the potential of spotlighting and ranking labour practices through information dissemination, Sabel, O'Rourke, and Fung sidestep the question of the legitimacy of replacing the state in private ordering in developing countries.

Critique

Export Processing Zones in Developing Countries

This analysis of codes of corporate conduct and the regulatory state in developing countries starts by paying attention to the places where the codes are particularly likely to be enforced. Place is a key element to understanding the significance of the rise in corporate self-regulatory initiatives. Corporate self-regulatory initiatives tend to apply in newly created spaces, where governmental regulation — notably in the field of labour, and often concerning freedom of association — either legally does not apply or systematically is not enforced, namely in EPZs based in developing countries.

Although the issue of the power of MNCs in their transborder activities has been the subject of scholarly concern for at least the last 30 years, little attention has been paid to the links between these self-regulatory initiatives and the particular places where they are meant to apply. Yet it is precisely by assessing the links that one can begin to understand the serious governance concern raised by these new initiatives.

Corporate self-regulatory schemes fit the EPZ model perfectly. Logically, an enterprise that establishes itself in an EPZ would seek to avoid entangling itself in a web of local regulation and enforcement. Indeed, the simplified, deregulated nature of these zones is designed to attract MNCs. This is not to suggest that all host countries explicitly exclude EPZs from the scope of all national labour legislation, although the tendency certainly has been noted (see, for example, Chapter 7). But the poor enforcement that may prevail throughout the host country is nonexistent within most EPZs. They are generally operated by EPZ authorities housed in ministries of trade or foreign affairs, outside the purview of comparatively weak ministries of labour. Physical entry is also frequently restricted, prohibiting access not only to trade unions, but also to labour ministry officials, including labour inspectors. Tellingly, in a report to the ILO on the social and labour practices of MNCs, a Turkish workers' organisation captured the EPZ experience as 'another country in the country' (ILO 2001).

The cumulative result of MNCs in many EPZs is the *de facto* deregulation of labour relations in these zones. Perched above these deregulated spaces is a new form of legality — self-regulation of labour standards by actors with annual revenues that far exceed the gross domestic product (GDP) of most developing countries in which they are based. This new form of legality emphasises not the fixed context of the domestic workplace, but rather the increasingly dynamic notion of the transborder product in the new international division of labour. When seen in the context of EPZs, therefore, self-regulatory initiatives fit logically within the framework of transnational consumerism through trade, in which Western consumers and MNCs seek to replace the state — and the social partners themselves (that is, local employers and workers) — as the new regulators.

The Specificity of Labour Regulation

Mechanisms such as ratcheting labour standards, therefore, are likely to supplant state-based regulatory initiatives in the labour domain, rather than simply co-exist with them. If the potential implementers of RLS are to be the new regulators of labour relations, then it is necessary for them to remain faithful to the principles that support labour regulations systems.[5] Yet from the start, the adaptation of a regulatory model that has been applied to products justifiably elicits a degree of scepticism when it is transferred to working people in the transnational production process in EPZs.

Corporate self-regulatory initiatives reflect the perception that in an increasingly interdependent economy, it is desirable to control the conditions under which imported goods are produced. These tendencies have developed from both protectionist and moral bases. Although the protectionism debate has been repeatedly rejected, at least from the moral perspective workers' rights advocacy strengthens the increasingly prevalent claim that it would be better not to have production at all than to produce under conditions that lead to the systematic violation of a limited set of fundamental human rights.[6] This approach has translated into attempts to establish effective labour conditions wherever there is some link between the worker and the final consumer product. However, by focussing on the product, codes of corporate conduct tend to treat labour conditions in a manner that parallels other production processes, notably processes that have an adverse impact on the environment.[7] Codes of corporate conduct are viewed simply on an administrative level, as regulatory links that connect workplace conditions to the global production chain.

At a fundamental level, though, labour is different for the simple reason that labour is about people. Through the productive process, people enter the market system in their capacity as a factor of production. However, their subordinated participation is expected to be other than that of a commodity merely bought and sold in relation to supply and demand. Labour law therefore mediates their access, infusing it with the dignity that the market alone cannot provide. To the extent, then, that labour law is where the roles of market ordering and public policy are traditionally mediated, the hard question for labour law, in light of changes in production and trade patterns, becomes the relative role of labour regulation in a transnational context (Langille 1996).

The dual purpose of labour law is to provide worker protection and worker agency through democratic participation (ILO 1919; Weiler 1990). These purposes have been accomplished in modern labour law through a state regulatory framework that allows groups with divergent interests to regulate relationships between themselves, while simultaneously establishing a floor of protective conditions. Profoundly redistributive, at a basic level labour law is the traditional terrain of domestic governments, often guided by international and regional standards that have been set through broadly participatory tripartite mechanisms at the ILO, and not infrequently with ILO technical assistance — drafting labour laws, helping to establish machinery such as inspection services, labour dispute mechanisms, tripartite advisory, and consultative bodies — to ensure that labour machinery is indeed functioning well. However, through collective action, independent workers' representatives negotiate the relationship between workers and management. In this manner, labour regulation establishes the channels that give workers a voice with which they can participate democratically in workplace decisions. Workers and management can then identify and focus on convergences of interest, while effectively negotiating, mediating, and, where necessary, arbitrating differences through largely state-provided dispute resolution systems.[8]

As with corporate self-regulatory initiatives, the RLS model changes the delivery of both key labour regulatory principles. RLS is characterised as a practical, ongoing approach to securing the former purpose, namely worker protection. RLS is meant to allow MNCs to compete in the articulation and adoption of protective standards, which they are also responsible for either implementing directly or — more likely — ensuring that they are implemented by their various chains of contractors along the production process, wherever they are situated on the globe. However, this approach accomplishes a limited degree of protection (Sabel, O'Rourke, and Fung do not even suggest that there would be a baseline starting point), while simultaneously multiplying the scope of management power, overreaching or supplanting state protective action required to mediate the effects of disproportionate bargaining power (even in collective relations), and preventing unacceptable bargains that might be arrived at through voluntary collective bargaining. Labour standards are quantified, but labour relations are overlooked.

Significantly, freedom of association is treated at best as an exception under the RLS framework. However, Sabel, O'Rourke, and Fung (2000) make the extremely important concession that 'organized workplaces might prove to be the most capable social monitors and agents of continuous improvement'. Yet it is precisely in relation to freedom of association that RLS would need to be most robust. In other words, in order for workers and unions to be able to provide the kind of information to be relied upon by the RLS, mechanisms are necessary to ensure that they can exercise their freedom of association.

Freedom of association is a cornerstone ILO principle, reaffirmed consistently in the 1919 Constitution, the 1944 Annex, and the Philadelphia Declaration, as well as in the core labour standards, the Freedom of Association and Right to Organize Convention, 1948 (No. 87), and the Right to Organize and Bargain Collectively Convention, 1949 (No. 98).[9] It was considered so central to the ILO that the obligation to comply was deemed to be a basic condition of membership, rather than required through ratification of specific conventions. The special status attributed to freedom of association has translated into extensively developed interpretations of this standard through context-sensitive reports and observations rendered by the ILO's supervisory system on freedom of association.[10]

It may not be surprising that many of the corporate codes of conduct and some co-ordinated self-regulatory initiatives ignore the topic altogether.[11] Initiatives that depend on consumer demand may invariably find that in countries such as Canada and the U.S., issues such as child labour tend to resonate better than freedom of association.[12] Even the most robust initiatives such as SA-8000 fail to provide the assurances that freedom of association principles will be made effective — on a day-to-day basis — without an industrial relations framework to sustain them. Little attention is paid to how freedom of association can meaningfully be promoted in relation to the new context and the limited abilities of individual states in developing countries. In the context of EPZs, this deficit is particularly stark.

The critique of the fixed-rule understanding of regulation and law proffered by Sabel, O'Rourke, and Fung (2000) shows a deep sensitivity to the material constraints faced by developing countries in establishing labour relations mechanisms, particularly inspection services and dispute resolution bodies. In the face of these constraints on state action, corporate self-regulatory initiatives draw on a degree of flexibility that states cannot as easily attain. The RLS model sheds light on the fact that many developing states are simply unable to legislate away major social and redistributional problems that are linked to labour standards violations. It opens the necessary discussion on how to draw upon the creative synergies of self-regulatory initiatives that have flourished, rather than dissipated, to buttress labour regulation.

But on the basis of this new regulatory realism, the RLS model potentially relegates the labour ministries in developing countries to the role of data collectors, and intergovernmental organisations become the sources of model legislation on transparency and monitoring of monitors. Meanwhile, the ability of monitors (even the high-profile monitors to which Sabel et al. allude) to curtail individual labour rights violations is itself deeply contested.[13] Independent monitors engaged in the task of ranking firms should not be expected to replace the panoply of inspection and dispute resolution bodies that support an industrial relations system; rather, the state mechanisms themselves need to be fortified through supportive technical co-operation activities. This kind of support, by international organisations such as the ILO through bilateral co-operation between states and potentially by the actors spurring on the code-related monitoring, would produce potentially more sustainable benefits for the export-oriented and local workplaces alike. RLS unfortunately sidesteps the central interaction of state law and private ordering in the labour relations context. It does not pay sufficient attention to the mediating role that needs to be played between MNCs and workers. RLS explicitly accepts that MNCs should be freed of the messy particulars of local regulation and enforcement channels, so long as they agree to provide information and accept to be monitored and ranked along a management-based competitiveness framework.

The preceding critique could be read as suggesting that state mechanisms are necessarily preferable to the regulation of labour relations over potentially more efficient and effective self-regulatory measures undertaken by the freely chosen, independent representatives of both workers and employers. Yet such a reading would miss the point that the opposition between state and private action is hardly stark in labour relations law and practice. Labour law is fundamentally pluralistic. As Harry Arthurs (1996, 3) puts it, 'the most distinctive feature of collective bargaining is not its nexus with the state; rather it is that collective bargaining relies upon employers and workers to generate and enforce the norms which govern workplace behaviour'. But, in contrast to the classic terrain of legal pluralist analysis, under lex mercatoria[14] — law developed by corporations to govern their own relationships — these corporate self-regulatory initiatives enable MNCs to extend their laws directly to workers along the global production chain, disregarding and even undermining any existing local enforcement efforts. Monitoring instead extends the ways that management would approach the product to the relationship of subordination of the workers in EPZs in developing countries.

Cosmopolitan Democratic Governance

Solving the problems that have spurred the attempt to address labour standards violations through codes of corporate conduct will call for concerted, complex transnational interactions that grapple with how to respect redistributional balances that labour regulation is meant to address, in a way that does not merely extend disproportionately great management power, harnessed by corporate do-gooders, into an invariably paternalistic form of consumer-driven *noblesse oblige*.[15]

Yet in their haste to join the bandwagon on self-regulatory and monitoring initiatives, some NGOs appear to have overlooked crucial questions about the role of the state in democratic governance in developing countries, as well as in industrialised countries. It is not at all evident that the majority of NGOs — particularly in Western countries — are even interested in displacing the state by themselves claiming the authority or assuming the responsibility associated with making and enforcing key public policy decisions. Yet, as Jill Murray (2001) has argued, the RLS proposal, if it worked at all, might create problems that are at the basis of why the state has the legitimacy to regulate on social matters in the first place, notably 'the disenfranchising from a public role in shaping social standards those unwilling or unable to purchase particular products' (42). This group ought to include those who make the products — the workers who should be at the core of hard and soft law approaches to labour regulation, nationally and transnationally. Perhaps the fact that they tend to be displacing 'somebody else's' state — notably in developing countries — seems to render the actions less objectionable, particularly when the state in developing countries is overwhelmingly perceived in the West to be undemocratic and corrupt.[16] This does not mean that these questions should not be raised; indeed, raising them is crucial. However, they should not be assumed to be part and parcel of the automatic justification for stepping in and overriding existing labour regulatory mechanisms in developing countries. Yet the privatised models of labour rights advocacy lead logically to this outcome. Sabel, O'Rourke, and Fung fully recognise this consequence, and explicitly endorse the displacement of the labour regulatory state in developing countries from their traditional task of regulating labour standards. This raises dilemmas beyond the pragmatic, which warrant greater theoretical conceptualisation.

It is true that the state in a re-regulatory environment on labour law need not — indeed, cannot, for the reasons articulated above — assume the same role that it was expected to assume in the past. Nor should the state be seen as needing to assume a purely domestic role; creating spaces at different governance levels is crucial in order to address the labour regulatory deficit and the *de facto* re-regulation of labour policy, particularly as it pertains to EPZs in many developing countries. Yet it should be recalled that with respect to labour regulation, representation by civil society and economic actors has been the norm, so much so that it has been a unique model of governance, combining hard and soft law, even at the level of international organisations through the tripartite ILO. Indeed, the state in transnational labour regulation should be increasingly understood along the lines of the classic but expanding industrial relations principle of tripartism. In other words, the state in labour regulation includes the local social partners; in most developing countries — unlike the *Wagner Act* tradition of Canada and the U.S. — they have played a major role in the ongoing process of establishing and enforcing ongoing labour relations policies and laws. Increasingly, attention has turned to how tripartite action in the labour policy field should be broadened to encompass different actors in the disaggregated state — trade ministries as well as labour ministries, MNCs in addition to local employers' organisations, NGOs that represent groups that have traditionally not been able to organise for a wide range of reasons (such as child labourers or migrant workers) in addition to trade unions. In any reconceptualisation of regulatory institutions to address transnational activities, attention needs to be paid to ensure that the democratic participation dimension is preserved and reinforced. This concern, relevant to the sectors of the economy in developing countries related to exports, is of course multiplied to the extent that what remains of the state mechanisms are all that is available to deal with the often greater (and less visible) difficulties in the rest of the economy (Bardhan 2001).

This is therefore largely a democratic participation critique of Sabel, O'Rourke, and Fung's work, proffered not to suggest that one should cling stubbornly to a traditional, fixed rule approach to labour regulation simply because it has theoretical justifications, but rather to emphasise how critical it is to ensure that new forms of labour regulation do themselves have robust labour relations justifications. In the absence of more theoretical attention given to democratic participation of the broad range of stakeholders in the transnational workplace — and not just or primarily Western MNCs, workers, and consumers — alternative regulatory proposals lose their validity.

Conclusion

Although this analysis has focussed on the critique of self-regulatory initiatives, this appraisal of proposals such as RLS ought not to be considered relentlessly critical. To the contrary, RLS and the plethora of self-regulatory initiatives, as well as the advocacy they have generated, have shown the potential to open up spaces for discussion — spaces that can be used to spotlight regulatory deficits and to promote cosmopolitan democratic participation by representative stakeholders in the labour context (see Blackett 2001). But ultimately consumer-driven re-regulatory initiatives should not be left to play the lead role in re-regulating labour law in developing countries. There is a role for the regulatory state in developing countries, and to enable developing countries to fulfil those crucial governance roles, some global redistributional policies — be they internationally conceived adjustment costs, technology transfers, broad technical assistance, or other measures meant to reduce the glaring North-South gap[17] — are essential.

Notes

1. The ILO has, however, drawn upon its supervisory mechanisms in the case of Myanmar's systematic forced labour violations in a manner that legitimises the use of sanctions against that country by member states (see Blackett 2002).
2. Perhaps surprisingly, though, Sabel and colleagues say nothing in their proposal about a role for the United Nations Global Compact. For an overview and discussion of the UN Global Compact, see Blackett (2001, 440–447).
3. Unfortunately, few details of how this public power could be marshalled within developing countries are provided.
4. The authors rely primarily on examples from environmental initiatives to suggest that the market provides ways to regulate against information overload (Fung, O'Rourke, and Sabel 2001b). However, the specificity of labour regulation is not contemplated in this form of comparison, as discussed below.
5. Sabel, O'Rourke, and Fung refer to several of the most well-known examples including the ISO-like Social Accountability International's SA8000, and seem to consider the Workers Rights Consortium to provide a particularly viable implementer of RLS. For a detailed discussion of these initiatives and some of the difficulties with their interface with domestic labour regulation, see Blackett (2001, 413–417).
6. On the subject of the rejection of protectionism, see, for example, the Singapore Ministerial Declaration (World Trade Organization 1996) and the Declaration on Fundamental Principles and Rights at Work and Its Follow-up (ILO 1998) Interestingly, the attempt of the World Trade Organization (WTO) to focus on trade and development at Doha in November 2001 has led to the exposure of the myriad of ways that industrialised countries retain highly protectionist policies, notably on agricultural policy, while exacting often painful liberalisation initiatives from developing countries. The failure of the WTO's Cancun ministerial meeting of September 2003 is widely attributed to this problem. For a valuable synthesis of the Cancun ministerial meeting, see 'Quatre dossiers ont concentré les désaccords et fait capoter la conférence' (2003).
7. See, for example, the analysis proffered by Fichter and Sydow (2002, 359), who say: 'Our focus is on the issue of production governance in networks and its (inter-) organizational practices in global commodity and production chains. ... We argue, that although unsatisfactory labor conditions are to be found in ... networks, this organizational form of economic activity is also a potentially effective means for organizing the governance of social responsibility in global commodity and production chains.'

8. Naturally, this model assumes a broadly democratic system or tradition within which to function. Indeed, it is a manifestation of workplace democracy, which thrives on the broader domestic context that facilitates democratic participation. It is precisely in contexts that are uncontrovertibly understood to be nondemocratic that soft approaches such as codes of corporate conduct seem particularly inappropriate, and more principled standpoints are necessary. The case of Myanmar is illustrative. Myanmar has been the subject of unequivocal international condemnation from various intergovernmental bodies. In the ILO context, a landmark report by independent legal experts found that the government had consistently violated the Forced Labour Convention No. 29 (1930) and the *jus cogens* norm against forced labour. Various ILO organs overwhelmingly approved two key resolutions, one of which recommended trade sanctions as the appropriate approach. For a fuller discussion, see Blackett (2002).
9. For a discussion of the centrality of the freedom of association principle, see Dunning (1998) and Swepston (1998). See also Blackett (1999).
10. Most notable is the special complaints-based machinery comprising a Committee on Freedom of Association, a Commission of Inquiry and a Fact-finding and Conciliation Commission, established following an Agreement with the United Nations Economic and Social Council (ECOSOC) in 1950 to address alleged violations of freedom of association. von Potobsky (1998) offers a detailed overview and assessment of the mechanisms.
11. See Diller (2000), noting that freedom of association is less commonly mentioned in U.S.-based codes than in European ones.
12. This point is reinforced by the fact that freedom of association tends to gain greater attention within the European context. A good example of this difference is the Clean Clothes Campaign <www.cleanclothes.org>.
13. Even those engaged in some of the most innovative practices on monitoring unequivocally affirm that monitoring is not enough (Van Heerden 2003).
14. For a discussion of *lex mercatoria* see Santos (1995), who discusses the various forms taken by the conflict between the 'transnational legal space' of *lex mercatoria* and the 'national state legal space'.
15. As Pranab Bardhan (2001) has stated, in asserting that initiatives on labour need also to invest in local community and workers beyond even the EPZs, 'this is no doubt much more expensive than implementing a simple RLS but massaging the conscience of the affluent consumers in an overwhelmingly poor world cannot be that cheap'.
16. Strikingly, Sabel, O'Rourke, and Fung (2001b) plainly argue as follows: 'We wholeheartedly endorse developing nations as full participants along with other parties, but disagree that the tropics (especially where governments are not responsive to their citizens) should possess a monopoly on labor standards.' Murray (2001, 63), however, correctly asserts that 'the RLS authors are committed to the principles of participation and empowerment of actors in civil society, yet the structures and processes they propose may turn out to be another version of top-down rule imposition of a neo-colonial kind'.
17. The failure of the recent WTO ministerial meeting at Cancun underscores the validity of this point.

References

Arthurs, Harry W. (1996). 'Labour Law without the State'. *University of Toronto Law Journal* vol. 46 (Winter), no. 1–45.

Bardhan, Pranab (2001). 'Some Up, Some Down: A Response to "Realizing Labor Standards"'. *Boston Review* February/March. <bostonreview.net/BR26.1/bardhan.html> (April 2004).

Blackett, Adelle (1999). 'Whither Social Clause? Human Rights, Trade Theory, and Treaty Interpretation'. *Columbia Human Rights Law Review* vol. 31, no. 1, p. 1.

Blackett, Adelle (2001). 'Global Governance, Legal Pluralism, and the Decentered State: A Labor Law Critique of Codes of Corporate Conduct'. *Indiana Journal of Global Legal Studies* vol. 8, no. 2, pp. 401–448.

Blackett, Adelle (2002). 'Mapping the Equilibrium Line: Fundamental Principles and Rights at Work and the Interpretive Universe of the World Trade Organization'. *Saskatchewan Law Review* vol. 65, no. 369, pp. 369–392.

Diller, Janelle M. (2000). 'Social Conduct in Transnational Enterprise Operations: The Role of the International Labour Organization'. In R. Blanpain, ed., *Multinational Enterprises and the Social Challenges of the XXIst Century: The ILO Declaration on Fundamental Principles at Work*. Kluwer Law International, The Hague.

Dunning, Harold (1998). 'The Origins of Convention No. 87 on Freedom of Association and the Right to Organize'. *International Labour Review* vol. 137, no. 2, p. 149.

Fichter, Michael and Jörg Sydow (2002). 'Using Networks towards Global Labor Standards? Organizing Social Responsibility in Global Production Chains'. *Industrielle Beziehungen — German Journal of Industrial Relations* vol. 9, no. 4, pp. 357–380.

Fung, Archon, Dara O'Rourke, and Charles Sabel (2001a). 'Fung, O'Rourke, and Sabel Respond'. *Boston Review* February/March. <www.bostonreview.net/BR26.1/fung2.html> (April 2004).

Fung, Archon, Dara O'Rourke, and Charles Sabel (2001b). 'Realizing Labor Standards: How Transparency, Competition, and Sanctions Could Improving Working Conditions Worldwide'. *Boston Review* February/March. <www.bostonreview.net/BR26.1/fung.html> (April 2004).

International Labour Organization (1919). 'Constitution'. <www.ilo.org/public/english/about/iloconst.htm> (April 2004).

International Labour Organization (1998). 'ILO Declaration on Fundamental Principles and Rights at Work'. <www.ilo.org/public/english/standards/decl/declaration/text> (April 2004).

International Labour Organization (2001). 'Follow-up and Promotion on the Tripartite Declaration of Principles Concerning Multinational Enterprises and Social Policy'. GB.280/MNE/1/1. <www.ilo.org> (April 2004).

Langille, Brian A. (1996). 'General Reflections on the Relationship of Trade and Labor (Or: Fair Trade Is Free Trade's Destiny)'. In J. Bhagwati and R. E. Hudec, eds., *Fair Trade and Harmonization: Prerequisites for Free Trade?*, pp. 231–265. MIT Press, Cambridge MA.

Murray, Jill (2001). 'The Sound of One Hand Clapping: The "Ratcheting Labour Standards" Proposal and International Labour Law'. *Australian Journal of Labour Law* vol. 14, no. 3, p. 42.

'Quatre dossiers ont concentré les désaccords et fait capoter la conférence'. (2003). *Le Monde*, 17 September.

Sabel, Charles, Dara O'Rourke, and Archon Fung (2000). 'Ratcheting Labor Standards: Regulations for Continuous Improvement in the Global Workplace'. KSG Working Paper Number 00-010; Columbia Law and Economic Working Paper Number 185; Columbia Law School, Public Law Research Paper Number 01-21. <ssrn.com/abstract=253833> (April 2004).

Santos, Boaventura de Sousa (1995). *Toward a New Common Sense: Law, Science, and Politics in the Paradigmatic Transition*. Routledge, New York.

Swepston, Lee (1998). 'Human Rights Law and Freedom of Association: Development Through ILO Supervision'. *International Labour Review* vol. 137, no. 2, p. 169.

Vallée, Guylaine, Gregor Murray, Michele Coutu, et al. (2003). *Les codes de conduite des entreprises multinationales canadiennes: aux confins de la régulation privée et des politiques publiques du travail*. Canada Law Commission, Ottawa.

Van Heerden, Auret (2003). 'Beyond Monitoring: What Codes of Conduct Should Do to Promote Labour Standards'. Paper presented at IIRA seminar on Codes of Labour Practices, Berlin,

9 September. <www.fu-berlin.de/iira2003/papers/special_seminars/S_1_Heerden.pdf> (April 2004).

von Potobsky, Geraldo (1998). 'Freedom of Association: The Impact of Convention No. 87 and ILO Action'. *International Labour Review* vol. 137, no. 2, pp. 195–222.

Weiler, Paul C. (1990). *Governing the Workplace: The Future of Labor and Employment Law*. Harvard University Press, Cambridge MA.

World Trade Organization (1996). 'Singapore Ministerial Declaration'. WT/MIN(96)/DEC/W. <www.wto.org/english/thewto_e/minist_e/min96_e/wtodec_e.htm> (April 2004).

Chapter 6

Standard Setting at the International Labour Organization: The Case of Precarious Employment

Leah F. Vosko[1]

Precarious employment is spreading internationally. Yet national governments have been slow to respond to its spread due to the rigid system of labour legislation, regulations, and policy modelled on a standard employment relationship (Fudge and Vosko 2001a, 2001b; Vosko 2000). In contrast, the International Labour Organization (ILO), the foremost international organisation operating in the field of labour, began to respond to the growth of precarious employment in the early 1980s, propelled primarily by representatives of organised labour, nongovernmental organisations (NGOs), and member states concerned about accelerating casualisation. From the early 1980s to the mid 1990s, the ILO introduced several new measures to address the rise of nonstandard forms of work threatening to deepen precarious employment in labour markets, entering uncharted areas of labour regulation and advancing important new policy frameworks for national and subnational governments to follow.

Between 1980 and 1995, the ILO adopted three new international labour standards — on Workers with Family Responsibilities (1981), Part-Time Work (1994), and Home Work (1996) — aimed at preventing growing pressure for flexibility from amounting to increased insecurity among workers. The creation of these standards is significant since each instrument advanced a regulatory framework for generating parity (that is, equal, equivalent, or comparable protections) among workers engaged in standard and nonstandard forms of employment and, therefore, each offered a model for national governments to follow in stemming the rise of precarious employment.

After the adoption of the Home Work Convention, however, the ILO changed course, shifting emphasis away from achieving parity, and thereby tacitly accepting the spread of precarious employment. This shift is apparent in the new international labour standard on Private Employment Agencies (1997), which abandons the goal of parity by calling only for 'adequate' protections for workers employed by private employment agencies and the ongoing discussion on the scope of the employment relationship, the tenor of which reflects a modest approach to re-regulation.

Paradoxically, however, at the very moment that the ILO is softening its approach to regulating nonstandard forms of employment, which leave workers highly vulnerable to dimensions of precarious employment, and altering the standard-setting structures by which it has traditionally operated, other international organisations are looking to the ILO for leadership in their efforts to develop hard and harder soft law instruments in pursuit of global governance.

This chapter sketches the ILO's standard-setting activities pertinent to precarious employment since 1980 as a means of evaluating the viability of the ILO model of setting standards. After defining a few core concepts, the chapter unfolds in four parts. Beginning with a brief history of ILO standard setting related to precarious employment, the first part describes and analyses key international labour standards on nonstandard forms of employment that were adopted between the early 1980s and the mid 1990s and were devised to provide guidance to governments aiming to curtail the spread of precarious employment. Through this inventory, it demonstrates that the international labour standards of this period reflected an attempt to prescribe comparable conditions for workers in standard and nonstandard employment relationships.

Building on this discussion, the second part examines the ILO's shift in course, beginning in the mid 1990s, from attempting to limit the spread of precarious employment to adopting conventions, recommendations, and increasingly softer promotional instruments aimed, at best, at regulating precariousness (Vosko 1998). This section devotes considerable attention to the substance of the new international labour standard set out in the Private Employment Agencies Convention and the 1997/98 stalemate over contract labour, which prompted broader inquiries into situations of workers needing protection. It reveals that the ILO effectively abandoned the goal of parity for workers employed by private employment agencies in the process of adopting both a new convention and a recommendation on this topic. Furthermore, prospects for the creation of not only hard law instruments but also soft hard-law instruments prescribing parity protections for workers in situations needing protection are bleak, especially for 'dependent workers' devoid of employment relationships.

The third part offers an explanation for the impasse over standard setting designed to curtail the spread of precarious employment. On the one hand, it attributes the deadlock to the absence of a sustained coalition between organised labour and labour NGOs inside the ILO, owing largely to the rise and spread of multiple forms of nonstandard employment and the erosion of the power of organised labour coming in its trail. On the other hand, it points to the emerging approaches to standard setting underway inside the ILO, which constrain the ILO's sphere of activity (and threaten to dilute hard law and hard soft-law measures in favour of promotional standards) and hence risk returning the organisation to its traditional emphasis on workers in standard employment relationships.

The fourth part concludes this inquiry by assessing the strengths and weaknesses of the ILO model as a template for global governance based on the findings of the case study.

This chapter addresses the shifting mix of soft laws or voluntary/promotional standards and the hard laws that have long operated in the ILO through a case study of standard setting on precarious employment. ILO standard setting represents an important area for analysis, because the ILO remains one of the few international organisations in which hard laws continue to operate, even though they are being re-evaluated and reshaped. In the ILO context, conventions are hard laws — they become formal, legally binding commitments once they are ratified by a member state. Recommendations, in contrast, which have traditionally supplemented conventions, are the soft laws of the ILO. After a country ratifies a convention, its associated recommendation can become more than a promotional instrument in practice, taking on the qualities of a hard soft law. Yet recommendations are increasingly becoming stand-alone ILO instruments, and softer forms of regulation, ranging from guidelines and protocols to codes of practice, increasingly find space in ILO activity.

Concepts and Definitions

Three substantive concepts frame this chapter: the standard employment relationship, nonstandard forms of employment, and precarious employment. The standard employment relationship is a continuous full-time, full-year employment where the worker has one employer and normally works on the employer's premises (Mückenberger 1989; Rodgers and Rodgers 1989; Schellenberg and Clark 1996; Vosko 2000). Its essential elements include an indeterminate employment contract, adequate social benefits that complete the social wage, reasonable hours of work, and employment that is frequently, but not necessarily, in a unionised sector. A high level of compensatory social policies, such as pension, unemployment insurance, and extended medical coverage, normally accompanies the standard employment relationship. The standard employment relationship became the normative model of (male) employment in industrialised countries after World War II.

In contrast to the standard employment relationship, which has features relatively easy to discern, nonstandard forms of employment are best defined by what they are not, rather than by what they are. In 1990, the Economic Council of Canada defined them simply as those forms of employment that differ from a full-time job. Nonstandard forms of employment range from own-account self-employment to part-time and temporary wage work to triangular employment relationships involving an agency, a client firm, and a worker. Moreover, workers in nonstandard forms of employment often engage in atypical work arrangements such as on-call work, multiple jobholding, and homework.[2] Nonstandard forms of employment are heterogeneous, yet they tend to entail a high degree of insecurity; workers in nonstandard forms of employment often have lower levels of social benefits and limited statutory entitlements compared to workers in standard employment relationships (Lipsett and Reesor 1997).

Although analysts often use the terms 'nonstandard' and 'precarious' synonymously when discussing employment, technically they should be differentiated. Not all part-time, temporary, or self-employed workers experience dimensions of precariousness.

Precarious employment, broadly conceived, encompasses those forms of employment normally involving atypical employment contracts, limited social benefits and statutory entitlements, job insecurity, low job tenure, low wages, and high risks of ill health (Vosko 2003; Vosko, Zukewich, and Cranford 2003). Still, because national governments as well as the ILO have historically modelled labour policy on the standard employment relationship, employers may resort to nonstandard forms of employment to escape regulation, contributing to the spread of precarious employment and, in practice, blurring the boundary between precarious employment and nonstandard forms of employment. In order to understand the ILO's historical and contemporary approach to precarious employment, it is thus critical to probe the nature and scope of international labour standards on workers with family responsibilities, part-time work, homework, and private employment agencies as well as to consider ongoing discussions on the scope of the employment relationship.

Toward Parity: Standard Setting on Precarious Employment, 1980–1996

Since its inception in 1919, the ILO has served as the international organisation responsible for setting workers' rights and protections through two sets of activities: setting standards, which involves crafting international labour conventions and recommendations designed to provide a framework for national labour legislation, regulations, and policies, and providing technical assistance or conducting activities devised to assist governments in implementing ILO standards at the national and subnational levels. The history of the ILO's standard setting and technical assistance is framed by efforts to institutionalise a standard employment relationship along with the Keynesian welfare state. From the time of the Labour Charter attached to the Treaty of Versailles in 1919 until the era of the 1944 Philadelphia Declaration and throughout the cold war, the ILO established a comprehensive set of international labour standards designed to strengthen the occupational welfare state through, for example, the creation of unemployment insurance and workers' compensation and to establish worker protections providing for weekly rest, minimum wages, protection from discrimination in employment, and so on. Consistent with these aims, the ILO has also pursued measures intended to extend social and labour protections to groups of workers, such as women and migrant workers, that do not conform to the male industrial norm. Through its technical assistance activities, the ILO has also devoted attention to conditions in the informal economies of industrialising and industrialised countries with the overriding aim of formalising and, more specifically, normalising the standard employment relationship.

Given the tenor of these decades of standard setting and technical assistance, officials inside the International Labour Office, member states, and representatives of organised labour began to express concerns about the spread of nonstandard forms of employment and the concomitant rise of precarious employment in the early 1980s. These concerns and the regulatory gap at the national level led the ILO to introduce three new international labour standards directed at workers with family responsibilities, part-time workers, and homeworkers respectively. Viewed collectively, these standards broke new ground because each aimed at generating parity between workers in standard employment relationships and workers in nonstandard employment relationships, while recognising, and in many instances endorsing, the growing plurality in employment relationships.

The goal of parity first surfaced in the Workers with Family Responsibilities Convention and its associated recommendation, which asserted that 'terms and conditions of employment, including social security coverage, of part-time and temporary workers should be ... equivalent to those of full-time and permanent workers' (ILO 1981, art. 21). The Part-Time Work Convention (1994) and its associated recommendation then took this sentiment forward. Still, although the parity goal itself shaped these instruments, parity provisions were largely relegated to non-binding recommendations — the soft law side of the ILO machinery — by the time the international labour standard on Home Work Convention was adopted in 1996. This standard is the weakest of these three instruments on the issue of parity, but it has symbolic value since core segments of the International Labour Code permitted the exclusion of homeworkers prior to its creation (Prugl 1999; United Nations Development Fund for Women 2000).

The preceding international labour standards set out to achieve parity in the four key areas of equality of opportunity and treatment, social security, work organisation and occupational health and safety, and freedom of association and the right to organise and bargain collectively. Framing these instruments, equality of opportunity and treatment is used to cultivate parity by advancing general principles of non-discrimination, on the one hand, and principles of equal remuneration, on the other hand. Regarding non-discrimination, the Part-Time Work Convention notes that 'measures shall be taken to ensure that part-time workers receive the same protection as that accorded to comparable full-time workers in respect of ... discrimination in employment and occupation' (ILO 1994a, art. 4). The Home Work Convention contains a provision requiring national governments to promote equal treatment between homeworkers and other wage earners, taking into account the special characteristics of homework (art. 4). Across all three standards, moreover, non-discrimination is used as a vehicle to prevent employers from resorting to nonstandard forms of employment for the purpose of escaping regulation.

The conventions on part-time work and homework also call for the promotion of equality of treatment in relation to remuneration. The Part-Time Work Convention states that:

> measures ... shall be taken to ensure that part-time workers do not, solely because they work part time, receive a basic wage which, calculated proportionately on an hourly, performance-related, or piece-rate basis, is lower than the basic wage of comparable full-time workers, calculated according to the same method (ILO 1994a, art. 5).

The Part-Time Work Recommendation underscores this sentiment, stipulating that part-time workers should 'benefit on an equitable basis from financial compensation ... received by comparable full-time workers' (ILO 1994b, art. 10). Taking this sentiment even further, the Recommendation on Home Work asserts that:

> For specified work paid by the piece, the rate of remuneration of a homeworker should be comparable to that received by a worker in the enterprise of the employer, or *if there is no such worker, in another enterprise in the branch of activity and region concerned* [emphasis added] (ILO 1996b, art. 15).

In both standards, parity is an explicit goal that the ILO recognises may demand diverse approaches and remedies, depending upon whether a worker works by the piece or for fewer hours than a standard work week.

The objective of parity takes sharpest expression in provisions related to social security in all three international labour standards. The Workers with Family Responsibilities Recommendation is unequivocal in its call for parity, stating that terms and conditions of employment, including social security coverage of part-time and temporary workers, should be 'equivalent to those of full-time and permanent workers' and indicating that workers (largely women) with family responsibilities are overrepresented in part-time and temporary work and thus disproportionately vulnerable to experiencing dimensions of precarious employment (ILO 1981, art. 21). Still, the Part-Time Work Convention contains the most comprehensivesocial security provisions: while it allows ratifying countries to exclude part-time workers with minimal weekly hours from the scope of social security schemes, except in regard to occupational injury benefits, this convention notes that 'thresholds ... shall be sufficiently low as not to exclude an unduly large percentage of part-time workers' (ILO 1994a, art. 8).[3] The Part-Time Work Recommendation also instructs governments to take multiple jobholders into account in statutory social security schemes by calculating benefits and entitlements based on their total hours of work, contributions and/or earnings accumulated from all sources (ILO 1994b, arts. 9, 13).[4]

The Home Work Convention acknowledges that equal treatment may be achieved in several ways, including extending existing social security provisions to homeworkers, adapting social security schemes to comparable workers, or developing special schemes geared to homeworkers paid by the piece (ILO 1996a, art. 25). Consistent with the international labour standard on part-time work, it also requires governments to promote equal treatment between homeworkers and other wage earners in relation to statutory social security schemes (ILO 1996a, art. 4). To protect wage levels and increase access to statutory social security, its associated recommendation also prescribes that, where intermediaries are involved, national governments determine the responsibilities of parties involved in subcontracting, temporary help work, and homework by the piece. Finally, the recommendations on part-time work and home work also call on governments to mandate written contracts informing workers of their specific conditions.[5]

A third area where the parity objective surfaces is in the area of work organisation and occupational health. The Part-Time Work Convention requires countries to ensure equal protection in the field of occupational health and safety (ILO 1994b, art. 4[b]). Similarly, recognising the health and safety implications where the home is the workplace, the Home Work Recommendation calls for the establishment of conditions under which certain types of work and the use of certain substances are prohibited in the home for reasons of safety and health; it also urges governments to compel employers to inform homeworkers of any hazards that are known to the employer or associated with the work assigned and, where appropriate, to provide them with necessary training (ILO 1996b, arts. 7, 20).

The fourth dimension of the parity goal is the requirement that ratifying governments fully extend the right to organise and bargain collectively to workers included in the Part-Time Work Convention (ILO 1994a) and in both the Convention and the Recommendation on Home Work (ILO 1996a, art. 4; 1996b, art. 11). Freedom of association and the right to organise and bargain collectively are central tools used by, for example, workers seeking prorated benefit schemes or legislation lowering thresholds for benefits. Thus, this dimension, too, reflects efforts to generate comparable conditions for workers in standard employment relationships and nonstandard forms of employment through the successful negotiation of international labour standards on workers with family responsibilities, part-time work, and homework.

From the early 1980s to the mid 1990s, the objective of parity framed international labour standards directed at extending statutory and employer-sponsored protections and benefits to workers in nonstandard forms of employment. In this period, the ILO's goal was to de-link nonstandard forms of employment and precarious employment in attempt to stimulate progressive changes in labour legislation, regulations, and policy at the national level.[6] Furthermore, the adoption of the Home Work Convention seemed, initially, to solidify this new direction as well as to accelerate efforts at progressive re-regulation. However, despite the successful negotiation of both a recommendation and a convention on homework and the broader attempt to generate parity in the instruments, negotiations were fraught with tensions.

It is beyond the scope of this chapter to examine how the Home Work Convention and its associated recommendation moved through the ILO standard-setting process. Suffice it to say that this instrument passed by a narrow margin in 1996 and only four countries — Albania, Finland, Ireland, and the Netherlands — have ratified the convention.[7] The explanation for the nonresponse from governments is that process proceeded without the participation of employers. The Employers' Group withdrew from negotiations over homework largely because of its rejection of the parity objective informing the discussion. Governments still negotiated the convention with organised labour due primarily to the determination of European Union governments and South Africa.

Yet the employers' withdrawal from these negotiations generated significant consequences, including a growing unwillingness among many member states to negotiate new international labour standards without a consensus in advance and employers' increasing resistance to the ILO negotiation process itself. Furthermore, organised labour expressed renewed discomfort (especially among representatives of industrial unions) with negotiating international labour standards in conjunction with even like-minded NGOs, due to their fears of quadrapartism or the introduction of a fourth social partner (NGOs) at the ILO that would water down organised labour's power at the ILO (Josselin 2001). The heterogeneous character of nonstandard forms of employment and the magnitude of the challenges of stemming precarious employment help explain this genuine fear. The adoption of the Home Work Convention was a decisive moment in ILO history precisely because organisations representing largely non-unionised homeworkers worked collectively to advise and support delegates of organised labour and the International Confederation of Free Trade Unions more broadly. They played a pivotal role not only in shaping the substance of this international labour standard but also securing support for a convention among member states (Vosko 2002; see also Prugl 1999).

Aggravated by low levels of ratification, the current impasse over creating new international standards — specifically, conventions or hard instruments — designed to curtail the spread of precarious employment is the most significant consequence of the deliberations over homework. The depth of this stalemate became apparent initially with the adoption of the Private Employment Agencies Convention (1997) and subsequently with the failure of negotiations toward a convention and recommendation on contract labour (1997, 1998). Backed by extensive country studies describing the occupational and industrial locations, demographic characteristics, and the forms of employment common to workers in situations needing protection, recent discussions on the scope of the employment relationship could alter this course, although prospects remain limited for hard or hard soft-law regulations targeting workers falling outside the ILO's corpus of standards.

Impasse: Standard Setting on Precarious Employment, 1996–Present

In 1997, the ILO adopted a new international labour standard on private employment agencies, ending the ILO's long-standing policy of prescribing either the strict regulation or the prohibition of private employment agencies and legitimising the triangular employment relationship involving a worker, an agency or intermediary, and a client firm (Vosko 1997). At the outset, ILO experts presented documentation revealing the unprecedented rise of temporary help work and called for a convention in this area (ILO 1994c). However, rather than addressing the rise of temporary help work through private employment agencies, the Private Employment Agencies Convention deals primarily with them as institutional actors, aiming to promote ethical conduct rather than encourage governments to regulate triangular employment relationships (ILO 1997d). The convention thus legitimises private employment agencies and, in so doing, temporary help work — a nonstandard form of employment where workers normally lack the full range of protection and benefits accorded to workers in standard employment relationships and, therefore, linked closely to the spread of precarious employment (ILO 2003b, ch. 3; Vosko 2000).

The substance of this convention illustrates the move away from parity inside the ILO in the late 1990s. This shift is apparent in the language of the convention: rather than calling for equal or comparable treatment, the Private Employment Agencies Convention directs national governments to provide 'adequate protection' to workers employed by private employment agencies. The convention touches on various dimensions of non-discrimination, social security, occupational health and safety, and freedom of association and the right to organise and bargain collectively, but it only mandates ratifying governments to extend adequate protection to workers employed by private employment agencies and fails to define 'adequate' or 'sufficient'. A draft convention on contract labour was also discussed at the International Labour Conference in 1997. Yet this instrument, as well as a proposal for a recommendation on the same subject, was ultimately rejected. On the surface, the failure of ILO constituents to negotiate a new standard appears to reflect a lack of consensus on the definition of contract labour, but the roots of its failure lie in its crumbling efforts at parity.

The Private Employment Agencies Convention and its associated recommendation each contain several provisions aimed at securing basic protections for workers employed by private agencies. In the area of equality of opportunity and treatment, for example, the convention directs ratifying countries 'to ensure that private employment agencies treat workers without discrimination' (ILO 1997d, art. 5). The recommendation extends this sentiment by proposing that governments prohibit, or by other means prevent, private employment agencies 'from drawing up and publishing vacancy notices or offers of employment in ways that directly or indirectly result in discrimination' (ILO 1997e, art. 9).[8] Yet none of these measures aims for parity between workers employed by private employment agencies (that is, temporary help workers) and workers in standard employment relationships or other wage earners. The convention also encourages ratifying countries to ensure adequate protection for workers working through private employment agencies yet only in relation to minimum wages. There is no mention of providing financial compensation to workers employed by private employment agencies on a basis comparable to other wage earners as in several of the pre-homework instruments (ILO 1997d, art. 11).

Adequate — rather than comparable — statutory social security benefits are also prescribed in the convention, justifying inferior levels of social protection for workers employed by private employment agencies and limiting the potential adoption of prorated benefits schemes among nations ratifying the convention (ILO 1997d, art. 11). The convention also requires member states to take necessary steps to ensure adequate protection in relation to maternity and parental benefits; hence, the language of the convention moves the International Labour Code away from equal treatment in this vital (and gendered) area of social protection. Provisions on freedom of association and collective bargaining are equally weak: rather than affirming workers' rights to freedom of association and to engage in collective bargaining, the convention requires countries to ensure that workers employed by private employment agencies 'are not denied the right to freedom of association and the right to bargain collectively' (ILO 1997d, art. 4). These rights are expressed in negative rather than positive terms. As well, requirements for written contracts are absent, as are detailed provisions governing the allocation of responsibility, especially secondary responsibility, between private employment agencies and their clients; this omission reveals constituents' failure to agree to use the convention as a vehicle for extending social protections to workers employed by private employment agencies.

The draft Convention on Contract Labour was considerably stronger than the international labour standard on private employment agencies that emerged in the same year (ILO 1997a). It covered several types of contract labour and its drafters aimed at parity in some measure. It also included an equal treatment clause, applied to a range of social protections, and directed ratifying countries to 'ensure that contract workers are afforded equal treatment with employees of the user enterprise or, as the case may be, with employees of the subcontractor or of the intermediary, for performing work which is essentially similar, under similar conditions and requiring similar qualifications' (ILO 1997a, para. 20). Additionally, it encouraged the assimilation of contract workers into standard employment relationships, mirroring measures contained in the Part-Time Work Convention, and facilitated the allocation of secondary responsibility when more than one employer or intermediary are present. There were several key omissions in the draft instrument, however. It made no mention of extending freedom of association and the right to bargain collectively or of training and injury benefits to contract labour, even though a commitment to parity shaped the draft instrument to a greater degree than the Private Employment Agencies Convention adopted that year.

Signalling a movement away from the parity objective, by 1998 efforts to establish a new international labour standard on contract labour were dashed. Rather than adopting a hard or even a hard soft-law instrument, ILO constituents adopted a resolution focussed on dependent workers in need of social protection. This resolution called on the ILO to hold meetings of experts and to take other measures necessary to complete the work initiated by the committee on contract labour. It also invited the ILO's Governing Body to place the issue on the agenda of the International Labour Conference within four years, with a view to the possible adoption of a convention and recommendation. Between 1998 and 2003, when the issue once again reached the floor of the International Labour Conference, the ILO had solicited 39 country studies to determine categories of workers in need of protection, held regional meetings, and hosted a meeting of experts to a report for the Governing Body.[9] In May 2000, the meeting of experts adopted a common statement, which served as the basis for both further study and the report on the scope of the employment relationship prepared for general discussion in 2003. The statement called for the adoption of a range of instruments on this topic, including conventions and recommendations. The parity goal was implicit in the report, whose drafters' main aim was to move the issue of precarious employment to the top of the ILO agenda once again.

ILO initiatives undertaken between 1998 and 2003 generated intense controversy. Tensions were sharp because the country studies had found myriad examples of what it labelled dependent workers in need of protection and highlighted the need to expand the discussion to all 'unprotected workers', a call that employers resisted vigorously.[10] Against this backdrop, in their report to the ILO Governing Body in November 2000 and March 2001, experts advanced three potential routes for the ILO in assisting workers in situations needing protection: put standard-setting activities on hold until more information is available, while recognising that this option could contribute to the acceleration of prevailing trends; draft a detailed standard, acknowledging that the diversity in national situations and differing views of workers and employers would remain obstacles to adoption and ratification; or construct a softer promotional instrument, weaker than a convention but distinct from a recommendation, encouraging states to formulate, implement, and adopt national policies to cover workers in situations needing protection (ILO 2000a, part 4, paras. 74–76).[11]

In response, the Governing Body opted to hold a general discussion on the scope of the employment relationship in 2003 and to prepare a report for discussion. The drafters of the report were mandated to consider the common statement of experts, country studies, and previous discussions on contract labour. Thus, the resulting report thus focussed on 'dependent workers' lacking protections normally attached to the employment relationship with attention to the following problems: unclear scope of law, disguised employment relationships, objectively ambiguous employment relationships, and the failure to comply with or enforce the law (ILO 2003b). It also gave special attention to triangular employment relationships since a lack of protection is pronounced in temporary help work even in the presence of an employment relationship.

After extensive debate, the conference recommended that future ILO action focus on disguised employment relationships and 'on the need for mechanisms to ensure that persons with an employment relationship have access to the protection that they are due at a national level' (ILO 2003a, para. 25). The situations of workers in ambiguous employment relationships and other dependent workers were set aside and the 'issue of triangular employment relationships was not resolved' (para. 25). The committee concluded that a recommendation represented an appropriate standard, although the discussion that took place in June 2003 illustrates that the adoption of this type of soft instrument is unlikely. Equally critical, the gradual yet consistent narrowing of scope limits prospects for parity. If an instrument is to evolve in the coming years, the employment situations addressed will be sufficiently limited that the persons covered will be *de facto* in the same situation as workers with employment relationships, excluding the expanding array of dependent workers devoid of employment relationships worldwide and thus highly vulnerable to precarious employment.

The issue of contract labour — now labelled 'The Scope of the Employment Relationship' — has been tabled temporarily. It will reappear on the agenda of the

International Labour Conference at the behest of the Governing Body in 2006, perhaps under yet a different name, focussing possibly on disguised employment relationships. Obstacles to the eventual adoption of a new standard remain profound: prospects for an alliance between organised labour and NGOs, backed by member states, are constrained by the lack of a unifying focus similar to that available in the negotiations of the Home Work Convention, especially given the problems with defining disguised employment. Developments at the broader level of standard setting could further thwart efforts. The current trend in the ILO amounts, at best, to encouraging member states to ensure adequate protections for workers in nonstandard forms of employment and workers in employment relationships that are disguised or otherwise ignored. Attempts at parity have essentially fallen off the agenda. At worst, it amounts to neglecting the rise of forms of nonstandard employment, such as triangular employment relationships and self-employment where workers exhibit qualities of dependence, and thereby tacitly accepting the growth of precarious employment.

Explaining the Impasse

The shift away from extending parity protections to nonstandard workers as a means of curtailing precarious employment is partly explained by two coincident developments — the limited power of organised labour since the adoption of the Home Work Convention due to the high level of non-unionised workers in nonstandard forms of employment worldwide and the ongoing review of ILO standard setting, which threatens to narrow the ILO's sphere of regulatory activity and reinforce the prevailing impasse.

In 1997, in the wake of the Home Work Convention, ILO director general Michel Hansenne delivered a report to the International Labour Conference titled 'The ILO, Standard Setting, and Globalization,' designed to renew standard setting in the organisation (ILO 1997c). The impetus was Hansenne's dual recognition that 'beyond a certain threshold, the production of new standards is subject to a law of diminishing returns' and that 'from a more political standpoint, the consensus kept alive by the cold war is wearing thin'. Reflecting on the events of the previous year, he noted further that

> *for the first time in the history of the Organization, a group refused at the last session of the Conference to take part in the discussion of a draft Convention on a question [Home Work]* which all three groups in the Governing Body had agreed to include on the Conference agenda. It would seem, in this particular case, that this position was taken because of the form of the instrument. Yet it appears that, generally speaking, *the pace of standard setting ... does not allow for the flexibility desired to take account of the degree of interest in potential subjects* [emphasis added].

Based on these observations and his view that ILO standard setting is increasingly critical in an era of globalisation,[12] the Director General proposed addressing the social dimensions of trade liberalisation through the introduction of a new ILO declaration on fundamental principles and rights at work and by updating the body of ILO standards. These two initiatives underpin the ILO's current direction.[13] Yet the tenor of the review of standard setting initiated by the Director General in 1997 and ongoing discussions within the Governing Body and the ILO Committee on Legal Issues and Labour Standards are also pivotal in the shift away from parity and in the ILO's growing reticence to create standards designed to curtail precarious employment.

In his examination of standard setting, Hansenne argued for greater emulation among states in labour policy. Additionally, he called for more focussed standards aimed at 'added value' — which he defined based on the number of workers potentially affected by the standard, whether or not the workers affected belong to lower economic strata and regardless of the severity of the problem in question. Additionally, he called attention to the importance of choosing the form of international labour standards more carefully. In advocating for the creation of international labour standards with added value, he contended that the absorptive capacity of states has greater limits in the face of globalisation and outlined what he perceived to be a growing misfit between the pace of ILO standard setting and the desire for flexibility among its three constituent groups, taking homework as an example. His solution: fewer, more focussed international labour standards with greater impact.

Citing an ILO report from 1964 that stated that 'there is no inherent virtue in a Convention', Hansenne also stressed that:

> There is no inherent virtue in a Convention … Although a Convention creates obligations, in certain fields, 'a standard which can be widely accepted as such may well be more effective in practice than obligations which are unlikely to be equally widely assumed'… the Recommendation should no longer be considered as a 'poor relation of the Convention' (ILO 1997c).

Hansenne's report thus called on governments to take seriously a softer approach and, at the same time, recognised the need to restore national follow-up procedures to recommendations and to ensure that harder conventions are not overly rigid. The Director General's report represents a call to limit the ILO's sphere of standard-setting activity in order to achieve a new consensus and to move away from hard law measures to soft law or more promotional standards.

'The ILO, Standard Setting, and Globalization' generated a great deal of discussion inside the ILO. In the context of the adoption of the Private Employment Agencies Convention and even in discussions on contract labour, it is credited for bringing employers back to the negotiating table. Yet developments in the Committee on Legal Issues and Labour Standards are not embraced by representatives of organised labour, many of whom question the soft law approach and limiting the ILO's sphere of standard setting more broadly. Yet despite the concerns raised by organised labour in a November 1997 meeting of the Governing Body, the International Organisation of Employers made its position clear: the Governing Body must achieve tripartite agreement on both the subject and the form of a new standard before a discussion reaches the floor of the International Labour Conference (ILO 1997b, para. 12).[14] The Report of the Governing Body on the issue of standard setting drew this conclusion based on the experience with the negotiation of the Home Work Convention:

> Recent experience at the Conference with controversial and complicated topics has underlined the importance of selecting a subject for standard setting *only when the necessary research and preparatory work has been carried out and the subject can reasonably be considered to be 'ripe' for standard setting* [emphasis added] (ILO 1998, para. 3).

In debates over the procedure for selecting subjects for standard setting, members of the Governing Body also used the case of homework to argue for a lengthier period of study before a standard is proposed. Based on this discussion, the Governing Body recommended that the International Labour Office examine more fully the potential scope of a given standard before proposing an item for the International Labour Conference:

> If it is difficult to specify the objective and the scope of the standards to which the subject under consideration could give rise *(situation illustrated by the question of private employment agencies)*, the Governing Body could request additional information concerning the constituents' level of interest in the subject as well as a more detailed report to enable it to ascertain ... whether, instead of being immediately selected with a view to the adoption of an instrument, this subject could, if selected, be the subject of a preliminary discussion prior to standard-setting action [emphasis added] (ILO 1998, para. 5[a]).

Here again, a new instrument pertinent to precarious employment is cited as an example of ILO constituents' failure to arrive at a common definition of the issues in question.

The ILO Committee on Legal Issues and International Labour Standards outlined a range of priorities to be taken into consideration in revising the standard-setting process based on the Governing Body's responses to Hansenne's report (ILO 2000b). Although discussions are ongoing, the committee is calling on ILO constituents to strive for added value and seek a balance between maintaining a consensus and a steady pace of standard setting. These developments amount to a narrower role for ILO standard setting — especially the negotiation of hard laws — than in the past, when standards aimed at curtailing precarious employment finally reached the ILO agenda and emphasising promotional instruments.

The ILO's reconsideration of the role of standard setting on substantive issues, therefore, offers a potential explanation for the organisation's declining commitment to the objective of parity. In light of its growing emphasis on consensus in devising standards and the call to narrow the field of international labour standards, a potential instrument on disguised employment relationships, for example, is unlikely to take the form of anything more than a recommendation. Given the minimal progress made in 2003, given the murky definition of a disguised employment relationship, and given especially that employers and several member states are on record as opposing standard setting in this area, the introduction of a hard law or even a hard soft-law instrument is unlikely.

Evaluating the ILO Model: Lessons from the Case of Precarious Employment

The case of ILO standard setting related to precarious employment highlights several key strengths of the ILO model as a template for global governance. Yet it also reveals profound weaknesses. The evolution of international labour standards aimed at curtailing the spread of precarious employment between 1980 and 1996 illustrates the potential role of international organisations in advancing issues difficult to raise at the national or subnational levels. International labour standards for workers with family responsibilities and part-time work, for example, emerged in a period when nation-states were just beginning to acknowledge the need to extend protection and benefits to workers in nonstandard forms of employment and, as a result, both instruments have served as important templates for national and subnational regulation. Developments over this period also reveal surprising openings in the ILO structure itself, specifically, the potential space that the ILO can provide for addressing concern specific to workers in precarious employment that are addressed inadequately by national and subnational legislation, regulations, and policies. In the case of the creation and adoption of the Home Work Convention, organisations such as HomeNet and trade unions such as the Self-Employed Women's Association of India took unprecedented advantage of this space and worked in alliance with organised labour (Vosko 2002). Paradoxically, employers' vehement opposition to the adoption and ratification of the Home Work Convention underscores the potential force of these hard law regulations and the constitutional obligations attached to ILO membership, where international commitments are growing in importance in the face of globalisation.

Still, the current obstacles to setting standards aimed at mitigating precarious employment highlight fundamental pitfalls in the ILO machinery. Since the mid 1990s, standard setting in this area has become a zero-sum game. If workers' and employers' representatives take opposing positions, their votes simply cancel each other out before an issue even surfaces on the ILO agenda, preventing important and timely matters from reaching the floor of the International Labour Conference and precluding the incorporation of proactive provisions in the international labour standards that are negotiated. This limitation leads to softer promotional standards, an ironic development since many intergovernmental organisations, including industrial or sector-specific organisations in sectors such as forestry, as well as some international financial institutions, are attempting to move beyond promotional standards in pursuit of global governance. In the case of precarious employment, some ILO constituents, including many workers' groups, increasingly argue that soft laws are preferable to the absence of standards altogether, especially when they contain strong substantive prescriptions for policy change, given weak national and subnational labour laws, legislation, and policies. This was evident at the International Labour Conference in 2003, when the Workers' Group began by calling for a convention but, ultimately, accepted a recommendation so long as it contained sound prescriptions for national policy development. Others see dangers in combined hard and hard soft-law instruments (such as the Private Employment Agencies Convention and its Recommendation) that legitimise nonstandard forms of employment where workers are vulnerable to insecurity, mandate the extension of weak protections and benefits to workers, and, hence, thwart efforts to curb the spread of precarious employment.

Notes

1. Field research for this chapter was undertaken at the International Labour Organization (ILO) in June 1997, July 2000, and July 2003. This research was made possible by the Social Sciences and Humanities Research Council of Canada, specifically a project on Workers and Social Cohesion, the ILO's Declaration division and representatives of various NGOs, emerging labour organisations of workers in the informal economy, International Trade Union Secretariats, and the International Confederation of Free Trade Unions. The author thanks the anonymous reviewers and Gerald Kernerman, for their comments on an earlier version of this chapter, and Arlene Dosen and Kim McIntyre, for their work as research assistants.
2. Notably, homework typically involves either the type of bilateral relationship common to the standard employment relationship or self-employment and work arrangements with the worker working offsite and paid by the piece.
3. The Part-Time Work Recommendation also directs governments to set progressively lower threshold requirements 'based on earnings or hours of work as a condition of coverage' in occupational social security schemes and, as appropriate, to grant part-time workers minimal or flat-rate benefits, especially in old age, sickness, invalid, and maternity benefits, as well as family allowances (ILO 1994b, arts. 6a, b).
4. In addition, it calls on member states to revise social security schemes that discourage part-time workers from entering a given occupation or field of work (ILO 1994b, art. 14).
5. In the case of homework, the call for written contracts is supplemented by provisions requiring that all written contracts contain the name and address of the employer and the intermediaries, the scale or rate of remuneration and the methods of calculation, and the types of work to be performed (ILO 1996b, art. 5[2]).
6. It is notable, therefore, that several efforts to stem precarious employment at the national level corresponded with standard-setting activities at the ILO. In Canada, for example, a federal task force on part-time work made recommendations similar to the ILO instruments, and the federal government, acknowledging the need to re-regulate the employment relationship, also conducted a review of the Canada Labour Code (Advisory Committee on the Changing Workplace 1997; Human Resources Canada 1995; Commission of Inquiry into Part-Time Work 1983). In the United States, in the same period, the Commission on the Future of Worker-Management Relations also probed similar questions (Dunlop 1994).
7. For a description of this process, see Vosko (2002).

8. It also calls on national governments to encourage private employment agencies to promote equality in employment through affirmative action programmes (ILO 1997e, art. 10).
9. For the country studies, national experts were asked to focus on four types of situations including subordinate work, 'triangular employment relationships', self-employment, and self-employment under conditions of dependence. In light of the ILO's concern to be sensitive to industry- and sector-based phenomena, they were also asked to pay particular attention to truck drivers in transport enterprises, construction workers, and salespeople, and to explore the grey area between formal and informal sectors and the situation of women workers.
10. 'Unprotected workers' is a confusing term. In its synthesis of the country studies, the ILO associated the term with 'a category of workers who appear to be excluded from the protection provided by the employment relationship, but who in fact carry out their work within the framework of concealed or disguised employment relationships'. It asserts further that, 'at the same time, there are objectively ambiguous situations, which are on the increase, which merit protection, since the workers involved are placed in situations of dependency, but in respect of which the scope of legislation may be too narrow, in that it does not allow an identification of the employer of the persons who should assume responsibility vis-à-vis the worker' (ILO 2000a, Conclusions, para. 102).
11. This final alternative is consistent with developments in the ILO Committee on Legal Issues and International Labour Standards, where the broad scope of the International Labour Code is increasingly under question.
12. In highlighting the growing respect for the ILO among international financial institutions, Hansenne cited a study by the Organisation for Economic Co-operation and Development (OECD) on labour standards and trade that demonstrates a correlation between the existence of fundamental labour rights in a country and economic security and stability (Torres 1996).
13. For a discussion of the history and evolution of the new Social Declaration, as well as its effects on standard setting pertaining to workers in precarious employment, see Vosko (2002).
14. All of the 117 speakers (from all three constituencies) also acknowledged the need to target standards more effectively to increase their impact (ILO 1997b, para. 14).

References

Advisory Committee on the Changing Workplace (1997). 'Collective Reflection on the Changing Workplace: Report of the Advisory Committee on the Changing Workplace'. <www.reflection.gc.ca/report_e.html> (April 2004).

Commission of Inquiry into Part-Time Work (1983). 'Part-Time Work in Canada: Report of the Commission of Inquiry into Part-Time Work'. Ministry of Supply and Services, Ottawa.

Dunlop, John (1994). 'The Changing Environment for Worker-Management Relations'. United States Department of Labor, Washington DC.

Fudge, Judy and Leah F. Vosko (2001a). 'By Whose Standard: Re-regulating the Canadian Labour Market'. *Economic and Industrial Democracy* vol. 22, no. 3, pp. 327–356.

Fudge, Judy and Leah F. Vosko (2001b). 'Gender, Segmentation, and the Standard Employment Relationship in Canadian Labour Law, Legislation, and Policy'. *Economic and Industrial Democracy* vol. 22, no. 2, pp. 281–310.

Human Resources Canada (1995). 'Seeking a Balance: Canada Labour Code Part I'. <labour.hrdc-drhc.gc.ca/labour/labstand> (April 2004).

International Labour Organization (1981). 'Workers with Family Responsibilities Recommendation'. R165. <www.ilo.org> (April 2004).

International Labour Organization (1994a). 'Part-Time Work Convention'. C175. <www.ilo.org> (April 2004).

International Labour Organization (1994b). 'Part-Time Work Recommendation'. R182. <www.ilo.org> (April 2004).

International Labour Organization (1994c). *The Role of Private Employment Agencies in the Functioning of Labour Markets*. International Labour Organization, Geneva.

International Labour Organization (1996a). 'Home Work Convention'. C177. <www.ilo.org> (April 2004).

International Labour Organization (1996b). 'Home Work Recommendation'. R184. <www.ilo.org> (April 2004).

International Labour Organization (1997a). 'Draft Convention on Contract Labour'. International Labour Organization, Geneva.

International Labour Organization (1997b). 'Follow-up on the Discussion of the Report of the Director-General to the 85th Session (1997) of the International Labour Conference'. GB.270/3/2. <www.ilo.org> (April 2004).

International Labour Organization (1997c). 'The ILO, Standard Setting, and Globalization: Report of the Director-General'. <www.ilo.org> (April 2004).

International Labour Organization (1997d). 'Private Employment Agencies Convention, 1997'. C181. <www.ilo.org> (April 2004).

International Labour Organization (1997e). 'Private Employment Agencies Recommendation, 1997'. R188. <www.ilo.org> (April 2004).

International Labour Organization (1998). 'Follow-Up on the Discussion of the Report of the Director-General to the 85th Session (1997) of the International Labour Conference (Part II)'. GB.273/4. <www.ilo.org> (April 2004).

International Labour Organization (2000a). 'Meeting of Experts on Workers in Situation Needing Protection'. Geneva, 15–19 May. <www.ilo.org> (April 2004).

International Labour Organization (2000b). 'Working Party on Policy Regarding the Revision of Standards. Second Item on the Agenda: Possible Improvements in the Standard-Setting Activities of the ILO'. GB.277/LILS/4. <www.ilo.org> (April 2004).

International Labour Organization (2003a). 'Fifth Item on the Agenda: The Scope of the Employment Relationship'. Report of the Committee on Employment Relationship, 91st Session. Geneva. <www.ilo.org> (April 2004).

International Labour Organization (2003b). 'The Scope of the Employment Relationship'. Report to the 91st Session of the International Labour Conference. Geneva. <www.ilo.org> (April 2004).

Josselin, Daphné (2001). 'Back to the Front Line? Trade Unions in a Global Age'. In D. Josselin and W. Wallace, eds., *Non-state Actors in World Politics*, pp. 169–188. Palgrave and St. Martin's Press, Houndsmills.

Lipsett, Brenda and Mark Reesor (1997). 'Flexible Work Arrangements: Evidence from the 1991 and 1995 Survey of Work Arrangements'. R-97-10E. <www.hrdc-drhc.gc.ca/sp-ps/arb-dgra/publications/research/abr-97-10e.shtml> (April 2004).

Mückenberger, Ulrich (1989). 'Non-standard Forms of Employment in the Federal Republic of Germany: The Role and Effectiveness of the State'. In G. Rodgers and J. Rodgers, eds., *Precarious Jobs in Labour Market Regulation: The Growth of Atypical Employment in Western Europe*, p. 267. International institute for Labour Studies, Geneva.

Prugl, Elisabeth (1999). *The Global Construction of Gender: Home-Based Work in the Political Economy of the 20th Century*. Columbia University Press, New York.

Rodgers, Gerry and Janine Rodgers, eds. (1989). *Precarious Jobs in Labour Market Regulation: The Growth of Atypical Employment in Western Europe*. International Institute for Labour Studies, Geneva.

Schellenberg, Grant and Christopher Clark (1996). 'Temporary Employment in Canada: Profiles, Patterns, and Policy Considerations'. Canadian Council on Social Development, Ottawa.

Torres, Raymond (1996). 'Labour Standards and Trade'. *OECD Observer*. <www1.oecd.org/publications/observer/202/010-012a.pdf> (April 2004).

United Nations Development Fund for Women (2000). 'Regional Policy Seminar on Home-Based Work'. 18–20 October, Kathmandu, Nepal.

Vosko, Leah F. (1997). 'Legitimizing the Triangular Employment Relationship: Emerging International Labour Standards from a Comparative Perspective'. *Comparative Labor Law and Policy Journal* Fall, pp. 43–77.

Vosko, Leah F. (1998). '"Regulating Precariousness": The Temporary Employment Relationship under the NAFTA and the EC Treaty'. *Industrial Relations* vol. 53, no. 1, pp. 123–153.

Vosko, Leah F. (2000). *Temporary Work: The Gendered Rise of a Precarious Employment Relationship*. University of Toronto Press, Toronto.

Vosko, Leah F. (2002). 'Decent Work: The Shifting Role of the ILO and the Struggle for Global Social Justice'. *Global Social Policy* April.

Vosko, Leah F. (2003). 'Precarious Employment in Canada: Taking Stock, Taking Action'. *Just Labour* vol. 3 (Fall), pp. 1–5.

Vosko, Leah F., Nancy Zukewich, and Cynthia Cranford (2003). 'Precarious Jobs: A New Typology of Employment'. *Perspectives on Labour and Income*, Catalogue No. 75-001-XIE, 16–26 October. Statistics Canada.

Chapter 7

Hard Law or Soft Law: India and International Labour Standards

C. S. Venkata Ratnam and Anil Verma

The International Labour Organization (ILO) has played a significant role in promoting international labour standards. A founding member, India has contributed to the codification of those standards. It has, in turn, benefited from them in developing its own legal and institutional framework for dealing with social and labour issues. In recent years, there have been efforts to link the standards to trade through the World Trade Organization (WTO) by introducing social clauses in international trade with sanctions for non-compliance. In parallel, voluntary initiatives such as the Clean Clothes Campaign and the Council on Economic Priorities Accreditation Agency's Social Accountability 8000 standard have begun to make progress in promoting social labelling, company codes, and consumer boycotts.

This chapter discusses the relevance of soft laws in reference to India. When there is international pressure on national governments for better compliance with international labour standards through appropriate legal frameworks and institutional mechanisms for strict enforcement — hard law — there is resistance on economic grounds and on political grounds (the issue of national sovereignty being among them). However, when one or more players in the marketplace want to apply similar pressure with legal backing, but using consumer power, the producers have little or no option but to agree to enforce international labour standards using initiatives referred to as voluntary. Yet, in reality, these initiatives are not voluntary, because of the coercive influence of the consumer's refusal to buy. However, although hard law wider, more universal coverage and impact, at least in the areas where it has jurisdiction, soft law — which has no statutory backing — has more limited coverage and impact.

International Labour Standards

The conventions adopted by the ILO constitute the international labour standards. Their purpose is to maintain certain basic minimum standards everywhere in the world. As a means of reconciling globalisation and social justice, international trade is being linked to a system of social clauses so that developing countries do not enjoy the benefits of trade liberalisation on the basis of comparatively low labour costs. There is a sound justification for international labour standards; their desirability is not in doubt. The controversy, however, is over the means of their enforcement, particularly concerning the attempted connection between certain core labour standards and international standards. Thus, the question is not whether international labour standards should be implemented, but whether the countries and companies that continue to pursue competitive advantages through violating fundamental rights should be punished through some sort of trade mechanism. Employers in several developed countries as well as unions, governments, and employers in several developing countries have resisted a formal link between labour standards and trade.

While producers in developed countries seem to seek to benefit by exploiting cheap labour in developing countries, workers in those countries believe their jobs may be affected by the comparative advantage of their labour. However, the three social partners in developing countries — labour and unions, management and employers, and government — consider linking trade in this manner to be an attempt by the governments and workers in developed countries to deprive the developing countries of the advantage of providing cheap labour. The social clause in a multilateral trade agreement can allow a country to ban the import of goods produced in another country in violation of a predefined international labour standard. In the words of Ernesto Zedillo, Mexico's former president, some protestors are determined to save developing countries from development.

The developed countries' arguments are based on the realisation that poverty anywhere is a danger to prosperity everywhere. There is also concern about the so-called race to the bottom encouraged by the notion that 'if you don't raise your standards, we may have to lower ours'. Social dumping could cause job losses in developed countries. Visa restrictions for software professionals from India entering the United States, for example, led Californian companies to consider shifting their operations to India if their easy access to skilled staff remained in danger.

However, given the abysmally low share of manufactured to value-added exports from developing countries to developed countries, there could be no net employment threat to developed countries to warrant any fear about a race to the bottom.

The conclusion of the Uruguay Round resulted in the General Agreement on Tariffs and Trade (GATT) giving way to the WTO in 1995. That year, the World Social Summit at Copenhagen committed itself 'to promoting the goal of full employment as a basic priority of our economic and social policies, and to enabling all men and women to attain secure and sustainable livelihoods through freely chosen productive employment and work' (World Summit for Social Development 1995). To this end, the summit reiterated the need to promote respect for seven ILO conventions covering equal remuneration and non-discrimination, freedom of association, the right to organise and bargain collectively, and the prohibition of forced and child labour. The Delhi Declaration of January 1995, issued at the Fifth Labour Ministers' Conference of Non-Aligned and Other Developing Countries, expressed deep concern about the efforts to link international trade and the enforcement of labour standards through the imposition of the social clause.

The Singapore Declaration of December 1996 at the WTO Ministerial Conference pledged a commitment to the observance of internationally recognised core labour standards, but conceded that the ILO was the competent body to determine and promote these standards. The trade ministers also agreed that labour standards should not be used for protectionist purposes and that the comparative advantage of low-wage developing countries should not be questioned.

In 1997, the International Labour Conference of the 113 member states of the Non-Aligned Movement and five observer states, including China, stressed the need to ratify core labour standards. The conference sought technical assistance and the promotion of economic development from the ILO and the world community for achieving higher standards of social justice. It further suggested that the ILO should provide help to developing countries facing protectionist pressures instead of implementing trade-related mechanisms such as social labelling or other measures that support the social clause. The ILO took the initiative in 1998 by adopting the Fundamental Principles Declaration (see Table 7-1) and in 1999 the Worst Forms of Child Labour Convention concerning immediate action. A number of developing countries still tried to push the WTO to

Table 7-1 International Labour Organization Declaration on Fundamental Principles and Its Follow-up

On 18 June 1998, at its 86th session, the International Labour Conference adopted the Declaration on Fundamental Principles. The declaration reiterates the binding nature of the Philadelphia Declaration of 1944 and requires compliance of the core conventions covering the following aspects, even by those countries that have not ratified the relevant conventions:

a) Freedom of association (Convention No. 87) and the effective recognition of the right to collective bargaining (Convention No. 98).
b) The elimination of all forms of forced or compulsory labour (Convention Nos. 29 and 105). In 1999, the ILO adopted Convention No. 182 concerning immediate action for the abolition of the worst forms of child labour, which became the eighth core convention.
c) The effective abolition of child labour (Convention No. 138).
d) The elimination of discrimination with respect of employment and occupation (Conventions Nos. 100 and 111).

All member countries of the ILO are required to submit a report on their progress in implementing the principles enshrined in the above conventions. The ILO, in turn, will prepare a report each year on one of the four categories of fundamental principles and rights. The report will be based on the information gathered and assessed according to established procedures. In the case of member states that have not ratified the fundamental conventions, the report will be based on the findings of the annual follow-up. In the case of member countries that have ratified the conventions, the report will be dealt in accordance with Article 22 of the ILO Constitution.

The above declaration does not, however, favour linking international labour standards with trade.

ensure respect for core labour standards and proposed establishing a working forum on trade, globalisation, development, and labour issues at the Seattle ministerial in 1999. Suspicions were roused when the American Federation of Labor and Congress of Industrial Organizations (AFL-CIO) pledged support to Al Gore in the U.S. presidential elections in 2000 and when U.S. president Bill Clinton said in an interview with a Seattle newspaper that trade sanctions should be used to enforce core labour standards.

In the developing and newly industrialising countries in the Asia Pacific region, the dominant notion is that labour standards improve only when economies and job opportunities grow. The link between economic growth and labour standards may be less than proportionate; that is, while labour standards may not improve at the same pace or rate as economies grow, the decline in labour standards would be more steep than the pace or decrease in economic growth. Some of the Asian countries that faced financial crises and chaos in 1997 experienced a dramatic deterioration in basic labour standards during the crises, as well as during the period of reform and recovery.

Although usually seen as an issue of developing and transition economies, the harmonisation of core labour standards within developing countries could be contentious and difficult (Block and Roberts 1998). In the case of the North American Free Trade Agreement (NAFTA), compliance with domestic regulations is required by the three member states — the U.S., Canada, and Mexico. However, this minimalist approach in NAFTA's separate agreement on labour rules out any scope for homogenising labour standards (Adams and Singh 1997).

Jo Marie Griesgraber and Bernhard Gunteur (1997) find that the reduction of tariff barriers on a world scale makes weak economies vulnerable to the power of multinational corporations (MNCs), and often leads to greater impoverishment of already poor countries and peoples. Lim Hoe (1995) stresses the need to protect workers' rights but considers that trade-related international labour standards may constitute hidden trade barriers. Keith Maskus and Jill Holman (1997) conclude that the potential economic impact of international labour standards, especially economic competitive advantage, could be counterproductive.

Much of the controversy about linking core labour standards and trade relates to the difficulties in harmonisation among countries at drastically different stages of economic development. In light of the fact that by 1993 there were already at least 68 regional trade agreements (Kirmani and Calika 1994), unless some integration were to take place among them, it is difficult to conceive of a global policy. The labour standards in the growing Asian economies have been the subject of major discussions that have dulled the shine of the past success brought by the export-led growth strategies in those same economies. In some, economic growth eventually did nonetheless lead to social development and improved labour standards. Although organisations such as the Organisation for Economic Co-operation and Development (OECD) and the International Confederation of Free Trade Unions (ICFTU) are keen to foster core labour standards, the Asia-Pacific Economic Cooperation (APEC) maintains that trade, investment, and development will bring improved labour standards.

Unfortunately, trade unions in many developing countries are still split along the cold war divisions of Eastern and Western federations of international trade unions trapped in an ideological dilemma. When global competition is based on cost-cutting strategies, production centres shift within the South, not necessarily only from North to South. If this goes unabated, labour standards fall rather than rise.

Thus fundamental differences persist within and among developed and developing countries. Employers in several developed countries — including two world bodies of employers, namely, the International Organisation of Employers (IOE) and the International Chamber of Commerce (ICC) — endorse the view that labour standards remain under ILO supervision. Governments do not necessarily reflect the interests of their country's workers. The ICFTU, comprising 221 affiliated unions that represent 150 million workers in 148 countries, supports the social cause, which links international trade with labour standards. In several Asian countries, particularly China and India, 172 trade unions support the social clause due to what they perceive to be enormous implications for low-paid, unskilled workers in labour-intensive export enterprises that provide these countries the comparative advantage of cheap labour. In Africa, the unions in some countries (South Africa, for instance) support the social clause, while a few others (such as Zimbabwe and Zambia) oppose it. Latin American countries are generally more amenable to linking trade and labour standards, and unions in Guatemala, Honduras, El Salvador, and Nicaragua have strategically used threats of trade sanctions by the U.S. to secure their own right to organise. Unions in several developing countries in Asia would do well to learn from such experiences. Instead, in many cases they have joined with governments and employers and continue to mistrust the real motives behind such a link. The dilemma of trade unions in many developing countries is that despite recognising the need to improve labour standards, they remain apprehensive that any link of labour to trade may wipe out some of the existing jobs.

Although trade unions in South Asia have declared that they were not content with the level of compliance with labour standards, several considered that linking labour standards to trade would hurt them more. Therefore, most have adopted a dual strategy: oppose such a link at the international level, but at the national level continue to pressure respective governments to improve labour standards. But this is not adequate. When the trade unions themselves are unable to realise the right to organise and bargain collectively or to mitigate the problems of wages that are below poverty level, inequality, child labour, and bonded labour, even after half a century of political independence, then it is useful to leverage the international support and assert, as does Juan Somavia, Director-General of ILO, that having 'jobs without rights is a sure trap into perpetual poverty'.

As Werner Sengenberger (2002, 92–94) states:

> Preventing sub-standard terms of employment and poor working conditions is the 'classic' justification of international labour standards ... [One of the] major impediments to advancing international labour standards relates to the sharp *shift in the balance of power* in both local

> and world labour markets at the expense of labour. This loss of power results from the weakening of trade unionism in many parts of the world. To a large extent, it is the outcome of globalization, which opened up new strategic options for employers, such as relocation. The mere threat of leaving existing locations is sufficient to change the bargaining power equation in their favour. In a large number of developing countries, trade unions have suffered from the policies of governments that attempted to gain competitive advantage by keeping EPZs [export processing zones] free from collective worker organisation. Higher average unemployment and faster structural change have caused loss of trade union membership and influence. Initiatives have been taken by national and international trade union organizations to counter the global trends through negotiations with companies and international organizations. The reach has been limited so far ... International labour standards linked to social objectives have to become priorities in national and international policy formulation ... The primary thrust for promoting international labour standards should be to provide *incentives* for various actors to comply with international labour standards. Such incentives, including preferential treatment in trade and investment policy for countries and enterprises respective international labour standards, should be formulated and provided on a multilateral basis. Punitive action, such as trade sanctions, should be used as a last resort in case of persistent violation of core labour standards.

From International Labour Standards to Corporate Codes

International labour standards have proven effective and useful in formulating national policies, legislation, and practices on diverse social and labour aspects. ILO conventions and recommendations cover a wide variety of issues: tripartite consultations at the national level and communications and grievance procedures; wages, working hours, leave, and holidays as well as occupational health and safety; non-discrimination as well as affirmative action for the aged, disabled, women, and other vulnerable groups; and labour markets, labour law, labour administration, human resources development, and so on. They are not binding in any of these matters, except when a state ratifies the concerned convention or reiterates the relevant principles in its own national legislation.

The U.S. government is known to use a variety of instruments and mechanisms to enforce international labour standards (Samford 1996). These include section 301 of the U.S. *Trade Act, 1974* (referred to as Super Clause 301), the Caribbean Basin Initiative, the European Union's Generalised System of Preferences, the extraterritorial application of the U.S. labour law, voluntary corporate codes of conduct, and the North American Agreement on Labour Cooperation (NAALC), which is the NAFTA's side agreement on labour.

Currently, several nongovernmental organisations (NGOs) and corporations are disseminating information about products and companies that do not involve child labour to consumers through what is now known as 'social labelling' (United States Department of Labor 1998). In addition to creating a production environment free of child labour, some have provided or funded schools for child labourers and employment or welfare benefits for their parents. Social labelling exists for exports in four industries, namely garments, leather products, coffee and tea, and hand-knotted carpets. Indeed, in the carpet industry, the main labelling programmes are Rugmark, Kaleen, STEP, and Care & Fair. Rugmark and Kaleen produce labels that are attached to individual carpets. STEP and Care & Fair are company certification programmes; participating companies use the labels for advertising and marketing purposes.

The Rugmark Foundation is a nonprofit, voluntary programme that provides market-driven incentives for carpet manufacturers to produce without child labour (Rugmark Foundation 2003). Established in India and Germany in 1994, it now operates in four countries, including Nepal and Pakistan, and provides independent, professional, and internationally credible monitoring as well as a certification system for carpets manufactured without use of child labour. Rugmark supports schools, rehabilitation, and other programmes for former child workers and their communities. As of December 2002, the Indian Rugmark Foundation had certified the export of about 2.68 million labelled carpets from 248 licensed exporters. The programme is enforced through an elaborate system of license approval, random inspections, and carpet tracking. It has conducted 123 436 unannounced visits to licensees' facilities, and has found 1379 children illegally working at looms and put them in seven schools that it has helped establish.

The Carpet Export Promotion Council (CEPC) was established by India's Ministry of Textiles as a quasi-government body, and set up the Kaleen labelling programme in India in 1995 (2003). It is funded partly through government grants and partly through subscription income from industry members. An independent agency conducts inspections and the government's development commissioner conducts periodic reviews. All registered exporters pay a fee for child welfare activities.

The STEP and Care & Fair programmes, initiated by Swiss and German NGOs respectively, require suppliers in India and elsewhere to honour their codes of conduct, which include eliminating child and bonded labourers, and to contribute to health and education programmes for carpet workers and their families (STEP Foundation 2003; Care & Fair for Children's Welfare 2003).

In the sports industry, Reebok International Ltd. uses labelling for soccer balls. By an agreement between the company and exporters in Pakistan, in association with the Federation of International Football Association and the ILO, balls produced for Reebok bear the label 'Guaranteed: Manufactured Without Child Labor' (Reebok International Ltd. 2003). The major elements of the programme are the reorganisation of all production at a new facility, where children below 15 years of age are prohibited to enter or perform work; external monitoring to ensure that children are not entering the workplace and that soccer ball panels are leaving the factory for stitching; and support for the education and vocational training of children in region of Sialkot, where the facility is located.

Another example is TransFair's fair trade label used in the coffee and tea industry (TransFair 2003). Its objectives are to ensure payment of a fair price that covers production costs plus a fair trade premium that flows to the actual producers, producers' participation in decisions regarding use of the fair trade premium, provision for advance payment, and the promotion of a long-term relationship between producers and importers.

The avowed aim of fair trade labelling in tea is to promote producers in the developing world at a disadvantage under the current trading conditions and help them achieve independence and equality. The higher price paid for the product, which mst be produced under conditions that comply with the core labour standards, goes directly to improving the living conditions of pickers, the plantation workers, and the smallholders.

Social labelling programmes are still voluntary and thrive on actual or potential consumer boycott threats. Awareness and acceptance of these programmes are growing, and the costs are often borne by the exporters and the importers themselves. Consumers in developed countries are also reportedly willing to pay a higher price for such labelled products, particularly those produced under certification programmes. As well, there is an ongoing effort to cover the subcontracting arrangements under these programmes.

In addition to labelling programmes, two kinds of codes of conduct are becoming popular in the countries of the North. These concern actual labour standards — what people are paid, how jobs are designed, the levels of safety, and so on — and formal policies that purport to shape corporate conduct in certain ways. Such polices refer to fundamental conditions that apply universally across workplaces. At the national level, the U.S. government announced a set of model business principles in 1995. These cover fair employment practices described in the 1998 ILO Declaration on Fundamental Principles: the avoidance of child and forced labour; the avoidance of discrimination based on race, gender, national origin, or religious beliefs; and respect for the right of association and the right to organise and bargain collectively (ILO 1998). Ironically, the U.S. has ratified only two of the eight core conventions of the ILO.

The IOE (1998) has expressed misgivings about labelling campaigns. The organisation points out that they deal with only a small amount of the child labour employed in export markets, without going to the heart of the problem. Furthermore, they act as a 'quick fix' for the problem of child labour by appealing to the 'feel good' sensibilities of consumers. A label does not guarantee the non-employment of child labour, and in fact may sometimes lead to a producer being penalised even though it has not employed child labourers.

The Case of India

As a founding member of the ILO, and one of the ten countries of chief industrial importance, India holds a non-elective seat in the organisation's Governing Body. It has ratified 39 of the 182 conventions (see Table 7-2).

The ILO has influenced India and India has influenced the ILO. A tripartite committee on ILO conventions oversees issues relating to international labour standards, proposals concerning new conventions and the ratification of old conventions, compliance with the provisions of the ratified conventions, and so forth.

For several years, the committee lay dormant, but it has recently become active once again. As in most other countries, the ILO conventions and recommendations have played a significant role in India's legal framework for wages, working conditions, welfare, social security, protection of the vulnerable members of society, human resources development, equality, and non-discrimination. India has also received technical co-operation and assistance in establishing several institutions and strengthening labour inspection and labour administration.

India's constitution and labour legislation uphold all the fundamental principles envisaged in the eight core international labour standards that constitute the fundamental principles (see Table 7-3). It has also ratified four of those conventions, and will also likely to ratify the Worst Forms of Child Labour Convention. India's record in this regard is better than that of U.S., which has not yet ratified six of the eight core conventions. Among the European countries, France and Germany have ratified all the core conventions. Although some South Asian countries have ratified more core conventions than India, it can be said with certainty that Indian record is no less, if not better, than its neighbouring countries (see Table 7-4).

India has not ratified Convention Nos. 87 or 98 concerning the rights to freedom of association and collective bargaining because of technical difficulties involving trade union rights for civil servants. This, however, is not a major hurdle, because the government can exempt certain services while ratifying the conventions under consideration. The real intention seems to be, as opined by Surendra Nath (1997), India's former Chief Labour Commissioner, that 'these conventions have not been ratified by India because the policy of the government has been to restrict freedom of association to only manual workers (by defining them as Workmen) and exclude supervisory and managerial workers. The other interest of the government is not to allow the right of collective bargaining even to industrial workers in government's departmental undertakings like the Railways, Post and Telecommunications, Central Public Works Department, etc. The government, on the basis of the Pay Commission's recommendations and not through collective bargaining, decides their pay, etc.'

There are certain restrictions on the formation of federations in the telecom sector.

In 1981, the right to collective bargaining was taken away from the insurance sector workers when it was found that collusion between the unions and the employers (public sector) was undermining the interests of the policy holders. Since then, insurance workers engage in consultations, but their pay revisions are made unilaterally by the concerned government department.

India's record in respecting the freedom of association and collective bargaining is much better than that of many other countries in the region that have ratified either one or both conventions. Mere ratification, however, cannot guarantee the achievement of the underlying objectives. It must be backed by a strong political will as well as a legislative framework and effective vigilance by the other social partners, such as NGOs and consumer groups. But these considerations cannot justify non-ratification. India ratified Convention No. 11 concerning the right of association for agricultural workers as far back as 1923. The Union Ministry of Labour in association with the National Labour Institute, set up by the ministry in the 1970s, has organised several workshops and rural camps to facilitate the organisation of rural workers. Still, given the huge social inequities in India, both in terms of caste and land distribution, there is very little unionisation of rural workers.

India still has a long way to go before it can comply with the core labour standards. Its large workforce in the informal sector would be content if even the minimum standards in national legislation — not international labour standards — were the maximum. For example, in the slate industry in one state, a union leader, an influential political party,

Table 7-2 International Labour Organization Conventions Ratified by the Government of India

Number, Title, and Year of Adoption		Date of Ratification
1.	Hours of Work (Industry) Convention, 1919	7 July 1921
2.	Unemployment Convention, 1919[a]	7 July 1921
4.	Night Work (Women) Convention, 1919	7 July 1921
5.	Minimum Age (Industry) Convention, 1919	9 September 1955
6.	Night Work of Young Persons (Industry) Convention, 1919	7 July 1921
11.	Right of Association (Agriculture) Convention, 1921	5 November 1923
14.	Weekly Rest (Industry) Convention, 1921	5 November 1923
15.	Minimum Age (Trimmers and Stokers) Convention, 1921	22 November 1922
16.	Medical Examination of Young Persons (Sea) Convention, 1921	20 November 1922
18.	Workmen's Compensation (Occupational Diseases) Convention, 1925	30 September 1927
19.	Equality of Treatment (Accident Compensation) Convention, 1925	30 September 1927
21.	Inspection of Emigrants Convention, 1926	14 January 1928
22.	Seamen's Articles of Agreement Convention, 1926	31 October 1932
26.	Minimum Wage-Fixing Machinery Convention, 1928	10 January 1955
27.	Marking of Weight (Packages Transported by Vessels) Convention, 1929	7 September 1931
29.	Forced Labour Convention, 1930	20 November 1954
32.	Protection Against Accidents (Dockers) Convention (Revised), 1934	13 January 1964
41.	Night Work (Women) Convention (Revised), 1934[b]	25 March 1938
42.	Workmen's Compensation (Occupational Diseases) Convention (Revised), 1934	13 January 1964
45.	Underground Work (Women) Convention, 1935	25 March 1938
80.	Final Articles Revision Convention, 1948	17 November 1947
81.	Labour Inspection Convention, 1947[c]	7 April 1949
88.	Employment Services Convention, 1948	24 June 1959
89.	Night Work (Women) Convention Revised, 1948	27 February 1950

and the district government officials wanted to help workers secure minimum wages. Duly encouraged, the workers went on strike for three days. The employers, who had multiple businesses, did not feel any pressure and therefore showed no eagerness to end the strike. On the fourth day, the workers agreed to return to work. The employers agreed to take the workers back, after further reducing their wages as a token of punishment for their temerity. In another case, a young and committed Indian Administrative Service officer in the state ministry of labour was not amused with the flourishing dollar-earning tobacco merchants getting away with paying below-minimum wages to their poor, non-unionised labour. He computed the difference in the wages

Table 7-2 International Labour Organization Conventions Ratified by the Government of India, cont'd

Number, Title, and Year of Adoption	Date of Ratification
90. Night Work of Young Persons (Industry) Convention (Revised), 1948	27 February 1950
100. Equal Remuneration Convention, 1951	25 September 1958
107. Indigenous and Tribal Population Convention, 1957	29 September 1958
111. Discrimination (Employment and Occupation) Convention, 1958	3 June 1960
116. Final Articles Revision Convention, 1961[d]	21 June 1962
118. Equality of Treatment (Social Security)[d]	19.08.1964
123. Minimum Age (Underground Work) Convention, 1965[e]	20 March 1975
115. Radiation Protection Convention, 1960	17 November 1975
141. Rural Workers' Organisation Convention, 1975	18 July 1977
144. Tripartite Consultation (International Labour Standards) Convention, 1976	27 February 1978
136. Benzene Convention, 1971	11 June 1991
160. Labour Statistics Convention, 1985[d]	1 April 1992
147. Merchant Shipping (Minimum Standards) Convention, 1976	26 September 1996
122. Employment Policy Convention, 1985	18 May 2000
105. Abolition of Forced Labour, 1957	18 May 2000

Notes

a Later denounced. The convention requires, *inter alia*, furnishing of statistics concerning unemployment every three months, which is considered not practicable.

b Convention denounced as a result of ratification of Convention No. 89.

c Excluding Part II.

d Article 8 of Part II.

e Branches (a) to (c), (e), and (g) covered.

Source: India Ministry of Labour, ILO Conventions Ratified by India <labour.nic.in/ilas/convention.htm> (April 2004).

Table 7-3 Components of the Social Clause and Indian Legislation

Social Clause Aspect	Indian Constitution/Legislation
Freedom of Association and Right to Collective Bargaining (Convention No. 87 and No. 98 respectively).	Freedom of association is guaranteed as a fundamental right in the Indian Constitution. The *Trade Union Act,* 1926, meets with part of the objectives of Convention Nos. 87 and 98. Both conventions, however, have not been ratified by the Government of India.
Forced Labour Convention, 1930, and Abolition of Forced Labour Convention, 1953 (Convention Nos. 29 and 105), which provide for progressive abolition of forced labour in all forms. Convention No. 182 concerning Immediate Action to End the Worst Forms of Child Labour.	Article 23 of the Constitution and the *Bonded Labour System (Abolition) Act,* 1976. India has ratified Convention Nos. 29 and 105. India is moving toward ratifying Convention No. 182.
Equal Remuneration Convention, 1951 (Convention No. 100) to eliminate sex-based discrimination in remuneration and provide for equal remuneration to both men and women for work of equal value. The four underlying bases for determination of work of equal value are skills, efforts, responsibility, and working conditions.	The Indian Constitution upholds the principle of equality between men and women. The *Equal Remuneration Act,* 1926, seeks to provide for equal remuneration to men and women.
Discrimination (Employment and Occupation) Convention, 1958 (Convention No. 111) covers any discrimination, exclusion, or preference that 'has the effect of nullifying or impairing equality of treatment' and that can be the result not only of legislation but also of existing factual positions or practices.	The Indian Constitution upholds equality, denounces discrimination, and encourages preferential treatment to disadvantaged groups in society. Convention No. 111 has been ratified.
Minimum Age Convention, 1973 (Convention No. 138) states that minimum wage for employment should ordinarily be 15 and raised to 18 in dangerous occupations.	The *Child Labour (Prohibition and Regulation) Act,* 1986, bans employment of children below the age of 14. In several laws, the minimum age is variously defined. Employment of children in certain heavy and hazardous industries (Schedule A, part 3) is prohibited by law and the government is taking steps to enforce it strictly. Several court judgements under public interest litigation actively support the prohibition and regulation of child labour.

payable according to the law and the wages actually paid to the workers, and arranged for the difference to be paid to the workers in his presence. After the labour commissioner and his staff left, the employers hired *goondas* who took the money back from the workers and warned them that if they reported the matter they would lose their jobs and their families could be in danger. In a third case, a government official reported that he had personally arranged for a bonded labourer to be freed from a landlord in a village in Bihar. The labourer was given money with which he bought some buffalo, which became ill and died. He sought employment in the village, but the landlord had warned the villagers not to hire him because he a troublemaker. Finally, the man was obliged to return to bondage, this time under even harsher conditions of employment.

These examples are not exceptions. They are common occurrences in many remote and rural areas. In the absence of a proper supervisory mechanism that fully understands and deals with the on-the-ground realities, efforts to improve labour standards can actually lead to a deterioration in labour standards.

In contrast, there are some highly organised and unionised sectors in India where collectively determined labour standards exceed the national and international labour standards in some aspects. During the 1920s and 1930s, the private-sector steel giant Tata Iron and Steel Company introduced a number of voluntary welfare measures long before such measures were incorporated either in the national legislation or international labour standards.

In the past, some of the legislative initiatives based on ILO conventions have had the opposite effect. For instance, measures to safeguard the interests of women through

Table 7-4 Ratification of Core Conventions in Select Countries

Country	Convention							
	29	87	98	100	105	111	138	182
Australia	•	•	•	•	•	•		
Bangladesh	•	•	•		•	•		
Canada		•		•	•	•		
China				•				
Germany	•	•	•	•	•	•	•	
France	•	•	•	•	•	•	•	
India	•			•		•		
Japan	•	•	•	•				
Nepal	•		•	•	•			
Pakistan	•	•	•		•	•		
Sri Lanka	•		•					
United Kingdom	•	•	•	•	•			
United States					•			•

maternity benefits and restricting their employment in night shifts in factories and underground mining have resulted in many employers hesitating to employ women. This reluctance, of course, should not lead to discontinuance of maternity benefits. Giving paternity leave would perhaps restore the balance and remove the incentive for employers to employ only males. The ILO's Night Work Convention was thus revised in the 1990s in order to facilitate the employment of women at night. The Government of India is pursuing, although rather half-heartedly, changes to some of the labour legislation. The changes would amend legislation to encourage the employment of women in night shifts in factories, particularly in the electronics industry and export processing zones.

India is actively pursuing projects to eradicate child labour in hazardous industries. The All India Organisation of Employers has undertaken a project in Jalandhar in Punjab to enter into an agreement similar to the one in Sialkot concerning the abolition of child labour in the manufacture of sports goods.

India has advocated the promotion of labour standards within the framework of the ILO's constitution. It has consistently opposed proposals to link labour standards and trade through social labelling and other measures. The 1995 Non-Aligned Movement meeting organised by India's labour minister in New Delhi adopted a resolution to this effect. India played an active role at the 1999 WTO ministerial meeting Seattle to prevent linking trade to labour and environmental issues. All three social partners — the government, employers' organisations, and a wide variety of national unions — are united against such a link for reasons that are articulated in the commitment of most developing countries to the ILO's pillars of voluntarism, tripartism, and free choice of social partners.[1] The mandatory imposition of labour standards, by whatever name they may be called, contravenes Article 19(3) of the ILO's constitution (ILO 1919). All social partners in India favour upgrading labour standards, but they oppose the connection to trade within the context of the WTO as it is currently constituted. They suspect doing so would result in artificial barriers against competition and, to paraphrase Zedillo again, save India and other developing countries from development. Furthermore, concern for improving labour standards should be more holistic and encompass the entire working class, rather than the microscopic minority engaged in production for exports.

Also, the superficiality of market players' commitment to labelling programmes is beginning to be exposed, with some buyers promising higher prices (whether the product is carpets in Nepal or garments and leather products in India) at the time of implementing voluntary initiatives but later reducing the prices on the pretext of intense competition.

Recently, there has been a tendency among some state governments to pursue competitive labour policies with a view to attracting investments and creating jobs. The Constitution Review Committee appointed by the Government of India in 2000 has been examining this aspect, given the need for a unified law on matters that fall within the ambit of the eight core labour standards under the 1998 ILO Declaration on Fundamental Principles.

A national consultation on international labour standards, which involved several NGOs and national trade union centres, made the following recommendations (Centre for Education and Communication 1996):

- Reject a link between labour and trade as related to the WTO.
- Uphold the principles of universal labour rights and the need for evolving structures to monitor the enforcement of labour rights.
- Set up a United Nations Labour Rights Commission as an alternative.
- Establish, at the national level, a powerful National Labour Rights Commission to monitor and enforce labour rights.

The 1998 Declaration on Fundamental Principles and Rights at Work and its follow-up substantially address the central issue (ILO1998). As long as the ILO takes care to make its instruments yield the desired results and resistance to the WTO continues to build, initiatives at the industry level through social labelling and at the community level through consumer boycotts will apply the kind of pressure that intergovernmental and international organisations are ill equipped to exert.

Conclusion

International labour standards provide the basis for national policy and legislation for developing universal minimum in terms of labour standards. Normatively, the rationale for compliance with international labour standards, particularly the core labour standards described in the Declaration on Fundamental Principles and Rights at Work, cannot be questioned. However, fundamental differences exist between developed and developing countries and among the principal social partners of labour and unions, management and employers, and government over apprehensions about protectionist tendencies. The arguments of the developing countries are based narrowly on the comparative economic advantage than on any long-term social advantage. The developed countries are overly concerned about possible adverse employment effects on their labour force and fear a race to bottom. The restrictions on free movement of workers, which benefits mainly the developing countries, has only exacerbated the fundamental North-South differences. In reality, it is no longer a North-South issue. Cost cutting is ultimately shifting the competition to South and among Southern countries and workers in developed as well as developing countries are losing in the process.

It is not only governments that have resisted international pressure at the macro level to link international labour standards with international trade. Unions and management in many developing countries have also held back. Often, national pride, sovereignty, and other emotional issues are raised in opposition. It is here that hard laws prove to be an elusive tool for achieving a universal minimum of labour standards. Against this backdrop, there has been some progress made in improving labour standards as a result of voluntary initiatives on the part of buyers and consumers to exert pressure on producers and suppliers, however modest those initiatives may be. When buyers and consumers exercise choice, producers and suppliers cannot question.

The problem, however, is that soft laws will have limited effect. Products from developing countries account for only a fraction of the world's total exports. The prices of bulk primary produce as well as manufactured products from developing countries have been falling. When the price falls, employers try to economise on the number of workers and the amount of remuneration paid to them. Most of the services from developing countries are not yet tradable, and there remain restrictions on the movement of people. All things considered, soft laws can cover only a small number of working people in developing countries. However, cost-cutting comparative advantages will make it difficult to ensure that the existing standards of labour are maintained in export-oriented businesses in developing countries. Beyond a point, if things are left only to market forces, practices can become amoral and averse to workers' rights and labour standards. Therefore, soft law is no substitute for hard law.

The international free trade union movement supports linking international trade to international labour standards. However, trade unions in some developing countries still oppose it. They need to learn from the experience of unions in Latin America and Africa, where the threat of sanctions through free trade and regional agreements was used to secure and improve workers rights. Moreover, those who propose the link of international trade and labour standards should focus not on only sanctions but also on incentives.

Note

1. For statements by different social partners, see Indian Industrial Relations Association and Friedrich Ebert Stiftung (1996).

References

Adams, Roy J. and Parbudyal Singh (1997). 'Early Experience with NAFTA's Labour Side Accord'. *Comparative Labor Law Journal* vol. 18, no. 2, p. 161.

Block, Richard N. and Karen Roberts (1998). 'Ranking Labour Standards across Countries: A Case Study of the United States and Canada'. 11th World Congress of the International Industrial Relations Association, 22–26 September. Bologna.

Care & Fair for Children's Welfare (2003). 'Catalog of Demands'. <www.care-fair.com/en/katalog.html> (April 2004).

Carpet Export Promotion Council (2003). 'The Magic of Indian Carpets'. <www.india-carpets.com/index.php> (April 2004).

Centre for Education and Communication (1996). *Labour, Environment, and Globalisation: Social Clause in Multilateral Trade Agreement, A Southern Response*. Centre for Education and Communication, New Delhi.

Griesgraber, Jo Marie and Bernhard G. Gunter (1997). 'World Trade: Toward Fair and Free Trade in the Twenty-First Century'. Pluto Press, London.

Hoe, Lim (1995). 'Human Rights, International Labour Standards, and Trade in a Globalised Economy: An Assessment of the Economic, Moral, and Political Dimensions'. Ph.D. diss. University of Birmingham, Birmingham.

Indian Industrial Relations Association and Friedrich Ebert Stiftung (1996). 'Social Clause in Trade: Trade Union Perspective'. New Delhi.

International Labour Organization (1919). 'Constitution'. <www.ilo.org/public/english/about/iloconst.htm> (April 2004).

International Labour Organization (1998). 'ILO Declaration on Fundamental Principles and Rights at Work'. <www.ilo.org/public/english/standards/decl/declaration/text> (April 2004).

International Organisation of Employers (1998). 'Employers' Handbook on Child Labour: A Guide for Taking Action'. In collaboration with the Bureau for Employers' Activities and the International Programme on the Elimination of Child Labour (IPEC) of the International Labour Office. <www.ioe-emp.org/ioe_emp/pdf/handbook_child_labour.PDF> (April 2004).

Kirmani, Naheed and Nur Calika (1994). *International Trade Policies: The Uruguay Round and Beyond*. International Monetary Fund, Washington DC.

Maskus, Keith E. and Jill A. Holman (1997). *Should Core Labor Standards Be Imposed through International Trade Policy?* World Bank, Development Research Group, Washington DC.

Nath, S. (1997). 'Labour Policy and Economic Reforms — India, 1991–1996: A Study in the Context of Industrial Restructuring'. Mimeo. New Delhi.

Reebok International Ltd. (2003). 'Our Business Practices: No Child Labor'. <www.reebok.com/x/us/humanRights/business/child_labor.html> (April 2004).

Rugmark Foundation (2003). 'Rugmark'. <www.rugmark.org> (April 2004).

Samford, Clay (1996). 'Strange Idea: The North American Agreement on Labour Cooperation and Other U.S. Approaches to Enforcing International Labour Standards'. Senior Honours Thesis. Cornell University, Ithaca.

Sengenberger, Werner (2002). 'Globalization and Social Progress: The Role and Impact of International Labour Standards'. Friedrich Ebert Stiftung, Geneva.

STEP Foundation (2003). 'Step by Step'. <www.step-foundation.ch/en/home.htm> (April 2004).

TransFair (2003). 'About TransFair'. <transfair.org> (April 2004).

United States Department of Labor (1998). 'By the Sweat and Toil of Children (Volume V): Efforts to Eliminate Child Labor'. Washington DC. <www.dol.gov/ilab/media/reports/iclp/sweat5/welcome.html> (April 2004).

World Summit for Social Development (1995). 'Copenhagen Declaration on Social Development — Part C: Commitments'. <www.un.org/esa/socdev/wssd/agreements/decpartc.htm> (April 2004).

Chapter 8

Trade Policy and Labour Standards: Objectives, Instruments, and Institutions

Michael J. Trebilcock[1]

The link between trade and labour has a long history (Charnovitz 1987; Leary 1996; Molatlhegi 2000; Salazar-Xirinachs and Martinez-Piva 2003). It has become one of the most contentious contemporary issues in trade and labour policy circles. When U.S. president Bill Clinton stated at the Seattle ministerial meetings of the World Trade Organization (WTO) in December 1999 that trade sanctions should be available under the WTO multilateral system against countries violating international labour standards, it provoked an intensely hostile reaction from developing countries. Indeed, it was a significant factor in the failure of WTO members to agree at that time on the launch of a new multilateral round. The subsequent ministerial declaration that launched a new round at Doha in 2000 confirmed the 1996 Singapore ministerial decision to remit all international labour standards issues to the International Labour Organization (ILO). However, the issue of a trade-labour link seems unlikely to go away. Regionally, the potential expansion of the North American Free Trade Agreement (NAFTA) into the Free Trade Agreement of the Americas (FTAA) will raise the scope and status of NAFTA's side agreement on labour in this context. And unilateral trade actions by states on account of labour practices prevailing in other states may well provoke trade disputes that will require adjudication by international trade dispute settlement bodies.

While it may well be the case that improved labour standards (and productivity) are in the best interests of developing countries themselves in promoting broad conception of development (Langille 2003), this chapter concentrates on the narrower issue of the potential interface between international trade policy and international labour standards in the WTO multilateral trading system. It proceeds by reviewing sequentially the choice of policy objectives, the choice of policy instruments, and the choice of institutional regime in structuring a link between trade policy and labour standards. On choice of objective, the chapter will argue for assimilating core labour standards into universal human rights, but not privileging core labour standards over other universal human rights. It rejects fair trade and race-to-the-bottom rationales for a link between trade policy and labour standards. On choice of instrument, it rejects conditioning the linkage of human rights–based trade sanctions on adverse trade effects in the sanctioning country. On choice of institutional arrangement, it favours according a pre-eminent institutional role to the ILO and United Nations human rights committees in making determinations of systematic and persistent violations of relevant universal human rights, leaving the WTO's Dispute Settlement Body with the determination of whether trade sanctions for violators have been applied in a non-discriminatory and consistent fashion and meet some basic standard of proportionality.

The Choice of Policy Objective

Out of both contemporary and historical discussions about the case for a link between trade policy and labour standards emerge several normative rationales for such a link. These rationales are often largely elided in debates, an occurrence that then greatly complicates the task of evaluating the appropriate choice of instrument and the choice of an institutional arrangement for vindicating the chosen policy objectives (Trebilcock and Howse 1999, ch. 16).

Unfair Competition

It is often argued that countries that sell into export markets goods produced by processes that fail to respect or comply with internationally recognised labour standards are engaging in an unfair form of competition; this competition deprives domestic producers in importing countries and producers and exporters in developing countries that comply with these standards of legitimate market share. The conduct of governments in countries where internationally recognised labour standards are not complied with are often accused of 'social dumping' or indirect or implicit and illicit subsidisation. These economic activities, entailing either lax labour laws or ineffective enforcement of nominally compliant laws (or both), are considered analogous to economic dumping and direct subsidisation that may attract antidumping duties or countervailing duties under Article VI of the General Agreement on Tariffs and Trade (GATT) and the current WTO agreement (Rodrik 1996; Blackett 1999).

Without further and careful specification, this rationale for linking international trade policy or sanctions with labour standards is largely incoherent. From the perspective of importing countries, generically lower labour costs in exporting countries enhance consumer welfare in importing countries, and by more than reductions in producer welfare in the latter. From the perspective of exporting countries, particularly developing countries, the latter rightly argue that in the early stages of industrialisation, entailing mass production of low-tech products such as textiles, clothing, footwear, processed agricultural products, low-cost and low-skilled labour is one of the principal sources of their comparative advantage; to deny them the ability to take advantage of this is to consign them forever to low value-added commodity production for developed country markets ('hewers of wood and drawers of water').

However, it is crucial not to overstate the significance of this source of comparative advantage. Data show almost a one-to-one relationship between labour productivity and labour costs in manufacturing in a wide range of developed and developing countries. About 90 percent of the international variation in real labour costs is associated with cross-country differences in labour productivity (Rodrik 1997; Flanagan 2002). Labour productivity, or total factor productivity, is a function of many factors, including public investments in education and training, health care, infrastructure, and law and order. It is thus a fallacy to assume that low wages are the main driving force behind today's global trade or foreign direct investment (FDI) flows. This relationship between labour productivity and labour costs explains why most international firms are not seeking to relocate to, for example, Bangladesh despite its low wages, and why most international trade and FDI flows are still dominated by developed countries as countries of origin and countries of destination.

Even stated in fairness rather than in welfare terms, if it is unfair for firms and workers in developed countries to be forced to compete with firms and workers in developing countries with access to low-paid, low-skilled labour, then it is equally unfair for developing countries to have to compete with firms and workers in developed countries that depend on highly skilled labour forces, highly developed infrastructure, large public investments in education and research and development, extensive healthcare systems, effective law and order, and superior institutions — in most cases reflecting collective or public investments on a scale that far exceeds the capacity of most developing countries. Consequently, this unfair competition argument, in and of itself, is totally indeterminate and carries high risks of the link between trade policy and labour standards being exploited for protectionist ends. In this respect, it is important to emphasise that the unfair competition argument focusses principally on the welfare implications of non-compliance with international labour standards for citizens or interests in importing countries.

The Race to the Bottom

Historically, a major motivation for promoting international labour standards in the first place and then subsequently promoting a link between trade policy and labour standards (as, for example, reflected in the preambles to the ILO Constitution and the Havana Charter) is that low labour standards (including low wages) in exporting countries may undermine higher labour standards in importing countries by precipitating a so-called 'race to the bottom' — a form of prisoner's dilemma — that can only be pre-empted by international agreement on minimum labour standards and their enforcement (see Alben 2001).

This argument is a variant on the unfair competition argument reviewed above. Instead of assuming that importing countries will maintain the *status quo* with respect to their more stringent labour standards and accept a loss of market share, it assumes that such countries will in fact progressively dilute them in order to avoid losing market share to imports from countries where these standards are not adhered to. Such a loss would result in a low-level equilibrium trap where all countries relax their labour standards to what many regard as sub-optimal levels, yet all simply retain their pre-existing share of trade or investment after the race to the bottom has run its course.

Despite the durability of this concern, for reasons given above — most notably that differences in conditions of employment largely reflect differences in productivity — there is little reason to suppose that liberal trade and investment regimes will precipitate a race to the bottom. Moreover, the empirical evidence provides no support for the claim that liberal international trade and investment regimes are leading developed countries to relax their core labour standards, or their labour standards generally, or that foreign direct investors are investing in countries with weak core labour standards. Indeed, the evidence suggests that with the notable exception of China, countries with such weak standards attract very little FDI either in general or specifically in the sectors where core labour standards are weak. Moreover, health conditions, civil liberties, and labour standards tend to be stronger, not weaker, in countries with more open trade policies (Flanagan 2002). Race-to-the-bottom concerns, like unfair competition concerns, largely emanate from the perceived welfare implications of non-compliance with international labour standards for citizens and interests in importing countries.

Core Labour Standards as Human Rights

Various core labour standards have been characterised as human rights in the United Nations Universal Declaration of Human Rights, the subsequent International Covenant on Civil and Political Rights, and the International Covenant on Economic, Social, and Cultural Rights. The ILO's 1998 Declaration of Fundamental Principles and Rights at Work enumerates a short list of core international labour standards, which are defined more fully in eight background covenants that are incorporated by reference, namely freedom of association and collective bargaining, the elimination of forced labour, the elimination of child labour, and the elimination of discrimination in employment. This list consistent with the characterisation of certain core labour standards or rights as human rights, especially those that guarantee basic freedom of choice in employment relations.

As Amartya Sen (1999) argues, the basic goals of development can be perceived in universal terms where individual well-being can be viewed as entailing certain basic freedoms irrespective of cultural context: freedom to engage in political criticism and association, freedom to engage in market transactions, freedom from the ravages of preventable or curable disease, freedom from the disabling effects of illiteracy and lack of basic education, and freedom from extreme material privation. According to Sen, these freedoms have both intrinsic and instrumental value. Importantly, in contrast to the unfair competition and race-to-the-bottom rationales for linking international trade policy and international labour standards, the human rights perspective focusses primarily on the citizens in exporting countries. The assumption underlying this concern with basic or universal human rights is that failure to respect them in any country does not reflect the will of the citizens but rather reflects the decisions of unrepresentative or repressive governments. Alternatively, that failure reflects the majoritarian oppression of minorities, such as children, women, or racial or religious minorities, or, alternatively again, reflects paternalistic concerns that citizens in other countries have made ill-informed or ill-advised choices to forego these basic rights.

The linkage of international trade policy, including trade or other economic sanctions, with core labour standards that reflect basic or universal human rights is a cogent one. When citizens in some countries observe gross or systematic abuses of human rights in other countries, the possible range of reactions open to them includes diplomatic protests, withdrawal of ambassadors, cancellation of air landing rights, trade sanctions or more comprehensive economic boycotts, or, at the limit, military intervention. Arguing that doing nothing is always or often the most appropriate response is inconsistent with the very notion of universal human rights. In extreme cases, such as war crimes, apartheid, genocide (in the case of Serbia, or the Holocaust in the case of Nazi Germany), it seems indefensible to exclude *a priori* economic sanctions from the menu of possible options. Whether it is the most appropriate option may, of course, be specific to the context and depend both on the seriousness of the abuses and the likely efficacy of the response — choice-of-instrument issues that are discussed next. But it is sufficient for the present purposes to restate the point that to the extent that core labour standards are appropriately characterised as basic or universal human rights, a link between trade policy and such labour standards is defensible and also arguably imperative, in contrast to the other two rationales for such a linkage — which, despite their much longer historical lineage, seem to be largely spurious and inconsistent with the central predicates of a liberal trading system. However, core labour standards viewed as basic or universal human rights, by promoting freedom of choice, are entirely consistent with a liberal trading regime that seeks to ensure other human freedoms, in particular the right of individuals to engage in market transactions with other individuals without discrimination on the basis of country of location (see Molatlhegi 2000, ch. 3; McCrudden and Davies 2000).

Having said this, the scope and definition of the class of human rights viewed as sufficiently universal as to warrant the potential imposition of trade sanctions for their violation are problematic in various respects. Even core labour standards are not susceptible to uncontentious understandings of their scope. Should child labour be defined only in terms of a minimum working age, or should some subset of exploitative child labour practices be targeted? What exact practices constitute discrimination in the workplace? What constitutes forced labour beyond slavery? When are freedom of association and the right to engage in collective bargaining fully respected, given that most countries deny or limit the right to strike in various contexts? Beyond core labour standards, while civil rights — for example, to be free from genocide, apartheid, torture, detention without trial, and so on — may be reasonably well understood and commonly subscribed to (at least in principle), political rights to engage in political association, criticism, or dissent, or to vote are much less widely recognised (Stirling 1996). Economic, social, and cultural rights are even less universally accepted ('The

Politics of Human Rights' 2001; Ignatieff 2001, 18–20). These issues have major implications for the choice-of-instrument and choice-of-institutional arrangements for structuring the link between trade policy labour standards.

The Choice of Instrument

Soft Law Instruments

Soft law instruments typically depend for their efficacy on exhortation, adverse publicity, social pressures for conformity, public shaming, or voluntary market reactions by consumers, employees, or investors. They do not usually attract formal legal or economic sanctions. The ILO's conventions on minimum international labour standards are the most prominent form of soft law in this context, in that they depend on ratification by individual member states and are subject only to investigation and reporting by ILO organs and the provision of technical assistance to enable countries to build capacity to implement them. Thus, compliance with ILO norms depends on a combination of public identification, embarrassment, and shaming (a mild stick) and technical assistance (a mild carrot). The ILO has been widely criticised by proponents of a trade-labour linkage for ineffective enforcement of its norms and, indeed, variable ratification of its conventions by many countries, including major developed countries such as the U.S. (which has, of course, been a prominent proponent of linking trade and labour). Many of the soft law, market-driven mechanisms described below have emerged in part out of frustration by nongovernmental organisations (NGOs) and other interest groups with the ineffectiveness of the ILO.

One class of soft law instruments entails a range of certification, labelling, and voluntary code of conduct mechanisms that purport to identify firms or products that conform to core international labour standards and hence respond to information market failures if consumers in importing countries derive private disutility from consuming goods produced in violation of core labour standards (Freeman 1994). The efficacy of these mechanisms turns largely on market reactions to the signals that they entail, principally by consumers and to a lesser extent by investors. These instruments are attractive in some respects in their focus on consumer (not producer) welfare in importing or consumer countries in that they depend on consumer preferences and a willingness to pay to vindicate those preferences; hence they are consistent with the normative predicate of liberal trade theory, which largely focusses on the potential for free trade to enhance consumer welfare. Typically, such mechanisms are either self-initiated by firms or industry associations or initiated by NGOs of various kinds, which negotiate them with firms or trade associations. The most ambitious initiative of this kind to date is the Global Compact, launched by UN secretary general Kofi Annan in 1999 and entailing voluntary corporate endorsements of nine principles, including core labour standards.

However, these mechanisms suffer from a number of limitations. Currently, they apparently apply to a small percentage of exports in a number of sectors where non-compliance with core labour standards is thought to be common, as with about 5 percent of exports in the textile and clothing industries. They vary widely in several dimensions, such as which core labour standards are recognised, how these standards are defined, if at all, and how effectively adherence to these standards is monitored, if at all (Diller 1999; Blackett 2001; Brown 1999).

In explaining the low, inconsistent, and often ineffective application of these mechanisms, a number of explanations suggest themselves. They, in turn, raise serious questions about attaching primacy to consumer preferences in importing countries in this context, despite the initial appeal of mechanisms that depend on consumer welfare as their reference point and the compatibility of this reference point with the predicates of free trade (see Boston Review 2003). First, consumers — even if fully informed about the conditions under which imports are being produced and violations of core international labour standards that particular modes of production may entail — in fact do not care enough about the intrinsic values reflected in these core labour standards (viewed as basic or universal human rights) to put their money where their mouths are. However, even if this were to be the case, the fact that the process of production and exchange may involve production or consumption externalities for other citizens in exporting or importing countries (a situation that is largely inherent in the notion of universal human rights) suggests that consumer preferences cannot be decisive in a human rights context.

A second explanation is that consumers in importing countries do care about these human rights values but are poorly informed about the conditions under which the goods they are consuming are produced in exporting countries. The cost of acquiring this information exceeds the value that they place on this information. In this respect, the voluntary and decentralised nature of the soft law mechanisms currently used in this context are an inadequate response to the information problems faced by consumers in importing countries.

Yet a further explanation for the low, inconsistent, and often ineffective application of these mechanisms is that consumers in importing countries do care about these intrinsic human rights values but confront serious problems with collective action, in that individual consumers who may be prepared to pay a premium for goods produced in conditions that meet core labour standards will be concerned that other consumers who share their concerns may opportunistically purchase lower priced goods and rely on others to bear the financial costs of vindicating their collective preferences. However, if every consumer suspects every other consumer of being likely to behave opportunistically — to free-ride on their sacrifices — an effective voluntary collective response may not emerge.

Beyond these reasons for not according primacy to consumer preferences in this context, a further serious limitation is associated with soft law mechanisms that link consumer responses to imports of offending goods, such as conflict diamonds or products made with child or forced labour. These mechanisms are unresponsive to violations of either core labour standards viewed as universal human rights or universal human rights defined more broadly that are occurring in non-traded goods sectors. For example, it is widely agreed that most child labour is not employed in export sectors (between 5 and 10 percent) but in domestic agriculture, services, retail, and the informal sector generally (Brown 2000; Morici 2001). Many abuses of civil and political human rights are not related in any direct way to traded goods sectors, such as civil rights abuses in Sudan and Burma, to take two current examples.

Hard Law Options

Hard law options typically entail formal legal or economic sanctions imposed by states either unilaterally or multilaterally. The first and most fundamental issue that arises in choosing such an instrument is the scope of the link between trade policy and labour standards. If core labour standards are conceptualised as universal human rights, how can these particular human rights be justifiably privileged over some subset of universally proclaimed universal human rights (subject to the definitional issues noted above)? Surely an act such as genocide, torture, or detention without trial warrants at least as serious concern from the international community — and at least as serious a set of legal or economic sanctions — as a violation of core international labour standards. To privilege core labour standards over these other human rights is quite overtly to elide the various normative rationales for intervention and to risk a protectionist rationale for trade or other economic sanctions.

Related to this point, a further issue arises relating to the scope of the link between trade policy and core labour standards: Why should trade or other economic sanctions be contingent on imports of offending goods? As in the discussion of market-driven soft law instruments and the limitations thereof, why should child labour or human rights violations in nontradable goods sectors demand any less concern, or any less severe sanctions, than such abuses in tradable goods sectors?

It follows from these two points that trade or other economic sanctions should not be confined to core labour standards, but should extend to at least a subset of universally accepted rights more generally, and that such sanctions should not be limited to imports of goods directly produced by the offending practices. This, in fact, suggests a very broad domain for linking trade and other economic sanctions with universal human rights. However, it also suggests some significant constraints on their invocation. In particular, the substantive rules governing their invocation as well as the procedures by which they can be invoked should be consistent with the human rights rationale for intervention and should exclude the unfair competition and race-to-the-bottom rationales for intervention.

In terms of substantive rules, this requirement assigns considerable significance to rules of non-discrimination and consistency. Here is a hypothetical example: the U.S. has no textile sector but has a significant clothing sector, and child labour is employed in producing exports for the U.S. market in India in both sectors; the U.S. would seek to impose trade sanctions only against clothing imports from India and not textile imports. Alternatively, even if the U.S. seeks to apply trade sanctions against both clothing and textile imports from India, given this supposed situation, it would not seek to do so against similar imports from Pakistan made with child labour for geopolitical or other reasons. The 1998 decisions of the WTO's Appellate Body in the disputes over shrimp and turtles, beef hormones, and Australian salmon, in other contexts, rightly emphasise the importance of non-discriminatory and consistent treatment of imports reflecting the ostensible rationale (in the context, a human rights rationale) for intervention, and not disguised protectionism. Unless a non-discrimination or consistency requirement is rigorously applied, trade sanctions based on hard law carry a high risk of constituting a disguised form of protectionism, with the human rights rationale only a mere pretext or cover for protectionist measures not motivated by human rights concerns but in fact by the other two illegitimate rationales for intervention discussed above. This is why it is important to resist arguments that differences in production and processing methods (PPMs) should render products unlike products for the purposes of Article III of the General Agreement on Tariffs and Trade (GATT). Instead, like products should be defined in economic terms, as the Appellate Body did in the case of asbestos in 2000, and should focus on substitutability by well-informed consumers. Justifications for inferior treatment of competitive imports can be remitted to Article XX (perhaps amended to include a universal human rights exception but subject to the non-discrimination conditions in the *chapeau*).

In defining the elements of a non-discrimination test, the purpose of the test must be clear: it should be designed to screen out cases of disguised protectionism in situations where trade sanctions have been invoked unilaterally, ostensibly on human rights grounds. But it should not require sanctioning countries to apply sanctions either to all countries in violation of core labour standards (or human rights) or to none at all — an all-or-nothing requirement that would likely make the perfect the enemy of the good. In other words, as a matter of international trade law there should be a negative duty not to discriminate for protectionist reasons, but there should be no positive duty to take affirmative action. While a sanctioning country could choose not to sanction all violations of core labour standards (or human rights) everywhere in the world, this form of trade sanction 'under-reach' should surely not be a legitimate concern of an international trade body, which cannot plausibly be transformed into a global human rights crusader. In contrast, the problem of sanction under-reach may be a legitimate concern of other international organisations (such as the ILO, or UN human rights committees), but this calls for action on their part (such as by adopting a regime analogous to the Convention on International Trade in Endangered Species, or CITES, which requires multilateral sanctions by members in particular cases). For the WTO, the principal concern is sanction 'over-reach', where the sanctioning country's actions in targeting some imports and not others seems principally explicable, on the basis that in the former case it has a domestic industry to protect and in the latter case it does not. Where it imposes trade sanctions in the latter case, they entail no benefits to the domestic producers of competitive products (which are non-existent), but does entail costs to domestic consumers; it can thus can be viewed as an action against material interest that can only be explained by the sanctioning country's genuine concern with violations of core labour standards (or human rights) in exporting countries (similarly in the case of bans on exports or FDI), where the government of the sanctioning state, by taking action, is seeking to solve collective action problems among its own consumers or citizens. Given this approach, in the hypothetical example above, the differential treatment of clothing and textile imports from India would be suspect, but the failure to sanction similar imports from Pakistan, for non-trade–related reasons, would not be.

In terms of procedural requirements for the invocation of trade or other economic sanction against violations of universal human rights in other countries, a number of options present themselves. First, a basic choice must be made between sticks and carrots, although this is often overlooked in debates over the link between trade and labour standards. Rather like the European Union's Generalised System of Preferences (GSP) regime, it is not difficult to imagine developed countries offering developing countries significant trade concessions (for example, accelerated implementation of the phase-out of the Multifibre Arrangement) if they commit themselves to an accelerated phase-out of offending labour or other human rights abuses. Unlike the Uruguay Round agreements, which for the most part entailed a single undertaking by member states, in this context such an arrangement might more appropriately take the form of a plurilateral agreement (like the Uruguay Round government procurement code). This plurilateral agreement would then be an option offered to developing countries and, for those that choose it, developed countries would bind their trade concessions (unlike GSP treatment). The commitments made by countries to observe core international labour standards and other universal human rights would have to be reasonably precisely defined in terms of ILO or UN conventions or covenants, so that violations could be rendered reasonably justiciable through an appropriate international dispute settlement process (perhaps vested in the WTO, but not necessarily or exclusively, as explored below). However, one of the limitations of the carrot option, as pointed out by Howard Chang (1997), is that it creates moral hazard problems: countries could persist in violations or engage in more egregious violations in order to attract larger concessions (or carrots).

A second option would be to require all countries that are parties to a regional or multilateral arrangement like the WTO to commit themselves to enforcing their own existing labour laws effectively (as under NAFTA's side agreement on labour), with enforcement provided through supranational or international dispute settlement processes and penalties. The limitations of this option are obvious enough. First, it addresses only violations of core international labour standards and not other universal human rights. Second, it assumes that member countries have already enacted substantive laws that reflect these standards and that the only problem is ineffective enforcement, which, in many cases, may not be the central problem.

A third option is to allow petitions for trade sanctions initiated by private parties under domestic law analogous to antidumping duties and countervailing duties. This option is clearly objectionable, because it espouses in its most naked form the two illegitimate rationales for a link between trade and labour standards and carries the highest risk of protectionist abuse.

A fourth option would be to allow unilateral state action against imports from offending countries (perhaps by amending the GATT's Article XX to include a exception for core international labour standards and universal human rights, or by an expansive interpretation of the public morals exception already contained within Article XX; see Charnovitz 1998). Such action would be subject to such measures that satisfy the conditions contained in the *chapeau* of Article XX prohibiting arbitrary or unjustifiable discrimination between countries where the same conditions prevail and prohibiting disguised restrictions on trade in the application of such measures, rigorously enforced as argued above. This option is obviously more appealing than expanding the scope of current private contingent protection mechanisms, but risks the capture of states imposing such measures by domestic producer interests. Moreover, it casts the burden of proof in important respects on the targeted country after such a measure has been imposed to initiate a formal complaint and make out a *prima facie* case that the measure in question does not comply with the requirements of Article XX. Similarly, in the absence of an attempt to link a core labour standards/human rights exception under Article XX to relevant background international conventions or covenants, in this context issues of justiciability in international dispute settlement are likely to be acute and place more of a burden on such a dispute settlement process than it can reasonably bear. Indeed, in the absence of a horizontal linkage with other international organisations with central mandates in the labour standards or human rights domains (discussed below), the 'necessity' test that attaches to most of the exceptions currently enumerated under Article XX is likely to prove intractable in evaluating the justification for trade sanctions (see Brown 2000; Maskus and Holman 1997; Roessler 2000, ch. 6).

A fifth option would be for all member states that are party to either regional or multilateral trading arrangements to negotiate a comprehensive set of rules setting out commitments to observe core international labour standards and other universal human rights (perhaps incorporating by reference relevant international documents). Given this approach, which is similar in some respects to that entailed in the Uruguay Round Agreement on Trade-Related Aspects of Intellectual Property (the TRIPS Agreement, which incorporates by reference the Berne, Paris, and Rome conventions), trade sanctions would come at the end of the dispute settlement process and not at the beginning. In other words, a country complaining that another country was in violation of its core labour standards or other human rights commitment would have to demonstrate a breach of these commitments and would carry the burden of initiating and proving, at least *prima facie*, such violations; and only if the complaint is upheld by the dispute settlement body could retaliatory trade sanctions be authorised in the event that non-compliance continued. This approach has several virtues: first, trade sanctions cannot be imposed until there has been a multilateral judgement that a violation of relevant core international labour standards or other human rights has occurred, and the country seeking to impose such sanctions bears the initial burden of initiating a complaint and proving a *prima facie* case. Second, it has the virtue of any rule-based system of laying out the substantive ground rules with some precision in advance, and thus minimises the potential for protectionist abuse of the regime. Third, and related, it renders problems of justiciability more tractable.

The Choice of Institutional Regime

Having discussed the choice of objectives and the choice of instruments in the light of those objectives in shaping a link between trade and labour standards, the remaining question is the choice of institutions to administer this link in the context of choice of objectives and choice of instruments. Two important and related considerations are relevant here: First, institutional specialisation, as in many other domains, has many virtues in vindicating desired policy objectives (see Srinivasan 1996; see also Sykes 2003). Second, because human rights, not trade effects, motivate the trade-labour link proposed, vesting exclusive or even primary jurisdiction in an international trade body risks compromising the normative rationale for the linkage, for example by giving primacy to adverse trade effects in either importing or exporting countries.

In cases of the most egregious abuses of universal civil human rights, such as apartheid or genocide, it is difficult to imagine as plausible the vesting of this function in a trade organisation (such as the WTO). Rather, following current international practices, one would imagine that the appropriate international organ for authorising or perhaps requiring such sanctions is the UN Security Council. In other less egregious cases, there will still obviously be questions of institutional legitimacy and competence in vesting the administration of such a regime in a trade organisation. One option here entails a sharp and exclusive institutional division of labour. For example, with respect to core labour standards, the authorisation or requirement for the imposition of trade or other economic sanctions could be vested in the ILO by way of elaboration of its sanctioning power under article 33 of the ILO Constitution.[2] This would follow, by way of analogy, the example of regimes such as CITES, which requires signatory states to ban imports of endangered species or products thereof. However, critics of the ILO are sceptical of the willingness or capacity of the ILO to implement and administer effectively such a regime. Defenders of the ILO, on the other hand, may worry that the attachment of economic sanctions to the powers of the ILO may destabilise the organisation, causing states to withdraw from membership or withhold ratification of its conventions to an even greater extent than is the case at present.

Another option is to imagine some form of horizontal co-ordination among international agencies, whereby, for example, the ILO would be wholly or largely responsible for determinations of systematic and persistent violations of core labour standards, or UN committees on human rights systematic and persistent violations of other universal human rights (other than the most egregious abuses), and the WTO would be responsible for overseeing the implementation of sanctions and ensuring that arbitrary and unjustifiable forms of discrimination and disguised protectionism are avoided, as well as ensuring they are proportionality in the scale of the trade sanctions imposed.

A variant on this option would be to have international organisations such as the ILO or UN human rights committees nominate members to dispute settlement panels or the Appellate Body of the WTO in cases involving complaints of violations of core labour standards or other universal human rights, and at the same time to take steps to render WTO dispute settlement process both more transparent and more inclusive in terms of admissibility of *amicus curiae* briefs from interested members of civil society (see Trebilcock and Howse 1999, ch. 3).

The preferred choice would be for some form of horizontal co-ordination with specialised international agencies with expertise and legitimacy in the labour standards or human rights fields that would make determinations of systematic and persistent violations of relevant norms despite whatever carrots and sticks (assisting and shaming) that the agency typically first brings to bear on violators. Thus, the 'necessity' test under Article XX of the GATT would fall largely within the domain of those agencies, although such determinations could be precipitated by unilateral state trade action under one of the options reviewed above, followed by a complaint by the targeted country. This would then lead to a reference under the Dispute Settlement Understanding (specifically Article 13) by the WTO panel seized with the complaint to a relevant specialised international agency for findings on violations of relevant international norms (which the WTO panel should accept as presumptively dispositive of the referenced issues). Such a determination may also go some distance toward meeting the conditions of non-discrimination or disguised discrimination in the *chapeau* of Article XX of the GATT. With respect to the proportionality of the proposed trade response, the WTO for its part should again be influenced by the nature of the horizontal agency's findings as to the seriousness and persistence of violations. Because adverse trade effects are irrelevant in this proposed framework of analysis, trade sanctions cannot be quantified in these terms. Here, more imaginative fashioning of trade remedies is called for. For example, NAFTA's labour side agreement provides for a system of fines, ultimately enforceable by trade sanctions, if a member state is found by a specialised panel to have engaged in a systematic and persistent practice of not enforcing its own labour laws, with the fine payable to the offending country to enhance its labour law enforcement.[3] This form of sanction suggests one option. Another option could entail denial of access to the dispute settlement process of the WTO as a complainant for so long as a member state is non-compliant. Still another option would be suspending voting rights in the WTO while non-compliance persists. Crippling a non-compliant country economically (particularly a poor developing country) with trade sanctions should be reserved as the remedy of last resort.

Conclusion

Many of the arguments on the appropriate link between trade policy and labour standards presented here flow out of critical questions that must be answered at the outset of any such analysis relating to the normative rationale for any such linkage at all. The unfair competition and race-to-the-bottom rationales are uncompelling, even incoherent, and provide a thinly disguised cover for protectionism — particularly on the part of developed countries *vis-à-vis* imports from developing countries — and rightly arouse the antagonism and cynicism of developing countries, which already struggle under enough disadvantages in trade and other domains without sustaining yet one more encumbrance on their ability to develop. However, the human rights rationale for a link between trade and labour standards is much more convincing and has important implications for the scope of the link, as well as for the choice of instrument and the choice of institutional regime. While the issues that arise under this rationale with respect to choice of instrument and choice of institutional regime are far from straightforward or uncontentious, resolving these questions is rendered vastly more difficult if the foundational normative rationale for a link in the first place is not clearly determined (as many historical and contemporary debates on the link between trade and labour standards have not been).

Notes

1. This chapter draws on a longer paper by the author, 'International Trade and International Labour Standards: Choosing Objectives, Instruments, and Institutions' in S. Griller, ed., *International Economic Governance and Non-economic Concerns: New Challenges for the International Legal Order*, Springer, Vienna, 2003.
2. Article 33 has been invoked only once — against Burma in March 2000 regarding the use of forced labour, where the ILO authorised (but did not require) its members to apply economic sanctions against Burma. Six months after final approval of the action, no member countries had taken actions (Elliott 2000).
3. Under Annex 39 of NAFTA, any monetary enforcement assessment shall be no greater than US$1 million for the first year of the agreement and thereafter no greater than 0.007 percent of total trade between the parties during the most recent year for which data are available, and must be paid into a fund to improve or enhance labour law enforcement in the party complained against.

References

Alben, Elissa (2001). 'GATT and the Fair Wage: A Historical Perspective on the Labor-Trade Link'. *Columbia Law Review* vol. 101, no. 16, pp. 1410–1447.

Blackett, Adelle (1999). 'Whither Social Clause? Human Rights, Trade Theory, and Treaty Interpretation'. *Columbia Human Rights Law Review* vol. 31, no. 1, p. 1.

Blackett, Adelle (2001). 'Global Governance, Legal Pluralism, and the Decentered State: A Labor Law Critique of Codes of Corporate Conduct'. *Indiana Journal of Global Legal Studies* vol. 8, no. 2, pp. 401–448.

Boston Review (2003). 'Stepping Up Labour Standards'. New Democracy Forum. <bostonreview.net/ndf.html> (April 2004).

Brown, Drusilla (1999). 'Can Consumer Product Labels Deter Foreign Child Labour Exploitation?' Discussion Paper 99-19, Department of Economics, Tufts University.

Brown, Drusilla (2000). 'International Trade and Core Labour Standards: A Survey of the Recent Literature'. Labour Market and Social Policy Occasional Paper No. 43. Organisation for Economic Co-operation and Development.

Chang, Howard F. (1997). 'Carrots, Sticks, and International Externalities'. *International Review of Law and Economics* vol. 17, no. 3, pp. 301–324.

Charnovitz, Steve (1987). 'The Influence of International Labour Standards on the World Trade System: A Historical Overview'. *International Labour Review* vol. 126, no. 5, pp. 565–585.

Charnovitz, Steve (1998). 'The Moral Exception in Trade Policy'. *Virginia Journal of International Law* vol. 38, no. 4, pp. 689–745.

Diller, Janelle M. (1999). 'A Social Conscience in the Global Marketplace? Labour Dimensions of Codes of Conduct, Social Labelling, and Investor Initiatives'. *International Labour Review* vol. 138, no. 2, pp. 99–129.

Elliott, Kimberly Ann (2000). 'The ILO and Enforcement of Core Labour Standards'. Policy Brief No. 00-6. Institute for International Economics, Washington DC. <www.iie.com/publications/pb/pb00-6.pdf> (April 2004).

Flanagan, Robert J. (2002). 'Labor Standards and International Competitive Advantage'. Working paper, Stanford Business School.

Freeman, Richard (1994). 'A Hard-Headed Look at Labor Standards'. Symposium on International Labor Standards and Global Economic Integration, United States Department of Labor, Bureau of International Labor Affairs. Washington DC.

Ignatieff, Michael (2001). *Human Rights as Politics and Idolatry*. Princeton University Press, Princeton.

Langille, Brian A. (2003). 'Re-reading the Preamble to the 1919 ILO Constitution in Light of Recent Data on FDI and Worker Rights'. *Columbia Journal of Transnational Law* vol. 42, no. 1.

Leary, Virginia (1996). 'Workers' Rights and International Trade: The Social Clause'. In J. N. Bhagwati and R. E. Hudec, eds., *Fair Trade and Harmonization: Prerequisites for Free Trade?*, pp. 177–230. MIT Press, Cambridge, MA.

Maskus, Keith E. and Jill A. Holman (1997). *Should Core Labor Standards Be Imposed through International Trade Policy?* World Bank, Development Research Group, Washington DC.

McCrudden, Christopher and Anee Davies (2000). 'A Perspective on Trade and Labor Rights'. *Journal of International Economic Law* vol. 3, no. 1, pp. 43–62.

Molatlhegi, Baatlhodi (2000). 'Trade and Labour Interface in the Context of Economic Integration: The Case of the Southern African Development Community'. Ph.D. thesis. University of Toronto.

Morici, Peter (2001). 'Labor Standards in the Global Trading System'. Economic Strategy Institute.

'The Politics of Human Rights' (2001). *Economist*. 18 August.

Rodrik, Dani (1996). 'Labor Standards in International Trade: Do They Matter and What Do We Do About Them?' In R. Z. Lawrence, D. Rodrik and J. Whalley, eds., *Emerging Agenda for Global Trade: High Stakes for Developing Countries*. Overseas Development Council, Washington DC.

Rodrik, Dani (1997). *Has Globalization Gone Too Far?* Institute for International Economics, Washington DC.

Roessler, Frieder (2000). *The Legal Structure, Functions, and Limits of the World Trade Order: A Collection of Essays*. Cameron May, London.

Salazar-Xirinachs, José Manuel and Jorge Mario Martinez-Piva (2003). 'Trade, Labour Standards, and Global Governance: A Perspective from the Americas'. In S. Griller, ed., *International Economic Governance and Non-economic Concerns: New Challenges for the International Legal Order*. Springer, Vienna.

Sen, Amartya (1999). *Development as Freedom*. Knopf, New York.

Srinivasan, T. N. (1996). 'International Trade and Labour Standards from an Economic Perspective'. In P. van Dijck and G. Faber, eds., *Challenges to the New World Trade Organization*. Kluwer International, The Hague.

Stirling, Patricia (1996). 'Use of Trade Sanctions as an Enforcement Mechanism for Basic Human Rights: A Proposal for Addition to the World Trade Organization'. *American University Journal of International Law and Policy* vol. 11, no. 1, pp. 1–46.

Sykes, Alan (2003). 'International Trade and Human Rights: An Economic Perspective'. University of Chicago Law School, Law and Economics Working Paper No. 188.

Trebilcock, Michael J. and Robert Howse (1999). *The Regulation of International Trade*. 2nd ed. Routledge, New York.

PART III
CREATING CODES OF CORPORATE RESPONSIBILITY

Chapter 9

Corporate Social Responsibility and the Evolution of International Norms

Hevina S. Dashwood

Corporate social responsibility (CSR) has become a prominent issue on the global agenda. In response to the declining regulatory role of the state in a context of globalisation, expectations have been raised about how companies should conduct themselves in their natural and social environments.

The current prevailing view — that the only function of business is to maximise profit — continues to dominate in the business world. One position in the literature holds that business cannot adhere to corporate responsibility, because of the highly competitive nature of the global economy, rendering CSR and profit maximisation incompatible. A related argument contends that the interests of shareholders should be given primacy in considerations of a corporation's responsibilities.[1]

In the past 15 years, however, various pressures have led to a shift in thinking on the part of companies. Business leaders are increasingly attuned to corporate responsibility. There is a growing recognition of how a company's reputation affects its bottom line. Many companies, especially large ones with international operations, have adopted voluntary codes of conduct. Such codes fall within the realm of soft law, lacking the hard law status of mandatory national legislation or international treaty law.

Both within and beyond the business community, there is a growing ethical concern for the societal impact of business, especially as it relates to international operations in developing countries, where the worst practices tend to occur.[2] To take the prescriptive stance that companies should be socially responsible assumes that it is appropriate for companies (alongside governments) to be involved in social policy. An issue such as whether a company should be involved in building schools in the communities in which it operates is a contentious one, as it arguably places too much power in the hands of already powerful multinational corporations (MNCs). To argue, however, that there is no scope for voluntary initiatives, and that corporate responsibility should be strictly a state regulatory affair, is unhelpful. An either/or approach to state regulation (that is, hard law) and voluntary initiatives (namely soft law) runs the risk of being overtaken by events. What is actually occurring is a hybrid situation, with state regulation prevailing in some areas along with voluntary initiatives in other areas. In short, although voluntary initiatives are no substitute for government regulation, they can complement or reinforce hard law.

Hard law, whether domestic or international in nature, can be difficult to enforce if it does not reflect a general societal consensus about its legitimacy. This point holds true in domestic contexts, in the presence of enforcement mechanisms, and is especially salient in the case of international law, where few, if any, viable enforcement mechanisms exist. To the extent that hard law depends for its legitimacy on societal consensus, it is important to establish the emergence of societal norms in support of CSR. There is plenty of evidence to support a shift in societal norms, reflected both domestically and internationally, in support of greater corporate responsibility. It is harder to establish whether those norms have been internalised by corporations, but it can be argued that many companies are beginning to engage in dialogue and learning about the issue of corporate responsibility.

What are the factors driving companies toward conforming to societal expectations about acceptable corporate behaviour? There are many such drivers, but the process of globalisation has been a critical one in bringing corporate responsibility to the forefront of the global agenda. There is evidence of shifting norms in terms of what is considered to be acceptable behaviour for corporations. Second, these forces have altered how companies calculate their interests. Third, for some types of business, the roles and identities they have crafted for themselves have led them to view corporate responsibility as consistent with their objectives. In short, there is both an interest- and a norms-based explanation for the growing attention of companies to corporate responsibility.

The extent of this shift in norms resonates with the constructivist approach in international relations (Wendt 1992, 1999; Ruggie 1998, Introduction). An important example is the literature that seeks to explain how and why governments that abuse human rights, in the absence of the traditional inducement of power politics, should gradually move to respond to human rights concerns (Risse-Kappen, Ropp, and Sikkink 1999l; Keck and Sikkink 1998). Although this literature is largely concerned with how state behaviour is influenced by those shifting norms and the efforts of transnational advocacy networks, companies can also be reasonably expected to respond to shifts with regard to acceptable corporate behaviour. Nongovernmental organisations (NGOs) enjoy considerable legitimacy in the promotion of norms-based behaviour, but other private actors — even self-interested ones such as corporations — can also play a role.[3]

This chapter begins by addressing questions raised about how CSR should be conceptualised. This is important, not only for the purposes of clarification, but also because how companies respond to corporate responsibility depends on their concept of what that responsibility entails. It then discusses the historical context, which includes early efforts to rein in corporations, dating back to the 1970s, as well as the devolution of political authority in the era of globalisation, embedded in a neoliberal consensus, and the emergence of the antiglobalisation movement in the political space provided by the end of the cold war. The reasons for companies' acting upon corporate responsibility will be briefly explored, as will be the issue of whether CSR can be solely left to the voluntary or private sectors.

A Definition of Corporate Social Responsibility

There is no single definition of corporate responsibility. Even the term 'corporate social responsibility' is contentious. The debate revolves around a number of broad questions. What does corporate responsibility entail, and how should it be conceptualised? To whom should corporations be responsible? To what extent can corporations adhere to social responsibility? Who should take responsibility? Should codes be voluntary or legislated?

CSR can be broadly understood to be the obligations that companies must respect as basic principles in their operations toward human rights, labour standards, and the environment. This definition is set out in the Global Compact, an initiative launched by United Nations secretary general Kofi Annan as a set of nine basic principles to which MNCs should adhere (UN 1999, 'About the Global Compact'). For example, in the area of human rights, corporations are called upon to ensure that they are not complicit in human rights abuses. In the area of labour standards, corporations are to respect freedom of association, disavow discrimination in employment practices, and work toward abolishing child labour. With respect to the environment, corporations are asked to support a precautionary approach to environmental challenges, namely to take steps to ensure that serious or irreversible threats to the environment are avoided (United Nations Conference on Environment and Development [UNCED] 1992, Principle 15).

There is a difficult debate about the precise scope and nature of a corporation's responsibilities. It is very important to distinguish between what companies should be doing, and what they are in fact doing.

Many people working in the area of corporate responsibility exclude from the definition any issues that are already regulated through national laws, the logic being that being in the realm of what corporations are obliged to do by law moves beyond the realm of responsibility (Carroll 1999). However, it is wrong to assume that corporate responsibility lies solely in the realm of voluntary initiatives. The term 'voluntary' conjures up an image of a corporation choosing which charity to support. The term 'a corporation's responsibilities' connotes duties that, even if not legally required, entail obligations. The term 'responsibility' implies behaviour and practice that resonate with societal norms and values that extend over and above legal requirements. In short, abiding by existing law represents the baseline of good corporate behaviour.

In the context of international law, if corporations expect to be able to enjoy rights (for example, a stable investment climate), then they must also be prepared to accept corresponding duties. The question then becomes what should be the extent of those duties, and to whom do those duties relate.

How one answers that question depends, in part, on one's understanding about the degree of compatibility, or incompatibility, between motivations of profit maximisation, and the collective interests of society that result from co-operative behaviour. The shift in international norms toward a more expansive understanding of corporate responsibility can be gauged in light of remarks made by Milton Friedman (1970): a corporation's sole responsibility is toward its shareholders, which can be met through increasing its profits. Although such a restrictive view still has a firm foothold in the corporate world, most companies are prepared to accept that their responsibilities extend beyond their shareholders.

Such a restrictive view ignores the fact that businesses do not operate in a vacuum, but must operate within a framework of laws, rules, and conventions. Indeed, the free market system within which businesses exist functions according to a legal framework that, for example, protects private property rights. Furthermore, businesses function within a social framework, underpinned by shared norms and values to which they must adhere (Culpeper and Whiteman 1998, 22–24). These norms are reflected in laws against, for example, racial discrimination. Moreover, businesses operate in the natural environment, and have a huge impact on it, through their extraction of natural resources and through their operations that harm the environment. Thus corporations must adhere to corporate responsibility because their economic activities affect the social and ecological systems in which they are embedded (24).

While corporations might behave as though profit maximization were the only legitimate criterion for behaviour, in fact, the discourse of CSR has broadened substantially. Many businesses now speak of stakeholders to whom the firm has a responsibility, encompassing employees, customers, and communities, and extending to the natural environment in which they operate.

One company, BP Amoco, likens its responsibility toward stakeholders to ripples in a pond (CDCAC 2001, 17–19). When BP Amoco drops its stones in the global pond, it causes ripples. Closest to the centre of those concentric circles are the shareholders, followed by employees and customers. Further out are local communities, national communities, and, at the outermost circle, international communities. In this analogy, the type of responsibilities a corporation might have in each sphere varies according to its size, the nature of its operations, and the stakeholder in question.

Corporate responsibility is most commonly associated with large corporations. Small businesses, when taken cumulatively, have an immense impact, yet generally lack the flexibility and resources on their own to implement the elaborate policy and management frameworks associated with corporate responsibility. Furthermore, the type of activity a corporation engages in influences its approach to corporate responsibility. For example, companies in the retail clothing sector deal with different sorts of issues from companies in the mining sector.

It is unrealistic to expect businesses (whether large or small) to meet their social obligations if they cannot realise a profit. The key role of corporations is to generate wealth, not create social policy. Business is expected to produce goods and services at a profit, and consumers, in turn, are expected to buy these goods and services. To suggest that corporations should make a profit is not, however, the same thing as saying that CSR is incompatible with profit making. Nor is making a profit necessarily incompatible with the interests of society.

At the same time that companies are expanding the scope of their responsibilities, serious doubts remain whether companies should be involved in the social realm. Especially in the context of developing countries, there is concern that already powerful MNCs will become even more powerful in the face of a weak state. In developed countries, the issue of the extent of a corporation's responsibilities depends on society's comfort level with the devolution of state regulatory functions to the private sector. Stale debates of an either/or nature are unhelpful, as already mentioned, and in fact have been overtaken by events. A hybrid of public and private roles has emerged.

The Emergence of Corporate Social Responsibility on the Global Agenda

CSR has become an important issue on the global agenda, but the issue is not new. In the 1970s and 1980s, companies began to improve their practices with respect to the environment and worker health and safety. Later, in the 1990s, companies began to embrace the social dimensions of corporate responsibility.

At the national level, governments in the 1970s and 1980s in the industrialised world began to tighten up their environmental regulations. Often, this was in response to major environmental disasters, such as the 1989 *Exxon Valdez* oil spill or the 1984 chemical disaster in Bhopal, India.

At the international level, efforts were made in those years to draw attention to issues of corporate responsibility. Yet international attention tended to zero in on specific corporations and activities in specific countries, as with the focus on apartheid in South Africa and the popular protests against Nestle after its aggressive marketing of infant formula in the developing world. Alternatively, concern about the activities of MNCs considered only their home-based operations, predominantly in industrialised countries.

The power and influence of MNCs, and efforts to address that reality, have long made them a magnet for international attention. A significant early effort to rein them in was led in the late 1960s and early 1970s by the growing number of newly independent developing countries, known as the Group of 77 (G77). Under the institutional umbrella of the newly formed United Nations Conference on Trade and Development (UNCTAD), the G77 demanded economic justice in the form of resolutions at the United Nations General Assembly calling for a New International Economic Order (NIEO). Among the various demands was a call for an international treaty governing MNCs that would ensure greater economic benefits to the underdeveloped host country. In 1972, the United Nations Economic and Social Council (ECOSOC) sponsored a resolution requiring the monitoring of MNC activities. In 1974, the Commission on Transnational Corporations was established in order to negotiate a code of conduct for MNCs.

Negotiations continued during the 1970s. From the start, developed countries resisted the various recommendations, but a number of forces helped to keep the G77's demands on the agenda. One such force was the growing number of newly independent countries, which outnumbered the developed countries in the General

Assembly. Together with the strength in numbers was the precedent set by Organization of Petroelum Exporting Countries (OPEC), which had successfully stabilised oil output and prices, and thus served as both an inspiration for developing countries and a red flag for oil-dependent developed countries. The period of détente in the cold war was another important factor, allowing developing countries to keep their demands for a NIEO on the international bargaining table.

By 1981, most of these advantages afforded developing countries had been lost. Many developing countries found themselves in heavily in debt by the late 1970s, which eroded what little bargaining power they may have enjoyed. The cold war détente came to an abrupt end with the Soviet invasion of Afghanistan in 1979, which heated up old tensions. Once again, security concerns crowded out the NIEO. In key countries such as Britain, the United States, and Germany, leaders with a neoliberal world view came into power, which was the antithesis of the demands contained in the NIEO. Negotiations for a UN code to govern MNCs dragged on until 1992, with the governments of the key industrialised countries favouring minimal state regulation in accordance with free market principles (Haufler 2001, 15–17).

The forces that allowed the NIEO to appear on the international agenda can be compared to those that led to corporate responsibility gaining prominence as a global issue in the mid to late 1990s. The rise of the antiglobalisation movement in the 1990s, which helped to shine the international spotlight on corporate responsibility, occurred in the context of the political space afforded by the end of the cold war, and the resultant diminishing of security concerns. The events of 11 September 2001, of course, pushed security concerns to the top of the international agenda once again, and have narrowed the political space for such issues as corporate responsibility. However, widely publicised corporate scandals, such as the Enron and WorldCom affairs, have ensured a continuing place for corporate responsibility on the global agenda.

Today, MNCs' activities around the world are now under scrutiny. There is something qualitatively different about the way in which corporate responsibility is conceived of today; one can speak of the globalisation of the issue of corporate responsibility since the mid 1990s. The globalisation of the issue of corporate responsibility reflects, in turn, the larger process of globalisation itself (Cragg 2001).

In this respect, transnational advocacy networks have played a significant role in placing corporate responsibility on the international agenda once again. Their influence has been exerted in a variety of ways, facilitated by the information revolution and the global reach of telecommunications. Through the internet, they can reach a large audience, and through antiglobalisation demonstrations at such venues as the annual meetings of the World Trade Organization (WTO) or G8, they are able to capture the media's attention. The problem, however, is that much of this global activism arises from Northern-based NGOs. Furthermore, MNCs tend to listen more readily to Northern-based groups, because they are concerned about the threat of consumer boycott.

Approaches to corporate responsibility are embedded in the larger controversy over the implications and inevitability of globalisation. The efforts of NGOs have helped to energise the debate over corporate responsibility, within the larger context of globalisation.

One debate centres on the unequal distribution of the benefits of economic globalisation (Ruggie 2000). The forces of economic globalisation have frequently magnified the pre-existing inequalities between the industrialised North and the developing South. In this context, MNCs are seen to be one of the major beneficiaries of globalisation. Notwithstanding the failure of the Multilateral Agreement on Investment (MAI), the interests of MNCs are protected by other international agreements. The WTO-based Agreement on Trade-Related Investment Measures (TRIMs), for example, restricts the policy autonomy of governments in an area of primary domestic concern for developing countries. Poor countries are not well placed to offer advice when international standards are being set, and they lack the bureaucratic capacity to represent themselves adequately in international organisations such as the WTO.

A related debate about globalisation concerns the imbalance in global rule-making (Ruggie 2000). Again, this is a North-South issue, but relates too to Northern-based NGOs and social movements. They are concerned with the perceived lack of democratic accountability in the decision-making process within institutions such as the WTO, and the fear that too much power is being divested to MNCs, which are accountable only to their shareholders (and, as Enron demonstrates, even that can be a stretch!).

Aware of the need to secure support from developing countries to reach the goal of maintaining an open global economy, the WTO has attempted to address these concerns, as reflected in the Doha ministerial declaration adopted in 2001. The Doha round of trade talks has sought to address economic balances by promoting developing country access to developed country markets in key sectors such as agriculture. However, although developed countries have preached the gospel of trade liberalisation and deregulation to developing countries for two decades, they have been reluctant to practice what they preach in their own economies. Indeed, at the WTO ministerial meeting in Cancun in September 2003, a group of 20 developing countries — referred to as the G20, but not to be confused with the G20 finance ministers and central bank governors — drew the negotiations to a halt by demanding trade-distorting subsidies on agriculture in exchange for a broader agreement on free trade rules.

Debates persist about whether the process of globalisation can be controlled, or about whether it is indeed inevitable. Many suspiciously believe that the developed countries and powerful interests are controlling the process. Claims that globalisation is inevitable are seen as attempts to deflect attention from the need for MNCs to behave responsibly. The issue of CSR is caught up in the larger processes, trends, and debates over globalisation. Furthermore, the very process of globalisation has helped to bring the issue of CSR under intense international scrutiny.

An important political byproduct of globalisation has been the devolution of political authority away from national governments. This devolution has occurred downward, from the national government to the local and municipal levels, and upward to the international and transnational levels. Although debates continue to rage about whether this devolution entails the demise of the state, what has in fact transpired is a multilayered process of governance, with the state, still an important actor, relinquishing regulatory responsibilities to private agents.[4]

Trade and investment agreements by organisations such as the WTO tend to reinforce this trend. In a globalised context, states — whether voluntarily or not — have devolved some of their regulatory powers to the private sector. MNCs figure largely in this process, as their power in relation to the power of governments in the jurisdictions in which they operate is often substantial.

This process is magnified in countries of the developing world, where the annual revenues of MNCs outstrip the gross national product (GNP) of the poorest of the less developed countries. Furthermore, the wave of economic liberalisation that took place in the 1980s and 1990s in many developing countries resulted in competition among them to attract foreign investment (Culpeper and Whiteman 1998, 18–20). MNCs are clearly in the driver's seat, as they seek out the most attractive (generally low-cost) places to invest.

An example of the upward devolution of political authority is the growing trend toward the privatisation of international standard setting. Again, this is not an entirely new phenomenon. For instance, the International Organization for Standardization (ISO) was formed in 1947, with a view to standardising commonly traded goods and services. Other examples include the International Accounting Standards Committee (1973) and, more recently, the International Association of Insurance Supervisors (1994).

In a related trend, private actors have been gaining power within international regimes. In part, this is due to the fact that the knowledge or information required to set standards is often held in private corporations, rather than in governmental organisations. Given that managing the globalised economy is increasingly complex, technical experts acquire greater power through their monopoly of knowledge. The private sector now has a much greater role in the public domain (Haufler 2001).

The various processes associated with globalisation have developed within the political context of the neoliberal consensus shared by the major industrialised powers. Advocates of neoliberalism favour trade liberalisation, state deregulation, and responsible fiscal and monetary policies. In the developing world, governments have been urged to privatise and deregulate, and to liberalise trade. In the developed world, efforts to bring budget deficits under control have generally been associated with the erosion of the welfare state.

The neoliberal context of globalisation is important to consider, for the devolution of political authority is not so much an inevitable process but a reflection of a political consensus that the role of the state should be minimised and that a greater role should be accorded to the private sector. The issue of corporate responsibility is thereby wrapped up in the larger political debate about the appropriate role of the state in society and the economy.

In a context where the power of the state appears to have diminished, greater attention has been brought to bear on the need for corporations to behave more responsibly.

Whatever the argument for or against state regulation of corporations, corporations themselves must come around to accepting that they hold responsibilities toward society.

There is a vast array of international instruments covering various aspects of corporate responsibility, including labour standards through the International Labour Organization (ILO), human rights — beginning with the 1948 Universal Declaration of Human Rights — and the environment, through the United Nations Environment Programme (UNEP) as well as a host of environmental conventions. Furthermore, trade agreements such as the North American Free Trade Agreement (NAFTA) contain rules intended to promote labour and environmental standards. Bodies such as the Organisation for Economic Co-operation and Development (OECD) have sought to establish rules governing such practices as corruption (such as the OECD's guidelines for MNCs).

Most of these initiatives are independent, such as the OECD guidelines, the UN Global Compact, or private codes. They therefore lack coherence. Side agreements, such as NAFTA's side agreements on labour and the environment, have only a limited reach. The result has been a hodgepodge of international legal rules and norms with varying degrees of enforceability. Investors and traders operate in a poorly defined international legal framework, where the monitoring and enforcement of international rules are very uneven (Culpeper and Whiteman 1998, 31).

Evidence of the shift in norms can be found in the wide range of responses to the need to promote corporate responsibility. These responses include company codes of conduct, standard setting by industry associations, international codes of an aspirational and issue-specific nature, and global reporting initiatives. Private agents promoting corporate responsibility include business leaders, NGOs, religious organisations, and prominent individuals. They have produced a plethora of organisations, important examples of which include the UK-based Prince of Wales Business Leaders Forum, the Switzerland-based Caux Round Table, and Business for Social Responsibility and the Interfaith Center on Corporate Responsibility, both in the U.S. Supporting the wide range of company codes are management programmes that oversee the implementation of company codes, such as the Global Reporting Initiative, Social Accountability 8000, the ISO 14000 standards, and AccountAbility 1000.[5]

A number of important developments reflect the new prominence of CSR on the international agenda. Two such developments include the updated OECD Guidelines for Multinational Enterprises and Kofi Annan's Global Compact (OECD 2000; UN 1999). CSR was a prominent issue at the Kananaskis G8 Summit in 2002, as it was a few months later at the World Summit on Sustainable Development in Johannesburg, where the role of corporations in promoting sustainable development was explicitly encouraged in the final document (UN 2002, 4, 37–39). It remained on the agenda at the 2003 G8 Summit at Evian (G8 2003).

These initiatives are important, because they bring together the proliferating international rules and conventions to create a single set of international rules of the game. This facilitates global monitoring and the eventual enforcement of rules throughout the world.

The Global Compact, which calls upon MNCs to adhere to basic principles of responsible behaviour, is especially important (UN 1999). It presents a set of nine global principles in the areas of human rights, labour standards, and the environment against which the activities of MNCs can be measured. It thereby holds the promise of moving away from the old, *ad hoc* approach to establishing rules on corporate responsibility. Because the Global Compact is under the rubric of the UN, there is an organisational forum within which the activities of corporations can be monitored. CSR can remain on the global agenda, thereby allowing for a sustained effort to promote international standards of acceptable behaviour.

The Global Compact reflects the new types of partnerships among the UN, member states, MNCs, international labour, and NGOs. Although some NGOs remain sceptical of the likely impact of the Global Compact, the possibility of establishing a dialogue nonetheless exists (Kell 2001). Furthermore, by considering these principles, corporations may potentially come to see CSR as more than just damage control, when public scrutiny focussed on such ruinous activities as environmental destruction. The Global Compact can provide space for argumentative persuasion and learning to take place.

Moreover, the Global Compact contributes to the debate about how to define CSR. This is especially important, as it goes to the heart of the issue of the precise scope and nature of a corporation's responsibilities.

Influences on Business to Be Socially Responsible

The forces of globalisation described above have started to alter how companies calculate their interests. There are a wide variety of reasons for companies to embrace corporate responsibility.[6] While it could be said that there is an ethical underpinning to these reasons, they should not be confused with altruism. They are primarily the result of careful consideration of a company's risks, liabilities, and interests. In the process of making these choices, however, companies engage in a norms-based dialogue about what constitutes acceptable standards of corporate behaviour.

Pressure from NGOs has played a very important role in forcing businesses to consider corporate responsibility. NGOs have used market incentives effectively to induce corporations to become more publicly accountable. Corporations are very responsive to pressure that affects their bottom line, such as consumer pressure, investor pressure, and media exposure.

Consumers have successfully advocated for greater knowledge about whether the goods they buy are produced in accordance with fair labour standards, with minimal harm to the environment, and without abusing human rights. Consumer campaigns are no substitute for government regulation, but by singling out companies in a particular sector for public scrutiny, they can have a substantial impact (Hutchinson 2001). For example, in addition to exerting influence on a particular company, other companies in that industry might fear the same treatment and change their behaviour.

Companies targeted by consumers might themselves seek government regulation to level the playing field. Another market incentive is socially responsible investment by shareholders concerned about responsible behaviour.[7]

Some companies have come to view corporate responsibility as consistent with their objectives. For example, Mattel, the large American multinational toy manufacturer, has a comprehensive code of conduct backed up by a well thought-out implementation programme that includes extensive monitoring and systematic, independent audits of the company and its suppliers. Mattel is motivated by competitive concerns, but also by consumer concerns: the programme is designed to reduce the risk of selling toys to children manufactured under conditions at variance with international conventions and human rights declarations. Another example is Levi Strauss, a major player in the apparel industry, which set out terms of engagement for its suppliers to require them to respect human and labour rights. It also carries out independent evaluation and monitoring of the impact of those terms of engagement (Cragg 2000).

Business leaders are increasingly aware that the maximisation of profits may be undermined if they are not pursued in an ethical manner. It is now frequently argued that a company's record on CSR has an impact on the bottom line through the company's 'reputational capital' (CDCAC 2001, 27–30). Today, when a product's image may be as important as the product itself, a company's social identity may be as significant as its brand identity. Nike, for example, learned the hard way that a negative public image can affect sales.

According to the Conference Board of Canada, 'reputation is fast becoming recognized as a dynamic corporate asset affecting firm performance' (Janet Rostami, cited in CDCAC 2001, 28). Companies that are perceived to be socially and environmentally responsible, such as the Body Shop, may gain a competitive advantage over more traditional firms. According to various surveys, most Canadians say they consider the conditions of production when buying consumer goods. Most report that they are prepared to pay higher prices for ethically produced products, such as those produced by Mountain Equipment Co-op. These Canadian attitudes reflect attitudes in other industrialised countries. A number of studies have begun to show a positive correlation between profits and social responsibility (CDCAC 2001, 29–30).

Corporate Social Responsibility: Voluntary Initiative or State Regulation?

Should CSR be left to the voluntary sector, or should states play a role? This is, in fact, not an either/or proposition: both are necessary. Both voluntary and national regulatory efforts must be backed up by international agreements that reflect the emerging global consensus on CSR.

The Canadian government has demonstrated a strong preference for voluntary adherence to CSR. For example, it supports the International Code of Ethics for

Canadian Business, signed in 1997. Such an approach may be appropriate in stable political circumstances, but not in zones of conflict, where the likelihood of severe abuses of human rights is very high. The problem with it can be illustrated by Canada's response to Talisman and its activities in Sudan. The Canadian government failed to take adequate action, even in light of the report by Peter Harker, who found that Talisman's oil operations were directly implicated in human rights abuses perpetrated by the Khartoum government (Harker 2000).

The Canadian government seems persuaded by the policy of 'constructive engagement' endorsed by the powerful Canadian Council of Chief Executives (formerly the Business Council on National Issues), which argues that companies should engage in more business with nondemocratic countries because it believes that trade would be a positive catalyst for change. In the aftermath of the Harker report, the government urged Talisman to behave in a more socially responsible manner (see Chapter 12). The government's stance is reminiscent of its response in the 1970s and 1980s to corporate trade and investment in South Africa, when it relied purely on voluntary approaches.

Even in cases less extreme than that of Sudan, the Canadian government's procurement powers affords it a largely untapped opportunity to ensure adherence to minimal standards. It has a wide array of government-support programmes, policy initiatives, and subsidies for corporate activities engaging in international trade and investment, handled through a number of departments and agencies, such as the Export Development Canada, Industry Canada, the Canadian International Development Agency, and the Business Development Bank of Canada. There is almost no evidence of any effort by the Canadian government to assert corporate responsibility criteria through these government support programmes (Hutchinson 2001, 10–12).

The failure of national regulatory initiatives as well as voluntary initiatives, as reflected in the case of Talisman, suggests that neither one or the other can be relied upon, and points to the need for an international regulatory framework.[8] For such a framework to be effective, corporations must internalise a sense of obligation or responsibility. As with all legal rules, whether domestic or international, they can be difficult to enforce if there is insufficient consensus about their legitimacy (Gordon 2000). Voluntary initiatives are therefore a necessary, but insufficient, part of the process of building an international regulatory framework for CSR.

Notes

1. For a brief summary of the various approaches to corporate responsibility, see Canadian Democracy and Corporate Accountability Commission ([CDCAC] 2002, 2–3).
2. For an authoritative account of the ethical case for corporate responsibility, see Bird (1996).
3. For an excellent recent analysis of the potential of non-state approaches to corporate governance, see Webb (2002). See also Florini (2000).
4. For an excellent description of this process, see Scholte (2000).
5. For a comprehensive list with links to websites, see Business for Social Responsibility (2003).

6. For a concise summary of reasons, see Haufler (2002).
7. In recent years, a number of instruments have emerged to promote socially responsible investment, including the Dow Jones Sustainability Indexes and FTSE4Good.
8. According to a survey conducted by the CDCAC (2002, 7), 84 percent of Canadians support the view that the Canadian government should work toward establishing such an international regulatory framework.

References

Bird, Frederick (1996). *The Muted Conscience: Moral Silence and the Practice of Ethics in Business*. Quorum Books, Westport CT.

Business for Social Responsibility (2003). 'Overview of Corporate Social Responsibility'. Issue Briefs. <www.bsr.org> (April 2004).

Canadian Democracy and Corporate Accountability Commission (2001). 'An Overview of Issues'. Discussion Paper. <www.corporate-accountability.ca/pdfs/discussion_paper.pdf> (April 2004).

Canadian Democracy and Corporate Accountability Commission (2002). 'The New Balance Sheet: Corporate Profits and Responsibility in the 21st Century'. Final Report. <www.corporate-accountability.ca/pdfs/FullReport2002.pdf> (April 2004).

Carroll, Archie B. (1999). 'Corporate Social Responsibility: Evolution of a Definitional Construct'. *Business and Society* vol. 38, no. 3, pp. 268–295.

Cragg, Wesley (2000). 'Human Rights and Business Ethics: Fashioning a New Social Contract'. *Journal of Business Ethics* vol. 27, no. 1–2, pp. 205–214.

Cragg, Wesley (2001). 'Ethics Codes: The Regulatory Norms of a Globalized Society?' In A. Soeteman, ed., *Pluralism and the Law*, pp. 191–223. Kluwer Academic Publishers, Dordrecht.

Culpeper, Roy and Gail Whiteman (1998). 'The Corporate Stake in Social Responsibility'. In *Canadian Corporations and Social Responsibility: Canadian Development Report, 1998*, pp. 22–24. North-South Institute, Ottawa.

Florini, Ann, ed. (2000). *The Third Force: The Rise of Transnational Civil Society*. Japan Center for International Exchange and Washington Carnegie Endowment for International Peace, Tokyo.

Friedman, Milton (1970). 'The Social Responsibility of Business Is to Increase Its Profits'. *New York Times Magazine*. 13 September, pp. 32–33.

G8 (2003). 'Fostering Growth and Promoting a Responsible Market Economy: A G8 Declaration'. Evian, 2 June. <www.g8.utoronto.ca/summit/2003evian/growth_en.html> (April 2004).

Gordon, Kathryn (2000). 'Rules for the Global Economy: Synergies between Voluntary and Binding Approaches'. Prepared for the Organisation for Economic Co-operation and Development, Paris. <www.oecd.org/dataoecd/57/25/1922674.pdf> (April 2004).

Harker, Peter (2000). 'Human Security in Sudan: The Report of a Canadian Assessment Mission'. January. Department of Foreign Affairs and International Trade, Ottawa. <www.reliefweb.int/library/documents/cansudan2.pdf> (April 2004).

Haufler, Virginia (2001). *A Public Role for the Private Sector: Industry Self-Regulation in a Global Economy*. Carnegie Endowment for International Peace, Washington DC.

Haufler, Virginia (2002). 'Industry Regulation and Self-Regulation: The Case of Labour Standards'. In A. F. Cooper, J. English and R. Thakur, eds., *Enhancing Global Governance: Towards a New Diplomacy?*, pp. 167–171. United Nations University Press, Tokyo.

Hutchinson, Moira (2001). 'Canadian NGO Policy Views on Corporate Social Responsibilty and Corporate Accountability'. Overview paper prepared for an NGO-Government Meeting,

Canadian Council for International Co-operation. Ottawa. <www.ccic.ca/e/archives/crs_2001_canadian_ngo_policy_views_on_corp_resp.pdf> (April 2004).

Keck, Margaret E. and Kathryn Sikkink (1998). *Activists Beyond Borders: Advocacy Networks in International Politics*. Cornell University Press, Ithaca.

Kell, Georg (2001). 'Dilemmas in Competitiveness, Community, and Citizenship'. Paper presented at 'Business and Human Rights Seminar: Towards Universal Business Principles', London School of Economics and Political Science, 22 May.

Organisation for Economic Co-operation and Development (2000). 'The OECD Guidelines for Multinational Enterprises'. Organisation for Economic Co-operation and Development, Paris. <www.oecd.org/dataoecd/56/36/1922428.pdf> (April 2004).

Risse-Kappen, Thomas, Steve C. Ropp, and Kathryn Sikkink, eds. (1999). *The Power of Human Rights: International Norms and Domestic Change*. Cambridge University Press, New York.

Ruggie, John (2000). 'Remarks on the Global Compact to the NGO Community'. United Nations, Geneva, 13 October. <www.unglobalcompact.org> (April 2004).

Ruggie, John G. (1998). *Constructing the World Polity: Essays on International Institutionalization*. Routledge, London.

Scholte, Jan Aart (2000). *Globalization: A Critical Introduction*. Palgrave, Houndsmills.

United Nations (1999). 'The Global Compact'. <www.unglobalcompact.org> (April 2004).

United Nations (2002). 'Report of the World Summit on Sustainable Development'. A/CONF.199/20. 26 August–4 September. Johannesburg. <www.johannesburgsummit.org/html/documents/documents.html> (April 2004).

United Nations Conference on Environment and Development (1992). 'Rio Declaration on Environment and Development'. <www.unep.org> (April 2004).

Webb, Kernaghan, ed. (2002). *Voluntary Codes: Private Governance, the Public Interest, and Innovation*. Carleton Research Unit for Innovation, Science, and Environment, Carleton University, Ottawa.

Wendt, Alexander (1992). 'Anarchy Is What States Make of It: The Social Construction of Power Politics'. *International Organisation* vol. 46, no. 2, pp. 391–425.

Wendt, Alexander (1999). *Social Theory of International Politics*. Cambridge University Press, Cambridge.

Chapter 10

The Role of Nongovernmental Organisations and Social Movements in Developing Countries

John W. Foster

The debate over the impact of corporate behaviour is frequently reduced to a consideration of incremental change of individual or sectoral corporate actions or discussions of voluntary codes of various descriptions. Through three reflections on the recent history of civil society engagement with issues of corporate accountability, this chapter examines the shift in emphasis among civil society organisations toward more fundamental issues. While the balance of gross power has moved over the last 30 years toward corporate actors, nongovernmental organisations (NGOs) and social movements are altering the field of engagement and seeking to reassert democratic citizen power. The result is to put the posture and actions of state actors in question, both nationally and multilaterally. Yet this brief review of individual campaigns, codes, and sectoral structures, as well as defensive battles against privatisation, investor guarantees, and the Agreement on Trade-Related Intellectual Property (TRIPS), suggests that the social movement of 'globalisation from below' is faced with a profound political challenge. The playing field will continue to tilt in favour of corporate and private privilege and protection, unless democratic control is re-established, and done so on a transnational or international basis as well as at the municipal, state, tribal, and national levels.

The Vision

A Vision Born

The legal suit launched in the United States against the Talisman mining firm recalls the deep roots of the social movement fighting against exploitation and for economic reform: the victories of the anti-slavery movement over the traders of their day more than two centuries ago; the church hall meetings and press campaigns that ripped regal respectability from the naked imperialism of King Leopold's wholly owned, if taxpayer-financed, project in the Congo (Hochschild 1998); the fight for labour rights that led to the formation of the International Association of Labour Legislation as the twentieth century opened and the formation of the International Labour Organization (ILO) shortly thereafter. These all expand the frame of reference for nongovernmental and social movement engagement with corporate responsibility.

The first theme for reflection is the shift in the balance of forces in the debate over corporate responsibility in the broadest sense. A key moment was the experience of observers at the United Nations Conference on Trade and Development (UNCTAD) in Santiago de Chile in 1972 — perhaps the high-water mark of the vision of a New International Economic Order (NIEO). There were many elements in that vision, among them the vision of a world in which scientific and technical advances would be shared in North-South terms. There was, in particular, the decision to move ahead with negotiations for an international code of conduct for multinational corporations (MNCs), situated within the United Nations and supported by the UN Centre on Transnational Corporations; this centre had been created to monitor corporate behaviour within the context of the code and other UN norms. The demand for the elaboration and implementation of a mandatory international corporate code of conduct entered the vocabulary and objectives of a number of NGOs and social movements at that time.

Virtually coincidental with the Santiago UNCTAD, Canadian churches inaugurated the Task Force on the Churches and Corporate Responsibility, patterned on the U.S. Interfaith Center on Corporate Responsibility, related to the National Council of Churches, and on the more specific campaigns for disinvestment in apartheid South Africa that were then current in the United Kingdom. This initiative was thus born in a context when Canadian NGOs were campaigning against imperial corporate ventures in Mozambique, Angola, Brazil, and Chile, among others, and women's and other networks were moving toward the Nestle boycott campaign. The latter was one of the most remarkable international efforts in popular mobilisation to change corporate behaviour, in this case regarding the aggressive marketing of infant formula.

A Vision Betrayed

Thirty years later, the UN Centre on Transnational Corporations, as such, is no more. It has been a victim of attacks from U.S. corporate and anti-UN pressures. The idea of a mandatory international code of conduct appears from time to time on the wish lists of NGOs and social movements. The Chilean polity continues to recover from more than 20 years of dictatorship resulting from a coup encouraged and welcomed by a number of multinational corporations (MNCs) and the U.S. administration. The crimes of the Pinochet dictatorship have finally begun to face international and domestic justice as well as the bar of history. The actions of those who connived to bring it to power remain largely exempt, secret, or forgotten.

The contrast between the context of Santiago in 1972 and the current situation is startling. There is constant corporate pressure for codes and guarantees that will expand and enhance protections for private property. Although the Multilateral Agreement on Investment (MAI) failed, other elements push on the General Agreement on Trade in

Services (GATS): bilateral investment guarantees, Chapter 11's investor-state procedures in the North American Free Trade Agreement (NAFTA), and investor rights. The vision that motivated the NIEO — the idea that the concert of nations might collectively hold MNCs accountable — seems a faded will-o'-the-wisp.

During the 1990s, at international conference after international conference, elements of civil society came together to build on the human rights and environmental norms of the previous four decades and to construct visions of a sustainable, just, and participatory future. The integrated and interpenetrating complex of human rights, the movement for gender equality and reproductive rights, the new commitment to social development, and the phenomenal and multidimensional vision of the 1992 United Nations Conference on Environment and Development (UNCED) at Rio all provided the grounds for new standards of global behaviour. The convening of large numbers of nongovernmental representatives created newly effective international networks of advocates, whose work was facilitated by the rapid deployment of new communications technology (Foster and Anand 1999).

Coincident, however, with these developments, and with the end of the cold war, was a new generation of trade and investment agreements, which included the Canada–U.S. Free Trade Agreement, NAFTA, and the World Trade Organization (WTO), in addition to the ill-fated MAI. As well came the pursuit of structural adjustment and the implementation of the so-called Washington Consensus. These all added up to a different programme for global change, one with much greater enforcement clout and more resources for implementation. Private investment, privatisation of public assets, the protection of private intellectual property, and the enhancement of investors — that is, corporate rights — became central to the programme, and they continue to be. The 2002 UN Financing for Development Conference in Monterrey, Mexico, updated and consolidated a consensus with the encouragement and protection of foreign direct investment (FDI) as the primary engine of development. Civil society organisations, in addressing the event, stated very clearly that they did not share that consensus and that corporate power, among other challenges, needed to be addressed by the global community.

In this context, Kofi Annan, Secretary General of the United Nations, has sought a way to return to the discussion of corporate accountability, with his collaborative and voluntary project of the Global Compact (UN 1999). Meanwhile, social movements and NGOs continue to campaign. Specific victories are gained. But the relations between corporate and democratic state power have shifted significantly, and systemic change seems desperately far off.

The Terrain of Engagement

The second zone of reflection has to do with the terrain of engagement. The actors, initiatives, strategies, and tactics that have characterised the decades since 1972 are too diverse to be little more than sketched in this brief treatment.

Individual NGOs and, particularly, coalitions and networks play various roles: as gadflies, monitors, intervenors, investors; as organisers of campaigns, legal interventions, boycotts and standards, and seals of approval. There have been several successes. The conflict diamonds initiative has deservedly won attention, and moved diplomats and corporations. The moratorium on logging in the Great Bear Forest in British Columbia, followed by the overall agreement to preserve it, is a tribute to the artful exercise of economic power brought to bear by the international NGO Greenpeace. The current effort to regulate and limit the trade in tobacco brings together health, environmental, legal, and trade expertise in fresh ways. The Clean Clothes Campaign and the alliance among students, consumers, and *maquila* workers have broken new ground. And the global reach of the Canadian Mining Watch, with pitifully limited resources, almost spans the reach of Canadian mining corporations, which have substantially more wealth.

Where these efforts have developed into agreements, codes of behaviour, or standards of compliance, the thrust is toward concrete and identifiable goals and resources for clear, persistent, independent monitoring, and toward sanctions of an economic or non-economic nature.

At the global level, the response from NGOs and social movements to the Global Compact is worth noting. Clearly, there are organisations and networks that welcome the re-entry of the United Nations into the field of corporate accountability. There are organisations, such as international labour bodies, with a long history of working through multilateral institutions — for example, the ILO on a variety of themes, including the social dimensions of globalisation, and the WTO on the social clause. These bodies take a long and measured view of this theatre of interaction.

Reservations, however, abound, particularly among NGOs and the globalisation-from-below networks. One key concern is with the relative weakness and poverty of the United Nations in comparison with many corporate actors. With the UN seeking corporate financial backing from such titans as Bill Gates and Ted Turner, many are sceptical about its freedom of movement in addressing issues of corporate behaviour and its impact.

Related to the UN's relative weakness is the question of capacity to do anything more than be an add-on to corporate public relations efforts. The staff complement of the Global Compact is limited, and it lacks a mandate for effective monitoring, let alone a critical and credible review of corporate compliance with the various codes contained in the Global Compact.

This structural frailty leaves the Global Compact with the sole hope of encouraging voluntary changes in behaviour using an approach of high-level co-option or collaboration with corporate leadership. What does the world body have to offer, beyond photo ops with the Secretary General? The response is that it offers a sort of seal of approval to corporations that sign on to the Global Compact. This has led NGO critics to cite an additional danger that the UN 'brand' — the familiar logo and shade of blue — will be corrupted, and be found side by side with the Nike swoosh or the Shell shell.

Some NGOs and foundations have responded by trying to influence what they obviously feel is a naive or ill-considered initiative. The response of others is to monitor and report in detail on the behaviour of corporations that have signed on with the Global Compact and to press the United Nations secretariat to suspend those — such as Nike or Novartis or Rio Tinto — that violate even its broad standards.

The debate over the Global Compact and its obvious weaknesses highlights the limitations of a focus on codes, particularly voluntary codes, of conduct. The terrain chosen by many organisations and networks stretches beyond this particular meadow or swamp. Whether in response to the new generation of trade and investment agreements, in frustration with individual campaigns, in impatience with voluntary measures or scepticism about corporate public relations exercises in accountability, NGOS and social movements have moved more extensively into the field of public regulation and rule making, beyond the issue of individual corporate behaviour or voluntary codes. More recently, in the face of WTO and World Bank pressure on governments to privatise public assets, in such key areas as water provision and public service delivery, NGOs and social movements have concentrated attention on both the critical monitoring and the development of alternatives to privatisation (Social Watch 2003).

At the regional level, perhaps the most obvious example, and one that was an indirect achievement rather than a direct one, was the creation of both the North American Commission for Environmental Cooperation (CEC) and the North American Commission for Labor Cooperation (CLC) in side agreements to NAFTA (Kirton and Maclaren 2002). This is, perhaps, an issue of being careful what you ask Santa for, as what 'Santa' Bill Clinton provided was not necessarily what either environmentalists or labour activists desired. In regard to the CEC, concerned environmental groups have been much more likely to take action regarding corporate or state behaviour outside the commission's procedures than to spend energy and money going through them (Foster 2002b). Similarly, while labour groups have been willing to use the CLC for symbolic and political purposes, few would argue that a job has been saved or restored through its operations. Many labour and environmental activists, if given the choice, would trade the abolition of NAFTA's Chapter 11 for the CEC and the CLC.

At a global level, the International Confederation of Free Trade Unions (ICFTU) has campained for a social clause and for recognition of the ILO's core labour rights within the WTO, with concomitant enforcement provisions. This is perhaps the most far-reaching attempt to condition investor and corporate behaviour from hiring on up, including global regulation.

On the environmental side, a coalition of representatives from less developed countries and environmental groups had had relative success in defending the priority of environmental agreements over trade and investment accords and in maintaining the priority of the precautionary principle. This has at least temporarily parried the attempt to assert the priority of trade and corporate concerns and the WTO.

Remaining in question is the current and hard-fought battle between the priority of the human right to health and the TRIPS regime with regard to the provision of affordable and accessible essential drugs for HIV/AIDS in the context of genocidal levels of death and disease (Foster 2002a). Pressure by AIDS organisations, Oxfam International, and Médecins Sans Frontières has made this issue one of the key 'make or break' items on the WTO agenda at Doha. A global coalition of Southern and a few Northern countries, together with NGO networks, faced the entrenched and embarrassing position of Pierre Pettigrew, Canada's Minister of Trade, and a few other largely Northern colleagues. The result was a political declaration cited by NGOs as a significant victory. The Doha declaration committed the WTO to further work to provide for importation of essential drugs by countries lacking domestic production capacity. The negotiations ran far beyond the target date committed in Doha and resulted in a highly questionable compromise, just before the ill-fated WTO Cancun ministerial in September 2003. The response of key NGOs has been not only to criticise the overly complex procedures of the agreement but also to take advantage of whatever windows it opens, as well as to develop their own demonstration projects and collaborate with Southern governments in testing the agreement through action to implement.

These various initiatives and battles form part of the tapestry of contemporary social movements. These movements have been typified by former president of Mexico Ernesto Zedillo as 'globalophobes', but are more appropriately described as movements of globalisation from below or horizontal social globalisation.

A connective tissue runs through much of this work. It is represented by the renewal of attention to existing treaty agreements, in particular to the Convention on Economic, Social, and Cultural Rights and other elements of the International Bill of Rights. Some quite small but creative organisations, such as the International NGO Committee for Human Rights in Trade and Investment (INCHRITI) and the International Federation for Human Rights (FIDH), have taken the leadership. But much larger organisations such as Amnesty International have also begun to shift attention in this direction.

During the last years of the twentieth century and the first few years of the twenty-first, human rights networks and multilateral human rights bodies have been engaged in a conceptual catch-up exercise. This exercise has essentially been provoked by the aggressive assertion of corporate prerogatives and the enshrining of these prerogatives in the new generation of trade and investment agreements.

A salient example is a document, produced in 1997 under the aegis of the International Commission of Jurists, that sets out the Maastricht Guidelines on Violations of Economic, Social, and Cultural Rights (1998). The experts drawn together by the commission clearly laid responsibility for the impact on human rights of the actions of third parties, particularly corporations, at the door of the state parties. They also reminded state parties that they should not conclude agreements that undermine or conflict with their covenant obligations. The conceptual links and working out of implications begun at Maastricht have been slowly working their way through official and NGO bodies in the years since.

In this regard, the work of INCHRITI has been instrumental in stimulating the UN Committee on Economic, Social, and Cultural Rights, and the Subcommission and Commission on Human Rights to study globalisation and trade and investment agreements, and to remind states parties of their responsibilities to respect, protect, and defend human rights.

The pertinence of these efforts is worth stressing, particularly because during the 1990s, in the search for bearings amidst the welter of market and corporate promotion, many NGOs and social movements discovered or rediscovered the remarkable significance of the human rights instruments. On the environmental front, a parallel but slightly different engagement has been taking place in the negotiation of new agreements on biosafety, climate change, and other issues. NGOs have been asserting the priority of basic principles and environmental protections over the trade priorities expressed by the WTO, NAFTA, and their processes (Swenarchuk 2001).

Turning Point or Perpetual Tunnel

A final reflection is on the question of whether the world is reaching a turning point in this engagement. Is it in fact a turning point, or is it perhaps some sort of tunnel with little sign of an exit?

First, the tunnel should be examined. The sorts of protection symbolised by NAFTA's investor-state mechanism, combined with the plethora of restrictions and supervision of the democratic formation and exercise of policy on the part of the WTO and other multilateral economic institutions, represent a significant narrowing of the field of political choice and political activity at the national and subnational levels. They are arguably contributing to the alienation of voters and cynicism about political options.

The post–11 September war on terrorism has had a chilling effect on the activists of globalisation from below. Trade activists in the United States, confronted with the push to pass fast-track legislation in Congress in 2001, did not feel that large public demonstrations were advisable. Playing corporate hardball is all too familiar; examples include corporations' withdrawal of donations from universities adhering to the Clean Clothes Campaign, or moves to restrict the ability of jurisdictions at various levels to boycott purchases from particular origins or producers, or the pressures that have been exerted on South Africa, Brazil, and numerous smaller states over the issue of drug patents. It does not take much imagination to consider that an activist's campaign to embarrass a particular firm or injure a particular brand is a corporate manager's idea of economic terrorism. This sort of neo-feudalism or corporatism is not so distant as to be simply a mirage.

As for a turning point, this moment essentially has to do with the revivification of democratic power at several levels, particularly at the global level. This is the sort of trend and hope that the Socialist International might have represented a hundred years ago. It is the challenge to which initiatives such as the 'third way' have been faint responses, inspiring more cynicism than mobilisation. The development of the World Social Forum, partly as a result of the meetings of a business 'cupola' in Davos and New York, has been quite remarkable. From its tentative beginnings in 2001 in Porto Alegre, Brazil, the event — initiated by an alliance of European and Latin American social movements and intellectuals — celebrated its fourth meeting in 2004 in Mumbai with more than 100 000 participants. Its agenda is a challenging and diverse exploration of alternatives to a world viewed as in the deadly grip of transnational capital.

Conclusion: A Profound Political Challenge

This brief review of individual campaigns, codes, and sectoral structures as well as defensive battles against privatisation, investor guarantees, and TRIPS leads to the conclusion that the various elements of the movement of globalisation from below face a profound political challenge. The playing field will continue to tilt in favour of corporate and private privilege and protection, unless democratic control is reasserted, and unless that can be done on a transnational or international basis and not only at the municipal, state, tribal, or national level.

One harbinger of this option is the World Social Forum and the regional social forums in Europe, Asia, and Latin America, which involve many more participants than can travel to Porto Alegre or Mumbai. It speaks to the many sectors, networks, and people who do not see themselves or their hopes represented by the business and political elites gathered at the World Economic Forum in Davos or by the major multilateral economic institutions.

The World Social Forum is probably much closer to a platform or a network of platforms than it is to a strategy to repossess public institutions for democratic purposes. Such a strategy would have to include too many elements that touch on the corporate role, but that have much more to do with broader social and political purposes. One component is the issue of electoral and electoral financing reform in the homeland of so many major corporations, and begin to level the democratic playing field.

The events of 11 September and the ensuing military and security actions by the U.S. and its 'friends' have set back democratic rights quite abruptly. This has occurred both at a general or societal level in Western countries, including Canada, and, more particularly, in the experience of social movements and NGOs with security surveillance and responses to street actions and other forms of resistance.

In the current rather extreme circumstances, the fate of the globe should seem more central in the field of public vision than is usually the case. Yet the appeals of those movements — such as the environmental movement and the anti-debt Jubilee South network — which ask political leaders to break with business as usual and project a transformative vision, seem largely to fall on deaf ears among political leaders.

The breakdown of the WTO ministerial talks at Cancun in 2003 were a short sharp shock to the assumptions of the leading Northern trade negotiators. Their priorities for so-called Singapore issues, including further protections for investors, ran into raging small farmers in the streets, and the newly coherent and firm opposition of large developing country governments in the form of the so-called G20 with other priorities. Northern arrogance also contributed to the embarrassing failure of the meeting.

If Pierre Pettigrew, Canada's trade minister during the WTO Ministerial in Doha, had had a 'Damascus road moment' and stated that the human right to health transcends the corporate protections of the TRIPS regime, for Africans with HIV/AIDS as much as for Canadians fearing anthrax or smallpox, one might have seen a little glimmer of the light that so many social organisations and networks seek. However, such light was unknown, as were mass demonstrations in the streets of the Qatar capital. It took almost two years of desultory TRIPS negotiations and several million more African deaths to reach the belated compromise on 30 August 2003, on the eve of the ill-fated Cancun ministerial. The Canadian government took advantage of the window opened by the agreement to initiate changes in Canadian patent and food and drug legislation. The precise terms of the new legislation were still at issue well into 2004, with NGOs pressing for simple and speedy procedures.

The large drug transnationals that have enjoyed the rich protection of the TRIPS regime for several years resisted the changes agreed upon in August 2003. Mandatory protections for private profit seem quite acceptable, yet mandatory assertions of the priority of the right to health over those protections are viewed with alarm. However, the battle to strengthen the enforcement of human rights provisions over the privileging of private profit has gained momentum.

The essential democratic challenge arising from a variety of confrontations between popular organisations and governments perceived to be servants of narrow corporate interests comes at a time when civil liberties and democratic dissent face renewed strictures and threats from the state's response to terrorist acts. The climate is hardly conducive to optimism. However, the emerging debate over state security raises and reinforces the same fundamental questions posed by the movement for globalisation from below. Beyond, yet also containing the issues of corporate accountability, is the fundamental issue of the survival and purposes of democratic life.

References

Foster, John W. (2002a). 'Canada and International Health: A Time of Testing on AIDS'. In N. Hillmer and M. A. Molot, eds., *A Fading Power: Canada Among Nations 2002*, pp. 191–208. Oxford University Press, Toronto.

Foster, John W. (2002b). 'NAFTA at Seven'. In *The USA and Canada, 2002*. Europa Publications, London.

Foster, John W. and Anita Anand, eds. (1999). *Whose World Is It Anyway? Civil Society, the United Nations, and the Multilateral Future*. United Nations Association in Canada, Ottawa.

Hochschild, Adam (1998). *King Leopold's Ghost: A Story of Greed, Terror, and Heroism in Colonial Africa*. Houghton Mifflin, Boston.

Kirton, John J. and Virginia W. Maclaren, eds. (2002). *Linking Trade, Environment, and Social Cohesion: NAFTA Experiences, Global Challenges*. Ashgate, Aldershot.

'The Maastricht Guidelines on Violations of Economic, Social, and Cultural Rights' (1998). *Human Rights Quarterly* vol. 20, no. 3, pp. 691–704.

Social Watch (2003). 'The Poor and the Market'. Third World Institute, Montevideo. <www.socwatch.org.uy/en/informeImpreso> (April 2004).

Swenarchuk, Michelle (2001). 'Civilizing Globalization: Trade and Environment, Thirteen Years On'. Report No. 399. Canadian Environmental Law Association, Toronto. <www.cela.ca/international/399global.pdf> (April 2004).

United Nations (1999). 'The Global Compact'. <www.unglobalcompact.org> (April 2004).

Chapter 11

Multinational Corporations, Globalisation, and the Challenge of Self-Regulation

Wesley Cragg[1]

Ethics Codes as Regulatory Norms in the Global Marketplace

Ethics codes are not of recent vintage. They are, to the contrary, integral elements of attempts throughout human history to understand and define the nature and the character of the ethical responsibilities of human beings for their conduct and its impact on others. Historically, ethics codes have been formulated with a view to guiding the behaviour of individuals, groups, organisations, governments, societies, and, most recently, corporations.

What all ethics codes share is the belief that those whose behaviour a code is designed to guide have ethical responsibilities, obligations, rights, and duties. Where codes typically differ is in their understanding of the nature of those responsibilities, obligations, rights, and duties. They also tend to differ in their understanding of the grounding or justification of the ethical responsibilities they characteristically articulate.

Corporate ethics codes are a relatively recent addition to the pantheon of ethical codes of conduct. This is not surprising since the modern corporation is itself of relatively recent vintage. And although corporate codes predate contemporary globalisation, it is globalisation that has thrust them into the limelight in a new and urgent way.

Globalisation has influenced the emergence and importance of corporate codes of conduct in two ways. First, it has been accompanied by the emergence of powerful multinational corporations (MNCs), whose activities are increasingly significant for people around the world. Second, it would also appear to have weakened the capacity of nation-states to regulate the conduct of MNCs. Self-regulation, guided by ethics codes, has emerged as one option for filling what some have described as a resulting regulatory vacuum. The chapter that follows proposes to evaluate this option.[2] Are ethical codes of conduct capable in principle of taking on the task of regulating international commerce?

Regulation has no value in its own right. Indeed, one of the underlying assumptions of liberal democratic systems of government is that regulation is only justified where it can be shown to provide benefits that are more substantial or beneficial than the harms, both extrinsic and intrinsic, of any action that imposes external constraints on the exercise of personal or collective freedom of action. Within this framework, regulation has been understood to be justified where its goal is protection against unwanted harms and unwarranted infringements of personal freedom and the establishment of rules designed to ensure equitable access to opportunities for personal or collective development.

In this sense, regulation has as its goal and justification the protection and enhancement of the common good. There is a difficulty, however. What the pursuit of these common goods requires and justifies has proven historically to be both highly contentious and difficult to achieve. Success in this regard requires two things. The first requirement is institutions or practices that allow reliable articulation of consensus or decision procedures for resolving disagreements about what is required and justified by way of regulation of individual and collective conduct. The second and equally important requirement is mechanisms for determining whether regulatory standards that have been put in place are respected.

There is a widely shared belief entrenched most deeply in liberal democracies that democratic institutions and practices are the most effective and reliable ways of determining the formal regulatory frameworks of modern pluralistic societies. There is, furthermore, a good deal of agreement on at least some of the ways of translating this consensus into workable institutions and practices at the level of the nation-state. What has not emerged is any consensus on how to achieve similar goals in international settings. This fact is part of the reason for the widespread interest in the potential efficacy of ethics codes and voluntary self-regulation. Is this interest justified?

The answer to this question would appear to be yes. This is not because a single set of authoritative institutions or practices for setting ethical standards for business conduct has emerged, a subject discussed below. What has emerged internationally, however, is widespread agreement with respect to the kinds of principles that should govern many aspects of international commerce. These principles have emerged from multifaceted dialogue in a variety of forums. In some cases, the principles have been set out in the form of ethics codes or statements of values and principles by organisations representing business interests, such as the International Chamber of Commerce (ICC) and the Apparel Industry Partnership Code of Conduct in the United States. In some cases, the issuing body is an international nongovernmental organisation (NGO), for example Amnesty International and Pax Christi International (1998). Increasingly, international institutions representing national governments, such as the Organisation for Economic Co-operation and Development (OECD), the World Bank, and the International Monetary Fund (IMF), are also articulating standards of conduct for international commerce. A good example is the OECD Guidelines for Multinational Enterprises (OECD 2000). Codes of conduct are also being articulated in settings involving participation on the part of groups and organisations drawn from more than one sector. In some cases, standards are emerging from privately initiated informal dialogue, for example, the Caux Round Table Principles for Business. In other cases, the driving force has been a formal institution such as the United Nations.

The resulting codes, standards, and statements of principle have focussed in some cases on specific countries, such as the McBride principles for Northern Ireland (to ensure fair labour practices), the Sullivan principles for South Africa (a code of conduct for human rights for companies operating in South Africa), and the Miller principles for China. In some cases, the focus is an industrial sector, for instance the apparel and chemical industries. In other cases, the focus is an issue or area of concern, for example bribery, the environment, or labour rights. Yet in other cases, the focus is on products or commodities such as rugs, coffee, diamonds, or lumber. Some of the codes or general statements of values and principles are comprehensive in nature and designed to cover commercial activities across a broad range of concerns, issues, and business ventures, such as the International Code of Ethics for Canadian Business (Human Rights Research and Education Centre 2003) and the OECD Guidelines for Multinational Enterprises.

A review of these international statements of values and principles designed to set standards for business conduct reveals a number of things.[3] First, there is remarkable consensus across the range of statements and codes about those aspects of commercial activities that require regulation or standard setting. Second, the values espoused are remarkably similar, despite sharp differences in the degree to which standards are elaborated or detailed across the range of influential statements that are widely regarded as models or points of reference in determining ethical conduct. Third, and equally significant, in virtually all areas where it is widely agreed that setting standards is important, there are examples of detailed articulations of standards required by the more general statements of values and principles. Finally, in virtually all cases, analysis reveals a great deal of consensus with regard to the need to protect against unwanted harms for which protection is required, infringements of personal freedom and the fundamental human freedoms that require protection, the importance of ensuring equitable access to opportunities for personal and collective development, and the importance of environmental protection.

Consensus is arguably clearest with respect to human rights. The United Nations Universal Declaration of Human Rights, set out following World War II, is today widely accepted as setting the standard against which the conduct of nation-states and their leaders, corporations and their leaders, and organisations generally should be judged. Human rights protection, furthermore, is area in which detailed standards and benchmarks have been elaborated and publicly debated extensively and in detail (for example, see Amnesty International and Pax Christi 1998).

There is widespread agreement on the need for standards with respect to the treatment of employees. Agreement extends to the exploitation of child labour, acceptable health and safety standards, freedom of association, and respect for principles of widely recognised principles of non-discrimination with respect to gender, race, religion, and so on. Here, too, as in the case of human rights, appropriate labour standards have been explored in detail by internationally influential bodies and set out in international conventions that have attracted wide endorsement by national governments. The International Labour Organization Tripartite Declaration of

Principles concerning Multinational Enterprises and Social Policy, for example, sets out principles with respect to employment, training, conditions of work and life, and industrial relations emerging from international deliberations involving labour unions, national governments, and MNCs (International Labour Organization [ILO] 1977).

Virtually all internationally influential, standard-setting statements of principles, values, and codes of conduct include principles calling for responsible environmental protection and responsible environmental practices.[4] The theme of sustainable development is not explicit in all statements, although it is a frequently specified standard. For example, the first general policy set out in the OECD Guidelines for Multinational Enterprises calls on corporations to 'contribute to economic, social and environmental progress with a view to achieving sustainable development' (OECD 2000, 19).

An area covered either explicitly or implicitly by virtually all influential comprehensive statements is that of corporate governance. In addition to the areas already covered, there is widespread agreement on the values and principles that should guide corporations in dealing with their shareholders, consumers, suppliers, and other stakeholders. The principles set out typically appeal to respect for the values of honesty, transparency, fairness, due diligence, respect for the law, and rejection of all activities permitting, encouraging, or tolerating bribery and corruption. Here again, standards of conduct based on respect for these general principles have been articulated by a number of internationally influential organisations, for example, the OECD, the ICC, and the Commonwealth Association for Corporate Governance.

Finally, although least extensively or consistently articulated, is an under-riding concern for principles of distributive justice. The Code of Ethics on International Business for Christians, Muslims, and Jews identifies meeting human and material needs as the primary purpose of business ('An Interfaith Declaration: A Code of Ethics on International Business for Christians, Muslims, and Jews' 1993). The International Code of Ethics for Canadian Business assigns responsibility to business enterprises for ensuring that all stakeholders affected by their activities receive a fair share of the benefits that accrue from those activities (Human Rights Research and Education Centre 2003). As already indicated, the OECD guidelines recognise that corporations have an obligation to contribute to the economic, social, and environmental progress of the communities in which they operate. What is equally true, however, is that international standard setting in this area is highly controversial, and emerging standards either vague or highly contested. An example is the contentious and unresolved debate over the concept of a living wage that has been suggested by some as a benchmark for minimum levels of acceptable remuneration for employees in the developing world. At issue, not surprisingly, is who should determine what should count as a living wage and how it should be calculated. An international consensus has yet to emerge.

Three important lessons emerge from this synopsis. First, there is widespread agreement within and across each of government sector, the private sector, and the voluntary sector that MNCs do have responsibilities that extend beyond their shareholders to those who are affected by their activities — that is to say, their stakeholders. Second, there is a great deal of agreement with respect to the values and principles around which regulatory frameworks should be constructed. Third, these understandings would appear to be emerging from extensive, broadly based dialogue and debate within and across all voluntary sectors.

Operationalising Codes of Conduct: Three Unsatisfactory Responses

Unfortunately, while there would appear to be considerable consensus on the general principles and standards that should guide the conduct of MNCs in the pursuit of international business opportunities, there is a good deal less agreement on how those principles and values should be operationalised in specific cultural and national settings, particularly in developing countries and countries in transition. Three questions are central: How should the general principles, on which there is wide international consensus, be interpreted and applied by corporations in the various national and cultural settings in which they do business? Who should be involved in the interpretive enterprise? And how should responsibility be assigned for ensuring that those principles and standards are respected?

Some of the answers that have been already advanced, sometimes in very influential circles, can be shown to be fundamentally unsatisfactory. An exploration of the problematic character of the unsatisfactory answers will set the stage for a third and more satisfactory response.

Ethics Codes and the Private Sector

The first widely canvassed and advocated answer would assign primary — perhaps even sole — responsibility for answering each of these three questions to the private for-profit sector. For some, this is both the purpose and the function of the voluntary ethics codes movement in the private sector. Ethics codes answer each of the questions. They set out principles and standards of conduct. Their form and content are, in the last analysis, determined by the industry-wide associations or individual corporations whose activities they are intended to guide, and the responsibility for ensuring that the code is respected is a voluntary matter to be overseen by whomever accepts voluntarily to be guided by it.

It would probably be true to say that the vast majority of existing codes of conduct have this character. Company codes are the best example. They are for the most part created in house, often by a company's legal staff, and interpreted, implemented, and enforced internally. In some cases, the code of conduct is not a public document, and implementation and enforcement are treated as a confidential matter.

Industry-wide codes share some, but not all, of these characteristics. Typically, they are public documents. Typically, they are adopted after some form of public consultation. And typically, they are offered as purely voluntary guidelines to be implemented voluntarily and monitored internally. Thus, it is not at all uncommon for model codes, however collaborative in their development, to state explicitly that the model code in question 'is not intended for legislation'.[5]

It has been argued that the primary purpose and social responsibility of the modern investor-owned corporation are to maximise profits.[6] Nonetheless, almost all the codes developed over the past decade acknowledge either explicitly or implicitly an obligation to contribute to common public interests beyond simply the enhancement of the value of their shares in the interests of their shareholders. What is commonly not recognised by those who acknowledge that corporations' responsibilities extend beyond their obligations to shareholders is the underlying weakness of codes of ethics that set out the social responsibilities of firms unilaterally with respect to wider public interests.

The weakness lies in the fact that unilateral definitions of the common or public good or the public interest and what constitutes contributions to the common good are likely to be either arbitrary or paternalistic. Furthermore, they ignore conflicts of interest that can come into play as a corporation or industry seeks to respond to the sometimes sharply divergent interests of its shareholders and other stakeholders. Moreover, they ignore the fact that in most circumstances, people themselves are likely the best judges of their own interests, including those interests they hold in common with others, and should be acknowledged as such.

These criticisms are closely related. Descriptions of common goods and public interests take on an arbitrary character when those who share those common interests are not significantly involved in their definition and description. The process is paternalistic when shared interests are defined by self-selected individuals or organisations, such as private sector corporations that assume that they are in a better position to define, describe, or interpret shared interests than others or than the public itself. Where a unilateral conception of shared interests and common or public goods is not paternalistic, it is likely to be arbitrary, because it constitutes a determination of shared interests in the absence of the input of the people with whom the interests are said to be held in common.

Conflict of interest is also a distorting factor. It is all too easy for those who have something to gain or lose by the outcome to define the public interest in ways that enhance the realisation of their own interests, or at any rate impede that realisation as little as possible. At issue may be simply the interpretation of public information. Or it may be the extent to which, or the manner in which, information subject to unfavourable interpretation by those whose interests may be thereby threatened is made public. Or it may be something as fundamental as assigning the burden of proof in determining the existence of health or safety risks where the success or failure in marketing a product or the profitability of a substantial investment for research or product development is at stake.[7]

Perhaps the fundamental flaw in unilateral definitions of what is and what is not in the public interest is the implied lack of recognition of the principle that individuals themselves are normally best able to define and identify their own interests. This is not to suggest that people cannot be mistaken about what is in their interest. It is only to say that if they are mistaken, they are in the best position in the normal course of events to identify their mistakes.

It is not simply a matter of identifying as efficiently as possible what is the case. Rather, it is a matter of according a central place in the identification and characterisation of public goods to the principle of equality. The function of the principle of equality in this context is to ensure that no voices are privileged in determining what interests people hold in common. The purpose of imposing regulations on international commerce is, in fact, to protect and advance public or common interests. To leave responsibility for determining the form and content of those regulatory norms in the hands of just one segment of society — the private sector or, more particularly, MNCs — clearly conflicts with the underlying principle that no voices should be privileged in determining what people have in common and how best to protect and advance what they may be.

Ethics Codes: The Public Sector

If voluntary self-regulation is not an acceptable way of determining the contours of a corporation's social or ethical responsibilities, it is clear that assigning that responsibility to governments — whether democratically elected or not, in partnership with the private sector or not — is equally unsatisfactory. Many of the reasons have already been mentioned (Cragg 2000, 2001, 2002). In addition to those mentioned above, the reasons include the reluctance and inability of most governments to regulate activity beyond their own borders, the impact of international agreements designed to remove obstacles to the movement of goods and services, and regulatory competition. All of these have worked to inhibit the capacity and willingness of national governments to regulate commercial activity nationally and internationally.

An additional factor, however, is frequently overlooked. Regulation is typically thought to be justified where its goal is to provide protection against unwanted harms and unwarranted infringements of personal freedom and to put in place rules designed to ensure equitable access to opportunities for personal or collective development. These things are widely thought to be common goods. and protecting or advancing them is considered in the public interest. The problem with democratic regimes is that 'the publics' whose interests the regimes are justified in protecting and advancing are national publics. When national governments seek to extend their definitions of the public interest beyond their national geographical boundaries, their definitions of that interest, or what has been described as the common good, become unilateral and therefore subject to the criticisms of unilateral definitions of the public good already set out. Democratic governments can speak for and legitimately describe only the common interests of those who elect them.

This problem is exacerbated where corporations are concerned by conflict of interest. In a global economy, the economic interests of the citizens of particular nation-states are closely tied to the capacity of their economies to compete effectively in international markets. One of the functions of a modern government is to ensure that conditions are in place to allow this to happen. The result is to align the interests of governments with the interests of corporations whose activities are subject to their regulatory powers. This puts governments in a double bind. Not only are they not well placed to define the common interests of populations that fall outside their national boundaries, but they also have a vested interest in creating or maintaining conditions that ensure national economic stability and growth regardless of the impact of their policies on the welfare of people in other national jurisdictions.

The practical implications of these factors are significant. Three examples will illustrate the point. First, almost every modern legal system prohibits the bribery of a country's own public officials, whether legislators, judges, civil servants, and police, to use just a few examples. Until very recently, however, only one country prohibited the bribery of foreign public officials. Indeed, until very recently, a number of the most advanced industrial democracies supported and encouraged the bribery of foreign public officials by allowing the deduction of bribes for tax purposes often without supporting documentation.[8] Recent moves by OECD countries to criminalise bribery illustrate the forces at work. None of the governments involved, with the exception of the United States, was prepared to move unilaterally for fear of undermining the competitive position of its own firms. It is equally significant that a collective monitoring mechanism is integral to the OECD Anti-Bribery Convention. This mechanism is designed to ensure a level playing field and reflects the assumption that, without effective collective monitoring, the governments involved could not be expected or trusted to impose or enforce regulations that might reduce the capacity of their own companies to win contracts in the global marketplace.

A second example: Virtually all governments in the industrialised world have export development banks or agencies that provide financial support for companies wishing to win international contracts. With very few exceptions, very little attempt has been made by the governments to insist that companies applying for loans under these programmes meet regulatory standards in their international business dealings that the government would require of them were the activities to fall under their own jurisdiction. Indeed, until very recently, most export development agencies have not even required assurance that businesses receiving financial support activity respect the laws of the countries in which they proposed to do business. Efforts to set conditions on loans or to require public monitoring to ensure that national laws are respected (environmental protection regulations or anticorruption laws, for example) have been largely resisted.[9]

A third example: One of the most striking features of the public debate over the regulation of international business conduct is the resistance of the governments of the developing world to see regulations relating to such things as human and labour rights, environmental protection, and the conditions under which goods are manufactured integrated into international free trade agreements. It is quite clear that this stance is motivated by a complex set of factors. It is equally clear that one of those factors is the belief that the developing world cannot trust the governments in the developed world to apply such regulations fairly. Governments are expected to defend national interests by tilting economic regulation in ways that favour their own country's economic interests.

It was suggested earlier that current understandings, particularly in the private sector, of the nature and function of corporate governance are heavily influenced by the view that the goal and purpose of the modern corporation are to maximise profits, something that might be described as a dominant paradigm. It would be a mistake to suggest that perceptions of the nature and function of national governments in international relations are similarly governed by a dominant paradigm. However, it is fair to suggest that the view of contemporary governments, particularly among industrially advanced liberal democracies, has been heavily influenced in the post-war period by 'political realism', a view heavily influenced by the thinking of Thomas Hobbes and very influential throughout the cold war. This view implies that nation-states cannot afford to be governed by moral principles in international relations because there is no way of guaranteeing that other nation-states will do likewise. The function of the nation-state in international diplomacy is to advance the interests of that state as effectively as possible.[10]

While it would be a mistake, as suggested, to argue that this particular view dominates international diplomacy at the current time, it would also be a serious mistake to think that political realism has faded in significance since the end of the cold war (see, for example, Kaplan 2000, 77; Surjadinata 1998). For the moment, its significance lies in the way it picks up and amplifies the view that the role of the nation-state is to advance national interests and the way in which it parallels for the nation-state, motifs that are deeply embedded in the dominant corporate governance paradigm. The national interests of the contemporary nation-state in an era of globalisation are, in large part, economic. The multinational corporation is the bearer of those interests in the international commerce. Under the conditions of globalisation, protecting and advancing the interests of corporate activities emanating from one's own country is therefore a primary obligation of national governments.

The focus on national self-interest does not eviscerate any inclination toward principled behaviour in the international arena. However, as with MNCs, it does tie any such inclinations closely to considerations of enlightened self-interest, which, in turn, requires that any commitment to be guided by ethical principles be framed by instrumental justifications. Seen from this perspective, nation-states have some important similarities to MNCs under conditions of globalisation.

In sum, globalisation has seriously eroded the capacity of the nation-state to set sound ethical standards for multinational corporations operating from its territory.

The Role of Civil Society

Globalisation has fundamentally altered the capacity and the willingness of nation-states to regulate international commercial activity. However, the factors that have weakened both the will and the capacity of nation-states to use the law to regulate commercial activity in the pursuit of common or public goods have also provided motivation and tools for individuals and groups concerned with the harmful impacts of international trade, particularly in the developing world, to monitor, share information, and organise public critiques of ethically questionable corporate behaviour. If competition and the drive to maximise profits have pushed corporations to ignore widely acknowledged principles of business conduct, the conditions that have allowed corporations to seek out low-cost, under-regulated environments in which to produce goods and services have also created conditions for sharing information about those conditions rapidly and globally. The technological advances that have provided the foundations for economic globalisation have also created conditions for international co-operation and resource sharing on the part of third parties concerned about the social economic and environmental impacts of the unbridled drive to maximise profits in relatively unregulated economic environments. The result has been the gradual emergence of a third sector that is collectively becoming known as civil society.

The emergence of this third sector of voluntary organisations or NGOs has been powered by two elements. The first is a growing capacity on the part of third parties to influence the activities of MNCs through publicity campaigns, consumer boycotts, and political pressure on local, regional, and national governments and international bodies. The second is the humanitarian and ethically grounded character of civil society organisations. Indeed, the capacity to exercise political influence is connected directly to the capacity of civil society organisations to attract financial and volunteer support for values grounded on moral principles that their advocates are persuaded should frame international commercial activity. The high profile of corporate ethical codes of conduct that speak to the interests of stakeholders other than shareholders is one of the most visible signs of the emerging political influence of voluntary sector organisations committed to raising the standards of business conduct locally and internationally.

The moral influence of civil society organisations is a function of the principles they advocate. The political influence of those same organisations is directly tied to their capacity to persuade potential supporters of the merits of their cause seen from a principled or moral perspective. Hence, the effect of their involvement in the debate over regulation of corporate conduct is to inject a focus that is grounded on concerns that go well beyond the economic interests of individuals, corporations, or governments to issues dealing with human rights, environmental protection, working conditions, the fair distribution of the costs and benefits of economic activity, and, in general, the ethical environment of business. As such, they add an important element to the whole issue of regulation and the construction of a new social contract governing the transaction of business in global markets.

Ethics Codes as an Effective Tool of Corporate Regulation?

This discussion began with the suggestion that along with globalisation has emerged what has been described as a regulatory vacuum that has left MNCs free in many respects to set their own standards of conduct in the pursuit of profits in international markets. Is self-regulation guided by codes of ethics an adequate way of filling the regulatory vacuum created by globalisation?

If codes of ethics are to fill that regulatory vacuum, they must have the capacity to protect and advance the public interests that globalisation has undermined. Do they have the capacity or the potential capacity? The building blocks for constructing an answer to that question are now in place.

It is difficult to see how a system of voluntary self-regulation designed and implemented unilaterally could adequately resolve the conflicts between private and public goods that corporations can be expected to encounter in the pursuit of business objectives in intensely competitive commercial environments. Simply reinvigorating a model (what has been described elsewhere as the post-war social contract) that located the responsibility of setting and enforcing standards of business conduct exclusively or primarily with governments is not a promising alternative. As has been seen, globalisation has realigned the interests of governments with those of the corporate sector. In international markets, national governments are not neutral, unlike competing commercial interests. They do not or cannot speak for or represent all the stakeholders whose interests are put into play by international commercial activity. And their regulatory powers are attenuated by principles of sovereignty and their hostility to attempts to extend the reach of legislated regulation beyond national boundaries.

Neither can the role of regulator be assumed by civil society, which has no regulatory authority or formal powers of enforcement.

At first glance, this would appear to leave no options — a pessimistic conclusion that has attracted some support. However, there are other options. The final task here is to sketch their contours.

Identifying the Building Blocks

What is required if ethics codes are to emerge as effective tools for regulating corporate activity in global markets? What building blocks emerge from this discussion? A number suggest themselves.

First, it must be acknowledged that corporations have social responsibilities that go well beyond the simple maximisation of profit. These responsibilities now constitute an integral feature of business, government, and public rhetoric.

Second, there is a widely shared consensus on the ethical principles that should guide corporate conduct in global and local markets. This consensus is finding expression in international ethics codes and corporate social responsibility standards articulated variously by international institutions such as the United Nations, the OECD, interfaith parliaments, private sector, and NGO associations and coalitions.

Third, sophisticated, relatively inexpensive systems of communication, such as the internet but also international media such as CNN, have made it possible for people widely separated by distance and culture to share information on the activities of MNCs and governments. Examples include the experience of Shell in Nigeria and the international attention and critique focussed on the OECD's efforts to negotiate the Multilateral Agreement on Investment, both of which generated co-ordinated responses.

Fourth, there is a growing awareness on the part of MNCs that both risk and reputation management now require paying close attention to social and environmental concerns and issues.

Fifth, and last, an increasingly sophisticated system of civil society organisations, composed in part of local and international networks of NGOs, is developing the resources and the capacity to monitor, both formally and informally, the impacts of commercial activity on a wide range of stakeholders.

These are, of course, only building blocks. To be effective, they have to be woven together into working cross-sectoral networks with the capacity, motivation, and resources to apply the principles on which consensus has emerged to the wide variety of economic activities and cultural settings encountered by corporations encounter in the global marketplace.

From Principles to Practice

Slowly, governments and corporations are acknowledging that civil society will have to be involved if ethically sound and operationally effective international and local regulatory systems are to be put in place. Progress is slow but evident. There are many illustrations of this development. Roundtables are increasingly common and have achieved some successes.[11] The OECD Anti-Bribery Convention's monitoring system has a place for NGO input. NGO pressure has opened Free Trade Agreement of the Americas (FTAA) to a more transparent negotiation process and wider NGO participation. Individual corporate initiatives are beginning to involve active partnerships with advisory councils with strong links to civil society.

Increasingly, too, NGOs are involved in monitoring the effectiveness with which codes of conduct are being implemented by corporations. This involvement is taking many forms. Virtually all the emerging monitoring standards — AccountAbility 1000 and Social Accountability 8000, for example — require active involvement of stakeholders in the monitoring process. NGOs such as Amnesty International and Transparency International are being invited to partner with corporations such as Shell to monitor compliance with corporate ethics codes. Moreover, NGOs are being called in to conduct compliance reviews or audits for large MNCs and publish their results.

What even the most successful initiatives point to, however, is the clear need for co-operative code building and code implementation involving the private sector, government, and civil society working co-operatively in cross-sectoral networks. Corporations have ethical obligations encompassing social, economic, and environmental concerns that extend beyond legal compliance. Governments have the capacity to create legal incentives and enforce standards. Civil society has the capacity to reach across legal and cultural boundaries to provide leadership in articulating the substantive, morally grounded principles that define acceptable international standards of business conduct.

However, effective code building requires implementation and monitoring strategies and management systems designed to ensure that codes once accepted are effectively followed. Here, too, co-operation across the public, private, and civil society sectors is essential. Civil society in particular must be involved in assessing the effectiveness of code implementation programmes.

Once again, there are many models to draw on in designing co-operative collaboration across the private, public, and voluntary sectors. Roundtables involving the three sectors have been particularly effective in Canada. The formation of organisations dedicated to co-operative standard setting and compliance monitoring has resulted in the development of two monitoring systems, Social Accountability 8000 and AccountAbility 1000.

Only time will tell whether the results will adequately address the public interest concerns caused by the regulatory vacuum that has accompanied globalisation and the exploitation of that vacuum by multinational corporations in pursuit of profit maximisation. If they do not, confidence both in globalisation and in the modern investor-owned corporation that would appear to be a significant beneficiary of globalisation is bound to be undermined. If the conditions required to ensure that corporate activity supports the pursuit of widely acknowledged public goods are not fostered and encouraged by the corporate sector, public demand for profound changes in the way international commerce is regulated is bound to grow. It is from this perspective that suggestions from voluntary self-regulation and corporate codes of ethics play an important role in setting standards for corporate conduct in global markets.

Notes

1. This chapter is a sequel to three papers on related topics by the author: 'Human Rights and Business Ethics: Fashioning a New Social Contract', *Journal of Business Ethics* (2000), vol. 27, no. 1–2, pp. 205–214; 'Business Ethics and Stakeholder Theory', *Business Ethics Quarterly* (2002), vol. 12, no. 2, pp. 113–142; and 'Ethics Codes: The Regulatory Norms of a Globalized Society?' in A. Soeteman, ed., *Pluralism and the Law*, Kluwer Academic Publishers, Dordrecht (2001), pp. 191–223. The author is grateful for financial support of the Social Sciences and Humanities Council of Canada through the Strategic Grants project entitled 'Ethics Codes: The Regulatory Norms of a Globalized Society'.
2. These themes are explored in a good deal more detail in (Cragg 2000, 2001, 2002).
3. A compendium of international codes, websites, and related material is available from the Voluntary Codes Research Group (2003).
4. The wording here parallels closely the statement of Model Business Principles, released by the United States Department of Commerce (1996) in 1996 as a voluntary guideline for U.S. business.
5. The wording here is quoted from the Model Business Principles (United States Department of Commerce 1996).
6. The most notable proponent of this view is Milton Friedman (1970). However, it is a view that is widely endorsed by chief executive officers of large MNCs and other private sector commentators.

7. The interpretation of information can be significantly influenced by the point of view of various stakeholders, even where it is accepted by key players, including investor-owned companies, that there is an obligation to protect the public good, such as human health. A fascinating study that illustrates this phenomenon is contained in *Value Assumptions in Risk Assessment: A Case Study of the Alachlor Controversy* (Brunk, Haworth, and Lee 1991).
8. There are obvious reasons for this. To require supporting documentation would be to require self-incrimination, because the activities in question would be contrary to the laws of the countries in which they took place. Bribery is unavoidably a covert activity.
9. Export Development Canada is a case in point. Until very recently, it has resisted any suggestion that it should impose requirements that would require Canadian companies seeking loans and export insurance to meet bribery, human rights, or environmental standards that might undermine their competitive position abroad. It is important to note, however, that this is gradually changing. Increasingly, export development agencies are responding to pressure to screen requests for financial support or insurance using environmental and social criteria. Again, Export Development Canada is an example. It is developing increasingly sophisticated and comprehensive social and environment policies to guide decisions on requests for financial support. The contours of the debates over the issues discussed here were also explored thoroughly in the Australian Parliament in the context of an extraterritorial application of an Australian code of conduct for Australian companies operating internationally.
10. For an insightful discussion of the influence of Hobbes on political realism and the influence of political realism on the nature and function of the nation-state and international diplomacy, see Conrad Brunk's (1987) discussion of 'Realism, Deterrence, and the Nuclear Arms Race'.
11. Two Canadian examples are the Whitehorse initiative and the Banff-Bow Valley Round Table, both with an environmental focus, and both successful in achieving a wide-ranging consensus on the values and goals that should guide the use and development of important natural resources.

References

Amnesty International and Pax Christi (1998). 'Multinational Enterprises and Human Rights'. Utrecht.

Brunk, Conrad G. (1987). *Contemporary Moral Issues*. 2nd ed. McGraw Hill Ryerson, Toronto.

Brunk, Conrad G., Lawrence Haworth, and Brenda Lee (1991). *Value Assumptions in Risk Assessment: A Case Study of the Alachlor Controversy*. Wilfrid Laurier University Press, Waterloo, ON.

Cragg, Wesley (2000). 'Human Rights and Business Ethics: Fashioning a New Social Contract'. *Journal of Business Ethics* vol. 27, no. 1-2, pp. 205–214.

Cragg, Wesley (2001). 'Ethics Codes: The Regulatory Norms of a Globalized Society?' In A. Soeteman, ed., *Pluralism and the Law*, pp. 191–223. Kluwer Academic Publishers, Dordrecht.

Cragg, Wesley (2002). 'Business Ethics and Stakeholder Theory'. *Business Ethics Quarterly* vol. 12, no. 2, pp. 113–142.

Friedman, Milton (1970). 'The Social Responsibility of Business Is to Increase Its Profits'. *New York Times Magazine*. 13 September, pp. 32–33.

Human Rights Research and Education Centre (2003). 'International Code of Ethics for Canadian Business'. <www.cdp-hrc.uottawa.ca/globalization/busethics/codeint.html> (April 2004).

'An Interfaith Declaration: A Code of Ethics on International Business for Christians, Muslims, and Jews' (1993). <astro.temple.edu/~dialogue/Codes/cmj_codes.htm> (April 2004).

International Labour Organization (1977). 'Tripartite Declaration of Principles Concerning Multinational Enterprises and Social Policy'. <www.ilo.org/public/english/standards/norm/sources/mne.htm> (April 2004).

Kaplan, Robert D. (2000). *The Coming Anarchy: Shattering the Dreams of the Post Cold War*. Random House, New York.

Organisation for Economic Co-operation and Development (2000). 'The OECD Guidelines for Multinational Enterprises'. Organisation for Economic Co-operation and Development, Paris. <www.oecd.org/dataoecd/56/36/1922428.pdf> (April 2004).

Surjadinata, Kenneth U. (1998). 'Revisiting Corrupt Practices from a Market Perspective'. *Emory International Law Review* vol. 12, no. 2, pp. 1021–1090.

United States. Department of Commerce (1996). 'Model Business Principles'. <www.itcilo.it/english/actrav/telearn/global/ilo/guide/usmodel.htm> (April 2004).

Voluntary Codes Research Group (2003). 'Codes of Conduct and Instruments of Corporate Responsibility'. Compiled by 'Voluntary Codes: The Regulatory Norms of a Globalized Economy?', York University. <www.schulich.yorku.ca/ssb-extra/businessethics.nsf> (April 2004).

Chapter 12

Canadian Corporate Responsibility in Sudan: Why Canada Backed Down

Robert O. Matthews[1]

This chapter traces the emergence and evolution of the two strands of Canada's relationship with Sudan. The first of these strands, variously referred to as human security, humane internationalism, and humanitarian concerns, emerged in the early 1980s in response to the evolving humanitarian crisis in the Horn of Africa and was reflected in a number of issues, notably the provision of humanitarian assistance, the promotion of human rights, and support for the peace process.[2] The second strand involved Canadian investor interests overseas. The purchase of a one-quarter interest in Sudan's oil operations in southern Sudan by Talisman Energy, a Canadian corporation, confronted policy makers in Ottawa with a real dilemma: Were these two dimensions of Canada's relations with Sudan compatible with each other, or were they irreconcilable? Would Talisman's presence in Sudan provide the citizens of that country, as it claimed, with 'the promise of a better life' and serve further as a 'catalyst for peace' (Sheppard and Manhas 2000)?[3] Or, alternatively, would its participation in the development of the Greater Nile Project undermine the search for a peaceful settlement, intensify Khartoum's campaign of human rights abuses, and thus derail Ottawa's attempts at establishing a humane foreign policy?

How Ottawa answered these questions would determine, or at least should have determined, its course of action. If the two sets of policies were judged to be consistent with each other, Canada could then have continued to pursue its human security agenda while Talisman undertook to behave in a socially responsible manner. In the context of this book, self-regulated obligations would have been sufficient to bring about the necessary changes in Khartoum's behaviour. However, if the two strands proved to be irreconcilable — that is to say, Talisman was either unwilling or unable to influence the Government of Sudan constructively — Ottawa should have been prepared to impose hard law commitments on Talisman, enforceable, as threatened by Canada's foreign minister, Lloyd Axworthy, on 26 October 1999, by 'economic and trade restrictions' (Department of Foreign Affairs and International Trade (Canada) [DFAIT] 1999). Non-involvement rather than constructive engagement should have been the appropriate course for any private corporation to take.

Evidence was compiled by a wide range of actors, including Canadian and international nongovernmental organisations (NGOs), international organisations, and even the Canadian government itself; that evidence demonstrated a close, negative link between the oil sector, on the one hand, and respect for human rights and promotion of the peace process, on the other hand. Consequently, most Sudan observers expected that Ottawa would take the necessary steps to force Talisman out of Sudan. That it did not pursue such a strategy meant that Canada's relationship with Sudan became, much like Canadian policies were toward South Africa during the apartheid years, 'complex, ambiguous and contradictory' (Freeman 1997, 5). Indeed, the parallel between these two cases is striking.

Humanitarian Concerns: The First Strand

In the first decades following Sudan's independence in 1956, Canada had effectively no relations with Sudan. Trade between the two countries was virtually non-existent.[4] The Canadian private sector displayed little or no interest in Sudan, and because it was a non-core country in Canada's aid programme, the Canadian International Development Agency (CIDA) extended little development assistance to Khartoum.[5] Consequently, Canada did not establish a diplomatic mission in Khartoum, and any consular or diplomatic concerns that arose were handled from Cairo until 1986 and, since then, from Addis Ababa. Only at the end of July 2000 did Ottawa open up a new post in Khartoum, a reflection of Canada's growing interest in the peace process and in the human rights situation there, as well as of the need to provide services for a growing number of Canadian nationals either visiting or working in Sudan.

The 1980s marked the beginning of an effective relationship between Canada and Sudan. In response to public concern with the growing humanitarian crisis in the Horn of Africa in the fall of 1984, Joe Clark, who was at that time Canada's Secretary of State for External Affairs, appointed David MacDonald as Emergency Co-ordinator for the African famine. His assignment was to assess the situation in the many countries in Africa affected by drought and famine, including Sudan, and to propose steps that Canada could take to alleviate the human suffering. The Canadian government provided significant funding for the drought-affected countries in Africa, but its generosity was inspired and spurred on by the sympathetic response coming from Canadians in all walks of life (see MacDonald 1985). According to MacDonald, as many as two thirds of all Canadians made contributions, with funds or in kind, to the famine relief programme. And while Ethiopia was initially the primary focus of the media's attention, in time Sudan took on considerable importance in the public eye as well. This crisis in the mid 1980s, then, galvanized an interest within the humanitarian community in the human tragedy taking place in Sudan, thus creating an effective and vocal 'Sudan constituency' in Canada. Although the famine in the Sudan region began in 1983, it did not reach its peak until 1988 when the 'death rates among southern Sudanese famine victims were among the highest ever recorded anywhere in the world' (Keen 1994, 76).

It was in response to this prolonged drought and the resulting famine as well as to the re-emerging civil war that Canada became involved in Sudanese affairs. The policies adopted by Ottawa in extending assistance to Sudan, in expressing opposition to human rights violations by Khartoum, and in encouraging and supporting the peace process there all reflected a humanitarian component within Canadian foreign policy. Whether the concern was for the 'safety of peoples' caught up in a violent civil war and suffering from hunger and famine, or for the rights of those same people trampled on by an oppressive government or warring parties to a conflict, the Canadian government was, through its various policies, acknowledging its commitment to a humane foreign policy (Hampson 2002, ch. 2).[6] In his policy statement on the Sudan in October 1999, Lloyd Axworthy expressed it in this way: 'Canada's commitment to human security, particularly the protection of civilians in armed conflict, provides the basis for its involvement in Sudan' (DFAIT 1999). Commercial interests were not articulated as a basis for Canadian involvement. This concern for the welfare of the Sudanese found its first expression in CIDA's programmes of assistance to their country. Beginning in 1982–83, aid to Sudan jumped to more than CA$20 million and continued at or above that level well into the 1990s. Much of this aid involved either bilateral food aid or humanitarian assistance administered by international relief organisations or NGOs. Indeed, the largest bilateral project was designed to develop new farming techniques to improve the cultivation of dryland crops. And when Canada suspended government-to-government aid to Sudan in 1992 in response to its serious concerns over that country's human rights record, CIDA continued to contribute humanitarian relief in the form of food aid, medical supplies, seeds, and tools, and channelled that aid through United Nations agencies and NGOs operating under the umbrella of Operation Lifeline Sudan as well as through the International Committee of the Red Cross (ICRC). In the decade following 1990, Canada's contribution to humanitarian relief in Sudan amounted to CA$110 million.

As is so often the case in civil wars, civilians are the principal casualties. Nearly 2 million people in Sudan have been killed since civil war broke out again in 1983, while more than 4.5 million have been internally displaced (with as many as 1.5 million internally displaced persons living around Khartoum). As part of its concern for their safety, Canada has co-sponsored UN resolutions calling for 'unimpeded access to civilians, for their protection, and for the extension of humanitarian cease-fires in order to ensure the safe delivery of emergency supplies to civilians' (DFAIT 1999).

Canada's humanitarian concerns were expressed in another way, through its condemnation of the Sudan government's human rights record, especially since the overthrow of the El Mahdi government in June 1989. This concern was usually registered in the multilateral context of the UN — in the Commission on Human Rights (when Canada was a member) or in the General Assembly and its Third Committee — and reflected the growing number of Canadians who had come to be concerned over poverty and oppression in Sudan. This growing concern was expressed through policy papers or at consultations organised by the government, where Canadian NGOs had occasion to provide information to the government on the human rights situation in various countries throughout the world and to make recommendations about what Ottawa's stance should be. The Sudanese constituency is large, but no one has a more effective voice than the Inter-Church Coalition on Africa (ICAFF), reflecting the long-term and close links between Canadian churches and the Sudan Council of Churches and the New Sudan Council of Churches. ICAFF's testimony, along with other civil society organisations in Canada, has had some influence on Ottawa and the stance it takes on human rights questions at the UN.

It was in 1992 at the UN General Assembly that a resolution was first passed citing 'grave human rights violations in the Sudan, particularly summary executions, detentions without trial, forced displacement of persons and torture' (UN 1992). The resolution expressed alarm at 'the large number of internally displaced persons and victims of discrimination in the Sudan, including members of minorities who have been forcibly displaced in violation of their human rights and who are in need of relief assistance and of protection'. At the same session, the General Assembly called for the appointment of a special rapporteur to monitor the human rights situation in Sudan and to report back to the UN Commission on Human Rights. Canada supported that resolution and similar ones passed in subsequent years. Indeed, at the 50th session of the commission, the Canadian delegation expressed 'in the strongest terms its condemnation of all parties to this conflict who suppress dissent, engage in the forcible displacement of people, attack civilian targets, use food aid for political ends, and fan the flames of ethnic and religious tensions'. The Canadian statement went on to express concern about the personal attack the Sudanese delegation had made on Dr. Gaspar Biro, the first UN special rapporteur, and urged Sudan to co-operate with him, stating that 'personal attacks on him are tantamount to attacks on the Commission and are not acceptable' (Francis and Khenti 1994, 10). Again in 1998, in a speech to the Third Committee, Canada's representative made a stinging critique of the government of Sudan, arguing that 'most brutal and repressive human rights violations' were committed in times of war or in the aftermath of armed conflicts, and specifically singling out the Sudan for 'arbitrary arrests, religious intolerance and persecution, as well as slavery' (UN 1998).

In more recent years, however, UN resolutions on Sudan have been more moderate in tone. While they all identify a continuing need for the UN special rapporteur, they refer to the human rights violations committed by the opposition group, the Sudan People's Liberation Army (SPLA), and acknowledge the positive steps undertaken by the government of Sudan. In agreeing to co-sponsor with the European Union the resolution passed at the Third Committee and the General Assembly in 2000, for instance, Canada expressed its encouragement for the positive steps undertaken by the Sudanese government to resolve the question of abducted children, recognising that the situation was complex and that progress might not come quickly (UN 2000).

The link between the discovery and exploitation of oil, on the one hand, and increasing human rights abuses, on the other hand, was first made at the UN Commission on Human Rights in 1999. Leonardo Franco, who was the UN special rapporteur on Sudan at that time, noted in his report that 'human rights situation was worsening because of strategies implemented in relation to exploitation of oil resources' (UN 1999). He went on to point out that the 'government's aim to establish a safe zone around the oil wells was leading to new risks for civilians. The human rights situation was of particular concern with regard to delivery of humanitarian assistance and it was even worse because the cease-fire agreements were legally weak. Sudan had the highest number of internally displaced people. The oil exploitation could add to the displaced'. It was this same issue, as discussed below, that Axworthy took up in his policy statement later that year. Under pressure from Canadian and international NGOs, he, too, registered his concern that oil exploitation was contributing to the violation of international human rights and humanitarian law.

Support for the peace process was a logical extension of humanitarian and human rights concerns. Humanitarian relief for the casualties of war was an acceptable short-term policy response, but unless it was to become a permanent feature on the international agenda, governments would have to deal with the underlying cause of the human tragedy — the civil war. Perpetual investment in a humanitarian response could not be justified without an equivalent investment in the peace process, both politically and developmentally. Without an end to that war, there could be no progress in ensuring the safety and security of the peoples of southern Sudan.

It was this realisation that led Canada to support efforts being made in the Horn of Africa to find a peaceful solution to the civil war in Sudan. With the collapse of the Abuja negotiations held in Nigeria in 1991 and 1992, countries in the Horn of Africa decided to launch their own peace initiative, what soon became known as the Intergovernmental Authority on Development (IGAD) peace process.[7] At a meeting held in 1994, the Sudan committee of IGAD, made up of Ethiopia, Eritrea, Kenya, and Uganda, approved a declaration of principles, the most important and controversial of which were the principle of self-determination and the separation of religion and state. To give support to that peace process, Canada — along with Norway, the Netherlands, the United Kingdom, Italy, and the United States established the Friends of IGAD (later renamed the IGAD Partners Forum or IPF). Worried about the lack of progress in these peace negotiations, between 1998 and 2000 the IPF pushed to create the Sudan Peace Talks Secretariat, the purpose of which was, as Axworthy noted in July 1999, 'to carry out continuous and sustained mediation efforts to reach a peaceful settlement to the civil war ... Canada's financial support for the Sudan Peace Talks Secretariat is a concrete indication of the importance Canada attaches to resolving the situation in Sudan' (CIDA 1999).[8] He went on to add that 'now that the Peace Talks Secretariat has been established, the much needed technical and political support are now formally in place to permit real progress in the talks'. Later in the same year, Canada appointed Senator Lois Wilson as Canada's special envoy to the Sudan peace process, where she led Canada's participation in the IPF.

Funding for this secretariat, amounting to nearly CA$400 000, came from CIDA's Africa branch. CIDA, through its newly created Peacebuilding Unit within the

Multilateral Branch, also gave its support to civil society initiatives on peace and reconciliation to complement the flagging efforts being made at the government-to-government level to reach peace in the Sudan. In particular, Ottawa helped fund Track II and Track III efforts, initiated by Canadian, regional, and Sudanese NGOs and designed to link civil society with the official peace process, encourage 'dialogue among the diverse elements of Sudanese civil society and support initiatives for community reconciliation'.[9]

Beginning in the early 1980s and extending throughout the 1990s Canada developed and maintained a series of humanitarian policies and programmes designed to provide for the safety, security, and welfare of the Sudanese. Underlying these concerns lay an overwhelming desire to bring an end to Africa's longest lasting civil war. During this period, Canada's engagement in the Sudan was thus driven by an ethic of humane internationalism rather than by material interest.

Foreign Investment in Sudan: The Second Strand

On 17 August 1998, Calgary-based Talisman Energy announced a friendly takeover bid for Arakis Energy's 25 percent interest in the Sudan Greater Nile Petroleum Operating Company (GNPOC). Arakis had bought its concession from Chevron in 1990, but was forced to bring in three partners in 1996 — the China National Petroleum Co. (40 percent), the Malaysian National Oil Co. (30 percent), and the Sudanese National Oil Co. (5 percent) — to share financing costs. Even then, Arakis was unable to raise its share of the funding necessary to bring oil to market, and so it entered into this deal with Talisman.

In announcing the agreement to purchase Arakis, Talisman's chief executive officer, James Buckee, went to great lengths to reassure his staff and shareholders that Sudan was much more peaceful than the 'lurid and exaggerated' stories in the press portrayed it, that John Garang, the leader of the SPLA and the Sudan People's Liberation Movement (SPLM), was losing support among his own people, and that the government was genuinely committed to using at least one quarter of the oil revenues for development in the South and thus to achieving an enduring peace. Among the many meetings Buckee held while in Khartoum, one of the 'most important' in his view was with Riek Machar, a former ally of Garang in the SPLA and, by 1997, vice-president of Sudan. The peace accord struck that year, agreed to by Machar, made it possible for Khartoum to extend its authority into some rural areas of the concession and thus expand oil development and complete the pipeline to Port Sudan. Buckee concluded his remarks, innocently enough, with one last reassuring comment: 'Our presence and continuing development of this area will also serve to increase international awareness and scrutiny.'[10] Little did he realise how true his prediction would prove to be, even if the target of international scrutiny was as likely to be Talisman as the government of Sudan.

Indeed, had Buckee done his homework more assiduously, he might have learned that oil had been one of the root causes of the civil war's recurrence in 1983. In 1979, four years after it began exploration for oil in southern Sudan, Chevron had made a significant discovery. Khartoum saw in this oil a golden opportunity to reduce its debt, but under the terms of the Addis Ababa Agreement of 1972 all taxes and revenue resulting from the exploitation of oil were to flow to the Southern Regional Government in Juba. Instead of negotiating a new wealth-sharing agreement with the South, President Jaafar Nimeiri did everything he could to maintain control over the oil and to deprive the South of this new wealth. A refinery was planned for the North (Kosti) rather than the South (Bentiu). Then, in 1980, the government introduced a map redrawing the boundary lines so that the oil concession would be located in the North. Finally, in direct violation of the Addis Ababa Agreement, he pushed through a bill re-dividing the South into three separate regions, each of which was to have limited powers and few, if any, sources of revenue. Most important, all the revenues derived from oil were to go to the central government in the North. And so oil, not an issue in the first phase of the Sudanese civil war, did become one when the war broke out again in 1983, and one of the first targets of the SPLA was the Chevron operation.[11]

A Clash of Interests

Arakis evoked some public criticism of its role in the civil war. In several articles in *The Toronto Star*, for instance, Martin Cohn drew attention to the symbiotic relationship between the Canadian company and the Sudanese government. Arakis, he wrote, depended 'for its security on the protective cordon of troops operating from a military base beside the oil camp' (Cohn 1998), and the latter 'opens its hospital doors to military men as well as nomads ... [and] services broken military trucks, provides electricity lines to their barracks and even pipes in water to the army camps' (Cohn 1997). However, the attention paid to Arakis hardly matched the intensity of criticism faced by Talisman for its involvement in the development of oil in the Sudan. But as Canada's largest independent petroleum company, with its technological sophistication and its ready access to capital, Talisman was able to assist its new partners in bringing oil to market even earlier than had been originally expected. Even before the oil started to flow through the pipeline, Khartoum had been able, as noted one careful observer of Sudanese affairs, 'to borrow against these revenues' (Reeves 1999). With Khartoum thus assisted by Talisman to fund its conduct of the war, the cost of which was estimated to be US$1 million a day, it is no wonder that it would become the target of media attention and public criticism.

As a result of protests at Talisman's annual shareholders meetings, divestment campaigns, and lobbying efforts in Canada, the United States, and Great Britain, by the fall of 1999 opposition to Talisman's Sudan operation had grown into a broad-based public campaign. In the U.S., it drew support from both ends of the political spectrum, with the Afro-American community concerned about charges of human slavery and the Christian right worried about the persecution of Christians and animists in the South. In Canada, the campaign was spearheaded by a newly created umbrella organisation called the Sudan Inter-Agency Reference Group, made up of 23 Canadian churches and NGOs opposed to human rights abuses in Sudan and to Talisman's presence there. Underlying the positions taken by all these advocacy groups were two central points: that, even if only indirectly, Talisman was contributing to the massive displacement of people on and around the drilling site — a violation of Sudanese human rights — and that the millions in dollars in revenue going to Khartoum from the sale of oil increased the regime's capacity to wage war and thus reduced any incentive it might have had to negotiate in good faith in the IGAD peace process.

Talisman's response to these charges moved through a series of stages, starting with surprise, followed by denials, and ending with optimistic predictions for the future. Initially, the reaction in Calgary was one of putative incredulity, for after all a Canadian company had operated in the Sudanese oil patch for at least five years almost without incident. This surprised response was quickly followed by expressions of doubt as to the veracity of the claims made against the government of Sudan. Indeed, Buckee charged that Sudan had been 'unfairly blackened' by its critics, and that there was a 'huge machine generating false things about Sudan and it's very hard to separate fact from fiction' (Chase 1999). He argued that stories about Sudan's violent civil war were greatly exaggerated, because the country was 'becoming a source of relative regional stability' (Reeves 1999). And as for claims concerning Sudan's human rights record, Buckee openly stated: 'I don't believe accusations of genocide and I also think the accusations of slavery are grossly misrepresented and possibly manipulated' (Chase 1999). It was not long before Talisman had hired the services of a public relations firm, Hill and Knowlton, which soon had company spokespersons putting a more positive spin on their statements. Talisman began to argue that its presence in Sudan was a positive influence, imparting Western values and curbing or preventing human rights abuses. Its representatives began to claim that money flowing from oil would not be spent on purchasing military hardware but on 'roads and hospitals, especially in the south' (Chase 1999) and on 'improving life for the Sudanese' (Sheppard and Manhas 2000). They claimed further that isolating Sudan through a policy of sanctions would only harm the poorest members of society, while the development of oil offered the promise of prosperity and peace for everyone (Sheppard and Manhas 2000).

The critics of Talisman did not limit their attacks to the company; they turned their attention to Ottawa as well, suggesting that failing to take any action against Talisman and, even worse, remaining silent undermined its commitment to human security. As one of the major actors in the Canadian network opposing the Talisman investment put it in a personal communication to the author: 'How can the Minister [Lloyd Axworthy] talk about the rights of civilians and the need to provide for the security of civilians, when he stands idly by and lets Talisman ... put money into the hands of the Sudanese regime, a brutish regime if there ever was one, and one that will — make no mistake — use those resources to kill more civilians?' Canada supported the IGAD peace process and 'yet all the efforts and resources poured into IGAD by Canada and others are at risk of total negation, because Canada has failed to appreciate the role played by oil in the conflict'.

Under increasing pressure to do something or, at least to say something, that might resolve the contradiction between Canada's humanitarian concerns and Talisman's involvement in the production of oil in Sudan, Axworthy issued an important statement on 26 October 1999 (DFAIT 1999).[12] He expressed his deep concern for the possible adverse effects that the exploitation of oil might have on human rights and on the peace process, noting as well that questions surrounding the ownership of the oil and the distribution of any benefits derived from its sale were still unresolved. Axworthy stressed that the private sector in Sudan had an obligation to ensure that its operations fostered a climate conducive to the attainment of a just peace, and specifically sought assurances from Talisman that 'humanitarian and human rights laws are being upheld, and that oil extraction and export is not exacerbating the conflict in Sudan'. While admitting that there were well-documented violations of human rights and that Canada was willing to continue to support UN resolutions condemning Sudan for its human rights record, he appointed John Harker to lead an 'independent and unhindered Canadian mission to Sudan' to investigate charges of human rights abuses and to establish whether or not oil extraction was exacerbating the conflict. If the evidence brought back by the Harker mission supported the claims of Talisman's critics that Canada's policy of human security was effectively negated by Talisman's engagement in extraction of oil in Sudan, then, the Minister threatened, Ottawa 'may consider, if required, economic and trade restrictions'. In the meantime, Axworthy called upon Talisman to undertake a number of interim measures: to implement the International Code of Ethics for Canadian Business, to encourage the government of Sudan to invite independent experts to monitor the human rights situation in Sudan, to initiate discussions with Khartoum on ways of setting aside revenue from the sale of oil in a trust that could then be used for humanitarian relief and development throughout the country, and, finally, to urge the Sudanese government 'to recommit itself to a meaningful and accelerated peace process in the context of the IGAD [Declaration of Principles]'.

As Eric Reeves (1999) has pointed out, Harker was 'asked to look for something that Mr. Axworthy has already declared exists'. In that sense, the Harker mission was little more than a time-honoured instrument used by governments to buy themselves time. The only difference in this case was that Axworthy had committed himself and the Canadian government to taking firm action against Talisman in the event that Harker submitted a report that linked oil and Talisman with the deterioration of human rights and the exacerbation of the ongoing civil war. Senior officials in the foreign ministry apparently cautioned him against including that threat, but Axworthy's commitment to human security was so firm that he was prepared, if necessary, to contemplate the government use of such instruments as the *Export and Import Permits Act* or the *Special Economic Measures Act*. The die was cast, and as if to make absolutely certain what the issues at stake were, the mission's report was entitled: 'Human Security in Sudan: The Report of a Canadian Assessment Commission' (Harker 2000).

Was human security at stake in the Sudan? To what extent did Talisman's presence compromise Canada's commitment to human security there? These were the questions that the Harker fact-finding mission would hopefully answer. It was invited to examine three specific issues: the question of slavery or abduction and the role played by Khartoum in supporting this practice; the alleged link between oil development and human rights violations, in particular, the forced displacement of people around and on the oil site; and the impact of oil development on the peace process. Setting off in December 1999, Harker's team first visited Europe for talks with various UN agencies in Geneva and with the European Union in Brussels. It then travelled to Khartoum, where it held meetings with government officials, opposition representatives, civil society activists, and foreign diplomats, as well as with displaced southern Sudanese and UN officials looking after them. After spending several days at the Heglig operating base of the GNPOC, and visiting several communities in the vicinity, the team was able to gain access to areas in southern Sudan not controlled by government forces. There they visited several isolated communities, one of which did not even figure on a map, having been recently created after an attack on the town of Leer 'sent people fleeing into the bush and swamps' (Harker 2000, 28). And as the report states, the existence of this newly created community 'constitutes an indictment of all who can act to promote human security ... but do not do so. Inaction is not an option' (28).

On the first question the Harker mission was asked to address — whether the abduction of Dinka by Baggara tribesmen was sponsored deliberately by the government of Sudan or is due to its inability to provide for the human security of its own citizens — the report concluded that the abduction of as many as 15 000 Sudanese women and children was probably not a deliberate government policy. Rather, the report described it as the result of 'war-strategies in which abductions play a large part' and practices to which the government largely turns a blind eye (Harker 2000, 41). Harker called for an end to government support for armed raiders: 'the [government of Sudan] must find the will and the way to look at the practice clearly, and move firmly to eradicate it' (41). He then went on to commend the government for establishing the Committee on the Eradication of Abductions of Women and Children (CEAWC), but argued that as a step to ending the practice, it was insufficient. The European Union helped to fund the CEAWC and committed to monitor its progress. Canada, Harker suggested, might well join the EU in pressing the CEAWC into improving the effectiveness of its work.

The Harker mission was also asked to examine the problem of internal displacement and how it was affected by oil development. Adding to an earlier report by Leonardo Franco, the UN special rapporteur on Sudan, Harker discovered that in a military offensive beginning in the spring of 1999 and lasting nearly two months, the government of Sudan managed to depopulate the areas around the oil fields. Indeed, between April and July 1999, the population of one area, Ruweng County, fell by as much as 50 percent (Harker 2000, 11). The Harker report, therefore, concurred with Franco's conclusion that 'a swath of scorched earth/cleared territory' was being created around the oilfields (50). Talisman may have dismissed such occurrences as 'merely an inter-tribal problem', but Harker remained unconvinced, arguing instead that 'displacement is now, and has been for some time, because of oil. This is a commodity which is seen increasingly as, and may well be, the enemy of a people with nowhere to go where they can enjoy human security' (11).

The last and central question for the Harker mission was 'whether the current oil operations exacerbate and prolong the war, or advance the pursuit of peace' (Harker 2000, 64). Harker's answer was straightforward and unequivocal: 'The evidence we gathered, including the testimony of those directly involved, directs us to conclude that oil is exacerbating conflict in Sudan. The [Sudanese government's] drive to exploit oil deposits in Western Upper Nile and South Kordofan, which can only be done with foreign help, has intensified the conflict' (64). The conflict, he claimed, was made worse by the oil operations in at least four different ways. First, concerned about the security of the oilfields, the government adopted a strategy of displacement and pacification, as already observed, with the catastrophic depopulation of Ruweng County serving as the most recent example. Second, the development of oil resulted in the intensification of fighting not only between Khartoum and the SPLA but also among southern Sudanese as well. At the same time, however, efforts were being made to encourage reconciliation among ethnic communities in the south, efforts that resulted in the Wunlit Dinka-Nuer conference in February-March 1999 and, more recently, in an understanding between Garang and Machar. Third, the airstrips and road systems developed for the oil operations provided the government of Sudan with a strengthened and extended infrastructure from which to launch their military campaign. Even Talisman admitted that the Heglig airstrip was used on two occasions by government aircraft for military purposes, although Harker reported that it had been more or less constantly used since May 1999. Fourth, and last, the discovery and development of oil highlighted the importance of who controlled and distributed its revenues. Although Harker was reluctant to examine the question of oil revenues, arguing that hard information was unavailable, his conclusion seemed to suggest that the way in which oil revenues were distributed had become an obstacle to reaching even a cease fire. 'It's difficult to imagine a cease-fire while oil extraction continues, and almost impossible to do so if revenues keep flowing to the GNPOC partners and the [government of Sudan] as currently arranged' (67). That all oil revenues went to Khartoum could not but fuel the government's war effort and make Khartoum less interested in a negotiated peace. Without a wealth-sharing agreement between North and South, it was hard to imagine any kind of peace.

In a country where the government acquiesced in, if not supported, the practice of abduction or slavery, forcibly displaced communities from their homes to secure an area for oil exploration and exploitation, often refused humanitarian access to many of the areas under opposition control, and bombed civilian targets indiscriminately, human security was clearly at risk. And knowingly or not, Talisman contributed to that suffering.

As the Harker report says, in a somewhat understated way, 'Sudan is a place of extraordinary suffering and continuing human rights violations ... and the oil operations in which a Canadian company is involved add more suffering' (Harker 2000, 15).

Although the Harker mission's primary purpose was to ferret out evidence and information concerning the human rights situation in Sudan and the link between oil and the war there, the report did not hesitate to offer advice, if not recommendations, to the participants in their study. To the parties to the conflict — the government of Sudan and the SPLA — the report appealed for a cease fire and an end to all bans on humanitarian flights. Only in this way could there be relief for the suffering people of Sudan. But even such a simple measure as a cease fire was not likely within reach, argued Harker, if the oil continued to flow directly to Khartoum. Either oil production had to stop or the revenues from its sale should be deposited in a trust fund for later use in the development of the entire country. As the government was not likely to halt the extraction of oil voluntarily, a cease fire would only emerge as part of a larger peace settlement. For officials in Ottawa, the report very carefully picked its way through the delicate minefield surrounding the question of sanctions. Although it did not call on the Canadian government to impose immediate sanctions on Talisman, it did counsel the government to hold the threat of sanctions — not through the *Special Economic Measures Act*, but through action under the *Export and Import Permits Act* or the Area Control List — in reserve, to be used in the event that Talisman was unwilling or unable to overcome the adverse effects that oil operations were having on the observation of human rights and the peace process in Sudan. Furthermoe, to Talisman it offered some advice for as long as it remained in Sudan. The company was reminded that it too had obligations under international human rights and humanitarian law, that it should closely monitor the actions taken by the Sudanese authorities, and that, when necessary, it should protest against any actions it deemed inappropriate. In a telling phrase, Harker concluded, 'if the company is either unwilling or unable to constructively influence the [government of Sudan], perhaps it should not be in the Sudan at this time' (Harker 2000, 70). This theme will be pursued in the next section.

An Ambiguous Relationship

In a remarkably short reply, dated 14 February 2000, to the long and detailed analysis of the Harker report, Lloyd Axworthy sought to set the future direction for Canadian policy toward Sudan (DFAIT 2000b). Given the evidence compiled by the Harker mission, and remembering the threat that the Minister of Foreign Affairs had issued only three months earlier, most of his critics expected that he would either force Talisman to withdraw or, minimally, impose limited sanctions on Sudan itself. Indeed, a commitment to the goal of human security seemed to require such a step. Instead, the government of Canada chose to adopt a policy of constructive engagement, while urging Talisman to behave in a more socially responsible manner.

To promote the cause of human rights, Canada undertook to provide whatever support was necessary to ensure that the UN's Leonardo Franco could resume his valuable work on the displacement of civilians and other human rights violations. At the same time, Ottawa would take advantage of its forthcoming presidency of the UN Security Council in April to address the humanitarian impact of the civil war on the people of Sudan. The Minister also promised to provide financial assistance to the CEAWC, either directly or through the United Nations Children's Fund (UNICEF). As for Talisman's role in all this, the Canadian government — or, more precisely, the Minister — backed down from his threat, conveniently seizing on the fact that Harker had not recommended the immediate imposition of economic sanctions while ignoring that the report had explicitly called on the government to hold the threat of sanctions in reserve. Instead, Axworthy called on Talisman to use its significant influence in Sudan to make a positive contribution to the human rights situation and to work with NGOs to develop an effective mechanism for monitoring its operations in Sudan.

As for the cause of peace, the Minister renewed his support for the IGAD process and its Declaration of Principles, drawing attention to Senator Wilson's recent visit to the region, during which she had stressed the urgency of reaching a negotiated settlement to the conflict. Axworthy also announced that Canada would soon open up a new office in Khartoum in order to monitor human rights violations from close up, to ensure that Canada could make a more effective contribution to the peace process, and to provide consular services to the growing number of Canadians in Sudan. The overall message was clear: 'engagement and dialogue are the most appropriate tools for advancing the cause of peace in Sudan' (DFAIT 2000b).

In its reply to the Harker report and through its subsequent actions, the government of Canada seemed to argue that the two strands of its relationship with Sudan were not incompatible, and that if the government worked harder at promoting human security in Sudan and Talisman became a good corporate citizen, human suffering in Sudan would diminish and the prospects for peace would improve. But, in fact, nothing since February 2000 would suggest that this policy of positive engagement has led to any improvement in the lives of the Sudanese. Quite the contrary, the practice of abductions continues, the government strategy of displacement remains in place unchanged, and, although some progress has been made toward reaching a peace accord, the two parties still remain far apart on a number of critical issues. The testimony offered by several close observers of Sudan merely confirms what Harker and his team made patently clear: the oil operations in which Talisman was involved were bound to add to the suffering of Sudanese.

Support for this view comes from a wide range of sources. Human Rights Watch (2001), a well-respected international NGO, recorded a litany of human rights abuses perpetrated by the government of Sudan. Even though Khartoum set up CEAWC, its support for the effort to eliminate slavery or abduction in Sudan was put into question by its unwillingness to record the identity of the abductors, let alone to prosecute any. The government continued to pursue a strategy of displacement through aerial bombardment and scorched-earth tactics in an effort to create what Human Rights Watch called 'a cordon sanitaire around new oil fields'. The government's disregard for human life was also demonstrated by an increase in government attacks on relief, health and school facilities, 'apparently aiming to deter or shut down the UN-led humanitarian operation, Operation Lifeline Sudan'. As for oil revenues and the ends to which they are put, the government openly announced its intention to use its new oil revenues for purposes of defence, witnessed to by Sudan's defence spending, which, between 1998 and 2000, increased by almost 100 percent.

In another investigative report, this time sponsored by NGOs in Canada and Great Britain, Georgette Gagnon and John Ryle documented a dramatic increase in armed attacks on civilians in and around the oil region in Western Upper Nile, carried out by both the government of Sudan and rebel forces (Canadian Auto Workers Union et al. 2001). However, what was new in the period under study (May 2001) was the increased number of direct attacks on civilians by government armed forces. Made necessary by the collapse of the 1997 Khartoum peace agreement between the government and Riek Machar, these 'attacks [were] part of what appears to be a renewed Government of Sudan strategy to displace indigenous non-Arab inhabitants from specific rural areas of the oil region in order to clear and secure territory for oil development' (Canadian Auto Workers Union et al. 2001, 2).[13] This study thus reinforced what the Harker report had found, namely that oil companies were implicated in this counter-insurgency strategy. And even though Talisman acknowledged that the Heglig airstrip had been used by Sudanese government forces for nondefensive purposes on at least four occasions — not just twice as noted by Harker — Gagnon and Ryle reported that such incidents occurred much more frequently and continued to occur. Much as did Human Rights Watch, these investigators found that Operation Lifeline Sudan's access to airstrips in rebel-controlled areas was progressively confronted by government flight denials and by the danger of aerial bombardment from government aircraft. The evidence continued to support the view that oil development exacerbated the civil war and assisted the war aims of the government of Sudan.

A final perspective is provided by Gerhart Baum, the UN's special rapporteur on Sudan in 2001, in his interim report (UN 2001, 7). He recorded that while the government had distanced itself from the practice of abductions, it had 'not yet taken concrete measures to prevent new abductions and its inaction has the effect of encouraging the recurrence of the phenomenon' (7). His report was particularly critical of the oil operations and their negative effect on human rights and the ongoing conflict. Even though the oil companies provide infrastructure and social services to the local population in the areas where they operate, Baum maintained that their impact is still on balance negative and would 'continue to face international criticism until military warfare ends' (13).

What these three reports document was the continuing, even worsening threat to the human security of most Sudanese, especially those living in the Western Upper Nile and, more recently, Eastern Upper Nile region. Notwithstanding recent efforts in

2001 made by Canada and IPF to re-energize the peace process and to cajole Talisman into acting more socially responsibly, the situation for most Sudanese had not improved; indeed, it had become increasingly worse, and the prospects for peace seemed dim. The hard fact was that the government of Sudan was simply impervious to the kinds of influence that Canada and Talisman could bring to bear. As long as oil revenues continued to flow and fill government coffers with badly needed foreign exchange, Khartoum had no reason to change its behaviour and bargain in good faith.

Support for this argument came from an unlikely source, Talisman's first Corporate Social Responsibility Report (Talisman Energy 2000). In that report Talisman described in detail how it tried to comply with its responsibilities under the International Code of Ethics for Canadian Business, creating what it called 'operating principles' with established standards in six areas: human rights, community participation, employees' rights, environmental protection, business conduct, and stakeholder engagement. These principles were in turn linked with identifiable objectives, performance indicators, and specified target dates. For example, in the section on concern for human rights, the report outlined a number of objectives, the most relevant of which for these purposes were the development of a human rights training and awareness programme for managers and employees working within Sudan, the development and implementation of a programme to monitor and investigate human rights violations arising from oil operations, and the development of guidelines concerning the use of oilfield infrastructure to ensure that it was not used for military purposes. In all these cases, a performance indicator was established and deemed to have been met. In examining these three objectives, it is important to note that the authority for dealing effectively with each of them is quite different: Talisman alone was responsible for ensuring that its employees were treated in a manner consistent with human rights codes concerning labour-management relations and for correcting any practices that fell short of those standards. However, when it came to investigating human right violations arising from oil operations, or for ensuring that its infrastructure was not used for 'military purposes', the reach of Talisman's authority was limited. With the appointment of a field co-ordinator, Talisman had the capacity to investigate some incidences of human rights violations, although only with the government's approval when operating outside the oil concession itself, and presumably not at all when the violations occurred in rebel-controlled areas. And Talisman could seek the government's promise to use its airstrip at Heglig for defensive purposes only, but any promises made were quite empty.

The problem addressed here is one that Talisman in fact acknowledged. In defining the objectives of its Corporate Social Responsibility Report, Talisman distinguished among three categories: Talisman objectives, GNPOC objectives, and so-called advocacy objectives (Talisman Energy 2000, 9). They differ in the extent to which Talisman was able to influence their attainment. In the case of Talisman objectives, the achievement of those goals, such as the training programme for its managers and employees, could be reached by Talisman acting on its own. When it came to so-called GNPOC objectives, Talisman had to secure the support of its oil partners if these goals were to be achieved. And, in the case of the advocacy objectives, Talisman had to obtain the full backing of the Khartoum government if those goals were to be attained. A re-examination of the issues involving the displacement of people from around the oil concession and the uses made of the Heglig airstrip makes it clear that Talisman had little or no control over how these matters were dealt with. It could approach the Sudanese government and urge it to desist from using its airstrip for military purposes and to stop its policy of forcibly displacing people living near the oil operations, but in the end, as Talisman admitted, Khartoum alone could decide how it was going to behave. Indeed, at times Talisman argued that even to protest too loudly might be intervening in the sovereign affairs of the Sudanese government. Talisman's Corporate Social Responsibility Report in effect described how a Canadian company would go about cleaning up its operations but within the limits of its authority. What it did not and could not deal with were the broader issues of human rights violations arising from the oil operations, the use made of the oil companies' infrastructure, and the purposes for which oil revenues were employed.

The prior question, then, that needs to be addressed is whether a corporation such as Talisman should invest in the first place in a country that is deeply divided by civil war. In a fascinating passage in the Harker report, the views of southern Sudanese on the development of oil were recorded: 'many Southerners believe … that oil in Western Upper Nile is being extracted under the authority of a government which has no legitimacy; they insist that oil can be taken from the area only when the South is under the authority of a government they recognize' (Harker 2000, 13). In Sudan, therefore, where sovereignty is hotly contested and where the object being fought over has become oil itself, companies such as Talisman should be actively discouraged from investing. As Harker made clear in his final piece of advice to Talisman, if it cannot or, even worse, is unprepared to change Sudan's destructive ways, then the wisest course of action for Talisman 'at this time' is to withdraw (70). Canada's policy of constructive engagement must be replaced by one of active disengagement, as only in that way can it resolve the contradictions and ambiguities in its present relationship with Sudan.

Why Canada Backed Down

One final questions remains: How can the decision made by Ottawa not to press Talisman to withdraw from its investment in Sudan's oil operations be understood? In short, why did Lloyd Axworthy back down? One plausible answer is that after much reflection and consultation, not only with Talisman but also within Canada and with Canada's allies, especially in Europe, it became apparent that the best way to bring about positive change in Sudan was not through isolation, but through increased contact with Khartoum. This certainly is the purport of Axworthy's own explanation, offered on 14 February 2000: 'Canada continues to believe that engagement and dialogue are the most appropriate tools for advancing the cause of peace' (DFAIT 2000b). It also parallels the European

Union's bilateral dialogue with the government of Sudan, which was launched following a visit to Khartoum by a European Union delegation in November 1999 for the purpose of supporting a strengthened peace process and encouraging democratisation.

But this argument fails, because it does not take into account Axworthy's firm and apparently unwavering commitment to human security and his belief, later reaffirmed by the Harker report, that Talisman's presence in the Sudan was undermining Canada's earlier policies of promoting human rights and working toward a permanent peace in Sudan. His so-called conversion to constructive engagement was nothing more than an attempt to make the best of an embarrassing situation. Indeed, it was Axworthy's own idea, brought forward at the last moment, that Canada's Department of Foreign Affairs set up a new office in Khartoum to monitor human rights and to promote peace, an idea that was used quite effectively to deflect criticism directed at him. Having failed to get the Canadian government to take action on its own on this matter, Axworthy did try to place Sudan on the UN Security Council agenda in April 2000, but that initiative failed because there was no support for it among Canada's closest allies.[14] Constructive engagement was all that was left.

Another explanation refers to the unavailability of any legal levers for the Canadian government to use in its efforts to change the behaviour of Canadian corporations abroad (Forcese 2001).[15] The triggers that justified the use of sanctions under the *Special Economic Measures Act* have been so narrowly interpreted that this act could not be applied in the case of Talisman. In the absence of any such legislative instruments, Ottawa had little choice but to rely on moral suasion. This argument fails to consider, however, the availability of other legal instruments, such as the *Export and Import Permits Act*, which could be used to place certain exports to Sudan under scrutiny, and the Area Control List, which was applied successfully to Warsaw Pact countries during the cold war. Even though these measures would have targeted Sudan rather than Talisman, they would have made it clear, as the Harker report pointed out, 'that Canada does not regard a "business as usual" approach to be appropriate under the circumstances' (Harker 2000, 69). This argument does not explain either why the government did not set about modifying the *Special Economic Measures Act* and other sanctions legislation so that they could be used against Talisman at some later date.[16] That Ottawa took sanctions off the table was not because the legal case for using them was weak, but for reasons totally different.

In the end, what is left are essentially domestic factors to explain Axworthy's sudden reversal of policy.[17] In the first place, with the help of a public relations firm, Talisman organised and managed a lobbying campaign designed to build support for its position within the senior ranks of the Canadian civil service and in Cabinet, notably that its ability to have an impact on its oil partners and on the government of Sudan was severely limited and that it had already suffered a serious blow in the stock exchanges of North America that had resulted in such a significant reduction in its share price that it might become a takeover target. As a result, Talisman and the oil industry in Canada were able to exercise considerable influence within the governing circles of Ottawa. At the same time, there was mounting concern in the same circles that the imposition of sanctions on Talisman would set a precedent, sending out a message that Canadian companies should think twice before they invest or even trade with countries whose human rights records are abysmal.[18] In striking a balance between the interests of the corporate elite and the concerns of the 'Sudan constituency' advocating support for human security in Sudan, it seems clear that policy makers in Ottawa came down in favour of corporate interests. In effect, human security was sacrificed on the altar of private sector profit and the imperatives of a global capitalist economy.[19]

Conclusion

Talisman finally withdrew from Sudan four years after it had purchased Arakis's share in its Sudan's oil concessions. On 30 October 2002, Buckee announced the sale of Talisman's 25 percent share in GNOPC to India's state-owned Oil and Natural Gas Corporation. Certainly, Ottawa played no part in this decision. Axworthy had been the sole advocate of Talisman's disengagement from Sudan, and when he left politics, his successor, John Manley, was much more sympathetic to corporate interests. Indeed, Canada's interest and involvement in Sudan has waned considerably since then.

Even though Canada failed to take any effective action against Talisman following the Harker report, the Canadian corporation's future in Sudan was still very much in doubt. Two events occurred that challenge the legitimacy of Talisman's presence there. On 8 November 2001, a class action lawsuit was filed in United States District Court for the Southern District of New York by southern Sudanese against Talisman Energy and the government of Sudan. To quote from part of the introduction of the amended complaint, submitted on 25 February 2002, the defendants, it is alleged, 'collaborated in a joint strategy to deploy military forces in a brutal ethnic cleansing campaign against a civilian population based on their ethnicity and/or religion for the purpose of enhancing [the] Defendants' ability to explore and extract oil from areas of southern Sudan by creating a cordon sanitaire surrounding the oil concessions located there'. It goes on to argue that 'the armed campaign, which … resulted in massive civilian displacement, the burning of villages, churches and crops, and the extra judicial killing and enslavement of innocent civilians, is possible only through [the] Defendants' collaboration and the Government's utilization of infrastructure, such as roads and airfields, constructed and maintained by Talisman'.[20] Included in this complaint is a document that ties Talisman directly to the scorched-earth warfare tactics used in the Western Upper Nile concession area by the government of Sudan in May 1999, the results of which are described in vivid detail in the Harker report. In a directive originating in the Petroleum Security's central office in Khartoum and sent to its office in Heglig (Talisman's nerve centre), the armed forces of Sudan are, at 'the request of the Canadian company', directed to 'conduct cleaning up operations in all villages from Heglig to Pariang' (cited in Reeves 2002). This directive, marked 'very urgent', was issued on 7 May 1999, two days before the Ruweng offensive began. Thus, Talisman Energy, the 'Canadian company', had its 'request' successfully fulfilled in the resulting Ruweng campaign. As described in the Harker report, this operation in 'cleaning up' the area from Heglig to Pariang succeeded in clearing and securing the areas around Talisman's oil concession while at the same time inflicting unspeakable destruction and suffering on the peoples of southern Sudan. This lawsuit still remains before the court even though Talisman has sold its stake in Sudan's oil fields.

The events of 11 September 2001 and the war on terrorism have given the United States a sharpened interest in Sudan, not only in obtaining information from Khartoum about the al Qaeda network but also in strengthening regional efforts to resolve Sudan's civil war.[21] The flurry of activity emerging from the U.S. peace initiative in 2002 resulted in a number of agreements — the Machakos Protocol (July 2002), which traded off a referendum on self-determination in southern Sudan against the application of the Sharia in the north, a comprehensive ceasefire (October 2002), an addendum to that ceasefire (February 2003), and a settlement of the security issue (September 2003). But any agreement that brought an end to armed conflict in Sudan would have been welcomed by Talisman. By contrast, given the oil interests of the Bush administration, it was highly unlikely that Washington would have taken any actions harmful to Talisman. Not surprisingly, then, when the White House finally gave its approval to the *Sudan Peace Act* in October 2002, the clause that would have prevented any oil company operating in the Sudan from raising capital on U.S. stock exchanges — a clause that had given great concern to Jim Buckee — had been removed.

In the end, Talisman's own explanation remains: 'It was time to turn the page ... we felt the controversy detracted ... from the strength of our other assets' (Bloom and Nguyen 2002). In effect, the adverse publicity Talisman received from its presence in Sudan put too much pressure on its resources and its share price. The subsequent dramatic rise in Talisman shares would seem to support that view.

What is more difficult to accept is Talisman's claim that it was a force for good and, with its departure, the plight of Sudanese would further deteriorate. It is hard to imagine that their human security could get any worse, but it is probably true that things will not get any better. There is no reason to expect that India's state-owned company will be any more willing or able to moderate the government of Sudan's policies than was Talisman. Given the almost umbilical link that exists between oil operations and the continuing civil war in Sudan, the international community has a responsibility to develop hard laws to proscribe or regulate direct investment in future conflict zones.

Notes

1. The author would like to express his gratitude to Konstantin Kilibarda and Anita Krajnc for assistance in research and to Anita Krajnc, David Melvill, David Mozersky, Cranford Pratt, and Ian Spears for their very helpful comments on an earlier draft of this chapter.

2. As Canada's Minister for Foreign Affairs, Lloyd Axworthy championed human security as the central preoccupation of Canadian foreign policy (see, for example, Axworthy 1997). On the other hand, Cranford Pratt (1999), critical of 'the near-total absence of poverty reduction and equity from Axworthy's advocacy of human security', prefers instead to use the term 'humane internationalism'. See also Pratt (1989). These terms are used here interchangeably along with the more general notion of 'humanitarian concerns'.
3. It should be noted that at the time this article was published, its authors were the vice-president, legal and corporate projects, and senior advisor, corporate responsibility of Talisman, respectively.
4. Between 1970 and 1985, Canadian imports from the Sudan ranged from CA$12 000 (1985) to CA$1.03 million (1980), while Canadian exports to Sudan varied from a low of CA$1.437 million in 1970 to a high of CA$23 million in 1980. Aside from the sale of aircraft, most of Canada's exports consisted of food products, part of its humanitarian assistance to Sudan (Statistics Canada 2003a, 2003b).
5. Throughout the 1970s, Canada's annual aid package to Sudan usually fell below CA$100 000, reaching a high point in 1977 of CA$1.03 million. To put that in perspective, however, in that same year Canadian aid to Tanzania was more than CA$92 million, to Ghana almost CA$40 million, and to the Ivory Coast CA$28 million (see CIDA 1977).
6. Fen Hampson distinguishes among three different interpretations of the term 'human security', of which two seem to be particularly appropriate to the case of Sudan: the 'safety of peoples' and the 'rights of people'.
7. For a comprehensive account of the IGAD peace process, see El Affendi (2001).
8. Funding for these peace-related projects now exceeds CA$2 million.
9. See the 'Backgrounder on CIDA and the Sudan' in DFAIT (1999).
10. Statement made by James Buckee to the staff of Talisman on the occasion of the purchase of Arakis's share in the GNPOC.
11. For an account of how oil became a contentious issue in Sudan, see Lesch (1998, 48).
12. All the quotations in the following paragraph are taken from Axworthy's statement (DFAIT 1999).
13. On 6 January 2002, John Garang, Chairman of the SPLM/SPLA, and Riek Machar, Chairman of the Sudan People's Democratic Front/Defence Force, agreed to merge their two movements under the name of the SPLA, thus putting the final nail in the coffin of the Khartoum peace agreement.
14. Canada managed to raise the issue of Sudan at the UN in August of the same year. During that month the government of Sudan engaged in aerial bombing that targeted civilians and relief organisations. Drawing on an earlier resolution that focussed on the need to protect humanitarian workers, Canada intervened and asked for a briefing from the UN Secretariat on the status of humanitarian activities in the Sudan (DFAIT 2000a).
15. Craig Forcese argues strongly in favour of amending Canada's sanctions legislation so as to give the government another option in dealing with Canadian companies abroad.
16. Georgette Gagnon, Audrey Macklin, and Penelope Simons (2003, 7) recommend 'the legislated adoption of a mandatory code of conduct applicable to TNC [transnational corporation] activity in conflict zones'.
17. See Forcese (2001, 47–49) for a similar explanation.
18. It may not be accidental that Ottawa had begun plans for Team Canada's sixth mission to China.
19. Cranford Pratt's dominant class theory is particularly useful as a tool to analyse and explain the shift in Canada's relations with Sudan, as well as the relatively weak impact that the so-called Sudan constituency had on government policy. See, for example, his excellent article 'Dominant Class Theory and Canadian Foreign Policy: The Case of Counter-Consensus' (Pratt 1983/84).

20. The full text of the case against Talisman in the U.S. District Court for the Southern District of New York (Civil Action No. 01 CV 998 [AGS]) is available from the website of iAbolish at <www.iabolish.com/classaction> (April 2004).
21. For an excellent analysis of recent U.S. policy toward the Sudan civil war, see Johnson (2003, ch. 11).

References

Axworthy, Lloyd (1997). 'Canada and Human Security: The Need for Leadership'. *International Journal* vol. 52, no. 2, pp. 183–196.

Bloom, Richard and Lily Nguyen (2002). 'Talisman Shares Rise on Sudan Sale'. *The Globe and Mail*, 31 October.

Canadian Auto Workers Union, Steelworkers Humanity Fund, The Simons Foundation, et al. (2001). 'Report of an Investigation into Oil Development, Conflict, and Displacement in Western Upper Nile, Sudan'. October. <www.iabolish.com/act/camp/divestment/Rylefinal.pdf> (April 2004).

Canadian International Development Agency (1977). 'Annual Report'. Ottawa.

Canadian International Development Agency (1999). 'Canada Supports Sudan Peace Talks Secretariat'. News Release 99-39, 26 July. <www.acdi-cida.gc.ca/newsro-e.htm> (April 2004).

Chase, Steven (1999). 'Talisman Plays with Fire in Sudan'. *The Globe and Mail*, 9 October.

Cohn, Martin Regg (1997). 'Oiling the Wheels of Revolution: Two Canadian Companies Defy Rebel Threats to Deal in Sudan's Hinterland'. *The Toronto Star*, 20 April.

Cohn, Martin Regg (1998). 'Oil Gushes amid Slavery, Hunger'. *The Toronto Star*, 17 June.

Department of Foreign Affairs and International Trade (Canada) (1999). 'Canada Announces Support to Sudan Peace Process'. Press Release No. 232, 26 October. <webapps.dfait-maeci.gc.ca/minpub/default.asp> (April 2004).

Department of Foreign Affairs and International Trade (Canada) (2000a). 'Axworthy and Minna Condemn Escalation of Violence in Sudan'. Press Release No. 197, 10 August. <webapps.dfait-maeci.gc.ca/minpub/default.asp> (April 2004).

Department of Foreign Affairs and International Trade (Canada) (2000b). 'Axworthy Outlines New Initiatives to Further Peace in Sudan'. Press Release No. 26, 14 February. <webapps.dfait-maeci.gc.ca/minpub/default.asp> (April 2004).

El Affendi, Abdelwahab (2001). 'The Impasse in the IGAD Peace Process for Sudan: The Limits of Regional Peacemaking?' *African Affairs* vol. 100, no. 401, pp. 581–600.

Forcese, Craig (2001). '"Militarized Commerce" in Sudan's Oilfields: Lessons for Canadian Foreign Policy'. *Canadian Foreign Policy* vol. 8, no. 3, pp. 37–56.

Francis, Daisy and Akwatu Khenti (1994). 'Report on the 50th Session of the Commission on Human Rights'. Submitted to the Human Rights Committee of the Canadian Council of Churches.

Freeman, Linda (1997). *The Ambiguous Champion: Canada and South Africa in the Trudeau and Mulroney Years*. University of Toronto Press, Toronto.

Gagnon, Georgette, Audrey Macklin, and Penelope Simons (2003). 'Deconstructing Engagement: Corporate Self-Regulation in Conflict Zones — Implications for Human Rights and Canadian Public Policy'. Social Sciences and Humanities Research Council of Canada. <www.eldis.org/static/DOC11715.htm> (April 2004).

Hampson, Fen Osler (2002). *Madness in the Multitude: Human Security and World Disorder*. Oxford University Press, Toronto.

Harker, Peter (2000). 'Human Security in Sudan: The Report of a Canadian Assessment Mission'. January. Department of Foreign Affairs and International Trade, Ottawa. <www.reliefweb.int/library/documents/cansudan2.pdf> (April 2004).

Human Rights Watch (2001). 'World Report 2000'. <www.hrw.org/wr2k1> (April 2004).

Johnson, Douglas H. (2003). *The Root Causes of Sudan's Civil Wars.* International African Institute, Oxford.

Keen, David (1994). *The Benefits of Famine: A Political Economy of Famine and Relief in Southwestern Sudan, 1983–1989.* Princeton University Press, Princeton.

Lesch, Ann Mosely (1998). *The Sudan: Contested National Identities.* Indiana University Press, Bloomington.

MacDonald, David (1985). 'African Famine and Canada's Response'. Canadian Emergency Coordinator/African Famine, Hull.

Pratt, Cranford (1983/84). 'Dominant Class Theory and Canadian Foreign Policy: The Case of the Counter-Consensus'. *International Journal* vol. 39, no. 1, pp. 99–135.

Pratt, Cranford (1989). 'Humane Internationalism: Its Significance and Variants'. In C. Pratt, ed., *Internationalism under Strain: The North-South Policies of Canada, the Netherlands, Norway, and Sweden*, pp. 3–23. University of Toronto Press, Toronto.

Pratt, Cranford (1999). 'Competing Rationales for Canadian Development Assistance: Reducing Global Poverty, Enhancing Canadian Prosperity and Security, or Advancing Global Human Security'. *International Journal* vol. 54, no. 2, pp. 306–323.

Reeves, Eric (1999). 'Sudan at War: Canada Responds'. *The Globe and Mail*, 27 October.

Reeves, Eric (2002). 'Talisman Requested Ethnic Cleansing in Sudan'. <www.iabolish.com/news/global/2002/talisman02-27-02.htm> (April 2004).

Sheppard, Jacqueline and Reg Manhas (2000). 'Sudan Experience Offers Corporate-Responsibility Lessons, Opportunities'. *Oil and Gas Journal* vol. 98, no. 46, pp. 70–73.

Statistics Canada (2003a). 'Exports by Country'. Catalogue No. 65-003-XMB. Ottawa.

Statistics Canada (2003b). 'Imports by Country'. Catalogue No. 65-006-XMB. Ottawa.

Talisman Energy (2000). 'Corporate Social Responsibility Report 2000: Sudan Operations'. <www.talisman-energy.com/pdfs/csr2000_report.pdf> (April 2004).

United Nations (1992). 'The Situation in the Sudan'. A/RES/47/142, 18 December. <www.un.org/documents/ga/res/47/a47r142.htm> (April 2004).

United Nations (1998). 'Third Committee Concludes Consideration of Human Rights Questions, Situations'. Press Release GA/SHC/3503, 10 November. <www.un.org/News/Press/docs/1998/19981110.gash3503.html> (April 2004).

United Nations (1999). 'Social Committee Hears of Progress in Countries with Human Rights Problems, But International Involvement Still Needed'. Press Release GA/SHC/3548, 5 November. <www.un.org/News/Press/docs/1999/19991105.gashc3548.doc.html> (April 2004).

United Nations (2000). 'Third Committee Winds Up Work for Current Session with Call for Respect for Humanitarian Law, Human Rights'. Press Release GA/SHC/3624, 10 November. <www.un.org/News/Press/docs/2000/20001110.gashc3624.doc.html> (April 2004).

United Nations (2001). 'Situation of Human Rights in the Sudan'. Report No. A/56/336, 7 September.

PART IV
INTERNATIONAL INSTITUTIONS AND SOFT LAW

Chapter 13

The World Trade Organization, the North American Free Trade Agreement, and the Challenge of Sustainable Development

Thomas A. Hockin

In assessing the environmental dimension of the NAFTA regime, three points stand out. First, there are the idiosyncratic habits of mind of Canadians and Canadian foreign and trade policy makers when it comes to foreign and trade policy. These habits are either hopelessly naive and self-serving or constructive. But they were very much in evidence in a 'clinical' example of the Canadian species in Uruguay, Montreal, and the Tokyo negotiations of the General Agreement on Tariffs and Trade (GATT) and the World Trade Organization (WTO), as well as in the negotiations of the North American Free Trade Agreement (NAFTA) and its two side accords.

Second, these habits have direct effects on the Canadian approach to the WTO and to the negotiations of NAFTA as well as its side accords. But here it is important to emphasise the advance made by both NAFTA and the side accords over the WTO in regard to stepping closer to civil society concerns about the environment. Third, there is a residue of value in both accords for the WTO, as the WTO evolves and tries to avoid imploding.

Canadian Habits of Mind

The prism through which Canadians view the world is coloured by Canada's domestic experience. There are some distinctive features about Canada's domestic experience and habits of mind that affect this Canadian worldview.

Arnold Wolfers (1956) noted that, until the Vietnam War and perhaps after, English and American thought and history — compared to continental Europe's — led to a distinct philosophy of international relations in the twentieth century. Although Canada was not explicitly included in his analysis, at least one fundamental premise then relevant to the United States and Britain still applies to Canada. The experience of all three nations stood in vivid contrast to Europe's experience. Geographical separation from the European continent — for Britain, the Channel, for the U.S., and Canada,

4000 kilometres of Atlantic Ocean — led to what can best be called a 'voluntarist' tradition in each country (Hockin 1971). This is the tradition that helped impel the search for moral opportunity in international politics.

On the continent of Europe, the propinquity of nations and centuries of suspicion had made sheer survival the major goal of diplomatic activity. Even today, looking from afar at Europe (before the European Community was founded), or the Balkans or the Middle East, Canadians are generally prone to be troubled and impatient with their Hobbesian dilemmas. Canadians have been more attracted by the opportunities to forge moral principles into international relations, and to try to elevate these relations above the unsavoury business of balance-of-power or terrorist politics. For Canadians, it remains not unnatural to suspect that the usual cause of international conflict is that other unruly nations simply transpose their own internal ways of doing things — their internal tyranny, violence, and revolutionary upheaval — into the international arena.

The Canadian antidote then is clear: the rules of the international game must be changed to elicit more salutary modes of international behaviour. Given this outlook, Canadians may often be inclined to see in military force a tool of the foreign disease, a manifestation of the barren and sorry condition of balance-of-power politics. For example, in entering into negotiations for the establishment of the North Atlantic Treaty Organization (NATO) in 1948, Canadian prime minister Louis Saint Laurent regretfully acknowledged that communist 'aggression' could be assuredly deterred only by brute military force. Yet the element most interesting about Canada's NATO initiative, however, was its concomitant insistence through Article 2 that NATO should be 'no mere military alliance' but an economic and civilising alliance (Reid 1992). Canadian policy makers since World War I have never believed, naively or not, that a Hobbesian balance of military power is the only possible international system.

However, Canada has a Janus face. One face is what might be called 'English Canadian United Church' voluntarism. On issues close to home, it is the 'Scots banker' tradition that shows its face. On matters of economics, laws of the sea, and, above all, Canadian-American questions, Canada's intention seems to be the exploitation of the Hobbesian system for national advantage or at least national protection.

On political and military problems in many other international constellations outside of North America, Canada's foreign policy from 1945 to the present has been distinguished by its continual, remarkably patient, voluntarist plodding search for something to supplement, even transform, balance-of-power politics. This 'something' includes the United Nations, disarmament meetings, trade and aid commissions, peacekeeping activities, the International Control Commission, the Commonwealth, and even NATO. Both Liberal and Conservative governments have a record of considerable voluntarism, be it Mulroney or Diefenbaker Progressive Conservative governments' initiative for South Africa or the Chrétien Liberal government's action to create the landmine treaty.

To be tough about problems close at hand and voluntarist about others further away is to hold a bifurcated view of world affairs. Yet this voluntarist streak still seems to be the foundation of most of what has been creative in Canadian foreign and trade policy. It is based on a touching faith in the possible results of sensitivity, goodwill, good intentions, another advisory committee, good offices, organisational creativity, and hard work. If there is a Canadian prescription for the international system, it is to inculcate each action in the system with a similar voluntarist faith. Voluntarism's content can be considered illusory, naive, or hypocritical and its exposition as trite, but it should not be summarily dismissed. It is an approach that finds exponents in the Liberal, Conservative, and New Democratic parties, and it has survived a great number of shocks and disillusionments.

The Canadian voluntarist worldview is supplemented by Canada's federalist challenges. These have been met with those expressions of good intentions, postponed timetables, further meetings, purposeful ambiguity, good offices, advisory committees, and all sorts of creative schemes to hold the country together. They include anything to diffuse stark choices by adding more actors and bringing more people to the table. These Canadian political federalist finesses are then transposed to the international arena by Canadians. These proclivities are even more evident in their contrast with the more Hobbesian private sector financial world.

This patient emphasis on tactics and schemes for organisation maintenance flows from Canada's frequently tedious and convoluted federal system, its tensions, and the pat responses of most federal government for another committee, another study, another meeting. The U.S. has task forces; Canadians have royal commissions and federal-provincial consultations.

Effects on Canada's Approach to Trade Liberalisation

This voluntarist and federalist style in Canadian foreign policy was evident in work of Brian Mulroney's government in trade negotiations during its second mandate, including the negotiations that led to the creation of the WTO as part of the culmination of the Uruguay Round. The Mulroney government worked hard to help create the Cairns Group of countries to press the Americans and Europeans to control farm subsidies. This approach helped Canadians defuse American influence without angering the U.S. Adding more players to a committee is a great Canadian tradition in global matters. Organisational co-ordination is another Canadian habit. In a quintessentially Canadian judgement, Sylvia Ostry lamented how disappointed she was that the working groups on the functioning of the GATT system (FOGS), meant to co-ordinate policies among the GATT, the World Bank, and the International Monetary Fund (IMF), had not made much progress. In the Montreal phase of the GATT negotiations, John Crosbie championed the idea of creating more than another GATT agreement but a genuine 'world trade organisation'. This instinct may indeed be useful. It is certainly very Canadian.

The triumph of NAFTA was to give Canada an extra option against a protectionist U.S. Congress and U.S. administration by adding Mexico to the bilateral Canada–U.S.

Free Trade Agreement (FTA) and by getting panel review of administrative determinations. As the Japanese trade minister put it some years ago: 'So Canada got equality with the U.S. on disputes. The U.S. is ten times richer than Canada and you got equality. I would like to get that for Japan.' In NAFTA, much room remains for American litigiousness. But much room also remains for Canadian federalist attempts to diffuse sheer U.S. economic power with panel reviews, expert committees, more advice from scientists, and similar devices.

The North American Free Trade Agreement

Equally important is the moral opportunity for environmental protection in the original NAFTA under President George H. Bush, Prime Minister Brian Mulroney, and President Carlos Salinas of Mexico, even before the side agreements were added. This moral opportunity was not Canada's unique contribution. At a minimum, Canada strongly supported the entrance of this opportunity. Here the original NAFTA dealt with four central environmental concerns in innovative ways. The first concern was the enforcement of international environment agreements, some of which contained trade provisions that may be inconsistent with the contracting party obligations under the GATT. The second issue was the fear that a state's trade obligations may prevent it from applying strict environmental standards to imported goods or vice versa. The third issue arose from the fear that significant disparities in standards and enforcement could lead to the creation of pollution havens. The fourth issue related to concerns that international trade dispute settlement panels may lack the expertise to address environmental issues adequately.

As Christopher Thomas (1998) noted, unlike the GATT and FTA, which contained scant reference to the environment, NAFTA expressly recognised environmental objectives and issues. Here the Mulroney government started by setting in place possible 'intentional' soft law guidance in the preamble. This preamble expresses three environmental intentions, among 15 listed overall, for entering into the agreement:

1. to protect and conserve the environment,
2. to promote sustainable development, and
3. to strengthen the development and enforcement of environmental laws and regulations.

Where the text is unclear in particular instances, the three intentions are to provide interpretative guidance in NAFTA dispute settlement panels. Canadians love 'indicative momentum-building premises'. They just might develop into something with bite.

NAFTA also has provisions dealing with its relationship to various international environmental and conservation agreements. In this set of disparate — and generally unrelated — provisions, the Mulroney government chose not to detail a comprehensive set of rights and obligations that exhaustively defined the relationship between trade and environmental regulation. Rather, the terms addressed specific trade-related and intellectual property and investment-related environmental matters. Here Canadian pragmatism prevailed, as perhaps American and Mexican pragmatism did as well.

Article 104 of NAFTA outlines the pre–side accord question of enforcement. Neither the GATT nor the FTA expressly states its relation to international environmental agreements. Article 104 is intended to deal with a state that might unilaterally prescribe standards that the environmental policies of another state must meet and restrict access to its market for goods exported from the non-complying state. This concern does not arise where states freely negotiate the right to impose trade restrictions in specified circumstances. This could be seen as an invitation for eventual side accords on the environment to advance progress on this problem.

As to the concern that national standards that are higher than international standards may be susceptible to challenge as unnecessary obstacles to international trade, the NAFTA text expressly states: 'A level of sanitary or phytosanitary protection different from that which would be achieved by a measure based on a relevant international standard, guideline or recommendation shall not for that reason alone be presumed to be inconsistent with this Section' (NAFTA 1994, art. 713[2]).

With respect to technical standards, the NAFTA chapters were derived from the work done in the Uruguay Round. The provisions in chapters 7 and 9 were similar to those in the Uruguay Round's Draft Agreement on Technical Barriers to Trade and the Draft Decision by Contracting Parties on the Application of Sanitary and Phytosanitary Measures. The requirement for a scientific basis for such measures — another Canadian proclivity — is emphasised throughout the chapters.

The concern that disparate levels of environmental standards and enforcement may affect firms' competitiveness and, consequently, international trade and investment flows is dealt with in Article 1114 on Environmental Measures (NAFTA 1994). It states the following.

1. Nothing in this Chapter shall be construed to prevent a Party from adopting, maintaining or enforcing any measure otherwise consistent with this Chapter that it considers appropriate to ensure that investment activity in its territory is undertaken in a manner sensitive to environmental concerns.
2. The Parties recognize that it is inappropriate to encourage investment by relaxing domestic health, safety or environmental measures. Accordingly, a Party should not waive or otherwise derogate from, or offer to waive or otherwise derogate from, such measures as an encouragement for the establishment, acquisition, expansion or retention in its territory of an investment of an investor. If a Party considers that another Party has offered such an encouragement, it may request consultations with the other Party and the two Parties shall consult with a view to avoiding any such encouragement. These are 'pure moral imperative' intentional clauses, at least on behalf of the environment and safety. This expression was much broader than that of the Uruguay Round up to that time. The Uruguay

Round's Draft Agreement on Trade-Related Aspects of Investment Measures was concerned only with a narrow range of investment matters.

The spectre of a state relaxing its environmental standards to attract or retain investment evoked a quintessentially Canadian approach. In a case of 'alleged relaxation' there is a requirement for 'consultation'. But the article noted above was seen as a significant step in recognising at least the link between a disparity in national laws and investment flows (and, indirectly, trade flows). The NAFTA text itself did not go further, even though the subsequent side agreement did.

If an environmental issue of consideration is only peripheral to such rights and obligations, it may not, therefore, be relevant to the resolution of the dispute. This was illustrated in the GATT Panel Report (1987) — 'Taxes on Petroleum and Certain Imported Substances' — where the panel found that a U.S. pollution cleanup tax imposed on certain imported substances was a valid border tax adjustment. Moreover, in the first dispute arising under the FTA, Canada successfully invoked its rights and obligations as a coastal state under the Law of the Sea when defending its right to maintain a catch/landing requirement for Pacific Coast salmon in 1990. That was background for the work on NAFTA. In that instance, the panel was prepared to construe a party's trade law rights and obligations in the light of other international legal rights and obligations.

A number of other provisions address environmental concerns, such as NAFTA's Article 1106. Moreover, the chapter on intellectual property provides, in certain circumstances, that parties may exclude inventions from patentability if such an exclusion is necessary to protect human, animal, or plant life or health or to avoid serious prejudice to nature or the environment. Thus, NAFTA itself departs from the GATT and the FTA in making environmental considerations expressly relevant to a range of rights and obligations.

The North American Agreement on Environmental Cooperation

The negotiation of the NAFTA side agreements on environment and labour was prompted by a conditional endorsement of the NAFTA by then Governor Bill Clinton during a presidential campaign speech in October 1992. After extensive negotiations, on 13 August 1993, the governments of Canada, the U.S., and Mexico announced that they had concluded the North American Agreement on Environmental Cooperation (NAAEC). On 14 September 1993, it was released to the public.[1]

This side agreement is concerned, above all, with the integrity of domestic environmental law enforcement, which was, in the Canadian government's view, a natural follow-on to Article 1114 of NAFTA. Protection is to be accomplished through commitments relating to transparency, to private access to domestic remedies, and through tri-national disciplines on certain trade-related instances of 'persistent patterns' of non-enforcement of environmental law.

The guiding body for the Commission on Environmental Cooperation (CEC) is the Council, made up of the three countries' relevant ministers or their representatives. It is expected to operate by consensus and to develop recommendations on elaborating on the agreement and supervise the secretariat and its budget.

Article 2 sets out a number of general commitments whereby each party agrees to take certain actions with respect to its territory, to consider implementing any recommendation by the Council regarding the appropriate limits for specific pollutants, and to consider prohibiting the export to the territories of the other parties of a pesticide or toxic substance whose use is prohibited within the party's territory.

In spite of all this, as the Canadian government entered the negotiations in the final weeks in August 1993, there was a political concern that environmental groups might describe the accord as simply freezing the *status quo* and nothing more. However, Canadian and other ministers had discussed this concern with environmental groups and others. The result was Article 3, which established an important principle with respect to levels of protection (CEC 1993). It states:

> Recognizing the right of each Party to establish its own levels of domestic environmental protection ... each Party shall ensure that its laws and regulations provide for high levels of environmental protection and shall strive to continue to improve those laws and regulations.

This principle is key. It is expressed in this way so as to retain each party's ability to make sovereign decisions respecting the nature and content of its environmental law. At the same time, no party is free to reduce the efficacy of its environmental law, because it is internationally obliged, regardless of the policy choices it has made, to ensure that it provides for high levels of environmental protection. Moreover, the parties are bound by a similar obligation in NAFTA Article 906(1), which states that 'the Parties shall ... work jointly to enhance the ... protection of ... the environment'.

As in the original text of NAFTA, in light of the growth in complexity of regulation in the environment, the three North American countries established certain transparency requirements in the side agreement. They sought to ensure that the law-making process in each country was as transparent as possible. They thus agreed that each party would promptly publish its laws, regulations, procedures, and administrative rulings relating to the Agreement. Each party is required to publish, in advance, any measure it proposes to adopt, and to provide a reasonable opportunity for interested people, including the parties, to comment on it. Each party is also obliged to notify another interested party of any proposed environmental measure that might materially affect the operation of the agreement.

This soft law of transparency is what the ministers of all three NAFTA countries liked to call 'the power of sunlight to act as a disinfectant'. The world of investment regulation requires disclosure of its producers of mutual fund products, not just on regulation but also required disclosure on everything about the products such as fees objectives. There is a similarity here to the side accords. Moreover, in matters of trade and customs, transparency obligations governing the publication of laws, regulations, and administrative rulings have long been complemented by an obligation to administer such laws, regulations, and procedures uniformly, impartially, and reasonably and to provide for the prompt review of administrative rulings. As Thomas (1998) observed when the side accords were negotiated, 'This latter obligation is increasingly being extended into new [non–trade/customs] areas of governmental action'.

Beyond transparency, the hard law bite intended in the new obligations relates to the enforcement of certain domestic laws. With the aim of achieving high levels of environmental protection and compliance with its environmental law, each party is obliged under Article 5 to enforce its environmental law effectively through appropriate governmental action. This obligation is not just vague soft law. Although undefined, the phrase 'appropriate governmental action' is elaborated upon in an illustrative list (CEC 1993, art. 5[1]). Indeed, minimum standards are specified with respect to private access to remedies.

A visceral concern fanned by Ross Perot and others in the 1992 U.S. presidential election was transboundary pollution. Explicitly, the NAFTA negotiators provided that the Council consider and develop recommendations on the provision by a party, on a reciprocal basis, of rights and remedies before its courts and administrative agencies for persons in another party's territory who have suffered damage or injury caused by pollution originating in its territory, as if the damage or injury were suffered in its territory.

Those who elaborated these advances in transparency, enforcement, and ameliorative steps in transborder pollution in the accord can be proud of their work. Some of this edifice can indeed serve as something of a model for other bilateral or multilateral agreements, even if they may be premature at present for the delicate WTO, given the scant administrative and governmental apparatus in so many developing countries.

With respect to dispute resolution, the NAAEC distinguishes between 'persistent patterns' of failure by a party to enforce its environmental law effectively (to which the comprehensive dispute settlement mechanism in Part Five of the Agreement applies) and all other matters that can form the subject matter of a dispute. There was a debate among the three governments on how to find the correct words for a serial 'neglect to enforce' but not to open a Pandora's box of excessive numbers of cases, some of them frivolous. They settled on the threshold term of a 'persistent failure to enforce'. It is a bit vague, but also a bit novel.

The negotiators were also bedevilled by the key element of any functioning legal system: to protect a government's discretion to prosecute or not. This is a key element of a nation's sovereignty. It was addressed in Article 45, through an essentially negative definition, by requiring a reasonable exercise of discretion and *bona fide* discretion. The definition of environmental law did not attempt to cover codes of conduct or non-binding guidelines. One can legitimately ask if is this too soft. Should the law have been harder?

Another typical Canadian proclivity — setting up advisory committees — was expressed in abundance in the NAAEC, in the ample provisions to increase public participation in the agreement. The Joint Public Advisory Committee (JPAC) created by the NAAEC consisted of 15 members. Its purpose was to provide advice to the Council on any matter within the scope of the agreement and on its implementation and further elaboration. A cynic would say that in order to co-opt some voices gently to be part of the process, they should be put on an advisory committee, not an executive one. This is a very Canadian practice, both in the public sector and in corporate boardrooms. It is not all cynical, for it invites novel voices inside.

The NAAEC dispute settlement process under Article 14-15 for citizens' submissions is initiated by a request from a party for consultations regarding an alleged persistent pattern of failure to enforce domestic environmental law (Tollefson 2002; Alanis Ortega 2002; Wilson 2002). Subject to the Council's rules of procedure, a third party that considers it has a substantial interest in a matter may participate in the consultations. Here there is a blend of Canadian soft law good intentions, accelerated by U.S. tendencies to open several stages for lawyers and litigants.

There is also a tediously patient Canadian tendency to avoid closure and allow for several second or third or fourth chances to settle an argument. If the parties fail to resolve the matter within 60 days of the request for consultations, a special session of the Council may be requested by any one of the disputing parties. It may call on technical advisors, have recourse to good offices, conciliation, mediation, or other dispute resolution procedures, and make recommendations. If the Council is unable to resolve the matter, it may, by a two-thirds vote on the written request of any of the consulting parties, convene an arbitral panel. It may do so only where the alleged persistent pattern of failure to enforce environmental law relates to a situation involving workplaces, firms, companies, or sectors that produce goods or produce services traded between the territories of the parties. Where the panel makes an affirmative determination, it must include any recommendations for the resolution of the dispute. If there still is no compliance, a monetary assessment is to be applied.

If that is not paid, then trade sanctions can be used between the United States and Mexico. From the start, it was clear that this reference to trade sanctions was overkill, dangerous for the U.S. and Mexico, and totally unacceptable for Canada. It was inserted into the agreement because the U.S. Democratic Party felt it was the necessary 'red meat' to feed the protectionist U.S. Congress. However, because a failure to pay a fine would be tantamount to abrogating the whole agreement, this makes the insertion of trade sanctions even sillier, given the obvious manageability of any monetary assessment.

There was considerable debate among the three delegations on the proper proportionality for these fines, given the enormity of the U.S. gross domestic product compared to Mexico's. Annex 34 of the NAAEC sets out the requirements for the imposition of a monetary assessment. It was decided during the initial year that the side agreement was in force, such assessments are limited to no more than US$20 million or its equivalent in the currency of the party complained against. In subsequent years, assessments are limited to a maximum of 0.007 percent of the total trade in goods between the parties during the most recent year for which data are available. Monetary assessments are to be paid into a fund and spent, under the direction of the Council, to improve or enhance the environment or environmental law enforcement in the party complained against.

There is another way in which the NAAEC could be a model in these times of civil society protests against freer trade and alienation between free trade advocates and environmental, labour, and other groups. These are its provisions for significant public participation and transparency. With respect to public participation, the agreement allows the Council to seek advice from nongovernmental organisations (NGOs) or individuals, and for the Secretariat to draw on information submitted by NGOs or individuals in preparing a report or a factual record. This will never satisfy everyone. But it does speak to the cry from well-meaning protestors to be able to get inside the discussion somehow. The Mulroney government never received much attention for this, because of its headline-getting rejection of trade sanctions for Canada.

There is one aspect of the accord that is 'too Canadianised', although the Mexicans probably have also been cautious on this aspect. This is in the well-meaning but innumerable steps to follow if parties fail to resolve a matter. The process is tortuously slow: 60 days to consult, another special session of the Council, a call on technical advisors, 60 days again, then an arbitral panel, 180 days for its report, 30 days for comments, then an action plan. This process is too lengthy, too Canadian sensitive, too American litigious, too Mexican *mañana*.

But these cavils aside, the NAAEC is meant to address the integrity of domestic environmental law enforcement. The negotiators saw it as a complement to the trade-related environmental provisions in NAFTA by addressing environmental law issues that may have an effect on international trade and investment. This was to be accomplished through commitments related to levels of environmental protection, to transparency, and to private access to domestic instances of persistent patterns of non-enforcement of environmental law. The negotiators felt that an important institutional precedent was set by the establishment of an independent tri-national Secretariat. In addition to its watchdog and reporting functions, the ability of the Secretariat to receive submissions from private persons substantially enhances the agreement's enforcement regime.

Implications for the World Trade Organization

In evaluating the NAAEC architecture, it is important to recall that good multinational structures are not necessarily lacking if their elaborate provisions are seldom pushed to final arbitration. They can be justified because they encourage settlement and enforcement at early stages.

With this reminder, one can assess the value of this NAFTA-NAAEC model for the WTO and its future trade-environment regime. Here it is important to note how fragile the WTO really is (Hockin 2001). Charles Carlisle, former deputy negotiator for the U.S. on WTO matters, has written that 'the Uruguay Round was not about free trade but freer trade, yet it almost failed' (Carlisle 1996). Similar comments were made to the G7 heads of government in Tokyo in 1993, with the additional comments that the amount of work, understanding, nuances, and emerging soft law alone made it essential to ratify the Uruguay Round, in order to prevent a worldwide psychological and then financial collapse. Those in the investment community know how disastrous for world financial markets the WTO's collapse would have been. China's accession, while necessary, could be a further burden. And the dispute settlement mechanism, although central to the WTO's credibility, has produced highly unsatisfying and confidence-destroying responses from many of the losers to panel decisions.

It is also vital to add momentum to the ministerial meetings such as that in Doha in 2001. After the quadrilateral trade ministers' meeting on the eve of the Tokyo G7 Summit in 1993, the heads of government would not have been able to take credit for trade or show leadership on the issue at their summit if the four trade ministers had not been able to agree on some new 'zero for zero' tariffs and other advances to the Uruguay Round to add momentum to the often immobile negotiations. An agreement was reached at four in the morning the day before the Summit opened.

The challenges in November 2001 at Doha and in September 1986 in Punta Del Este, Uruguay, are strikingly similar. All GATT ministerial meetings come at unpropitious times. In Punta, it rained and blew outside all week. Equally gloomy was the atmosphere inside the room where the negotiations took place. This room was the resort's casino — an appropriate location to launch a round of multilateral trade negotiations. Nothing was to be agreed until all was agreed. Incremental advances on services, intellectual property, and tariffs were all hostage to finding an agreement on agriculture. Cabinet ministers such as New Zealand's Mike Moore, Australia's John Dawkins, and Canada's Thomas Hockin, in the Cairns Group wanted a commitment to reduce farm subsidies to zero over a period of years. So did the U.S.

The same agriculture issue loomed large over the Qatar meeting. This time it was overlaid by the chasm of developing country versus developed country, especially on pharmaceuticals. Behind lay the spectre of all the civil society protests and the concerns raised by these actors.

Momentum clearly started at the fourth WTO ministerial at Doha. On 14 November 2001 — the last day — members announced that negotiations would commence immediately on

> i) the relationship between existing WTO rules and specific trade obligations set out in multilateral environmental agreements (MEAs)...;
> ii) procedures for regular information exchange between MEA Secretariats and the relevant WTO committees, and the criteria for the granting of observer status;
> iii) the reduction or, as appropriate, elimination of tariff and non-tariff barriers to environmental goods and services (WTO 2001).

This task given to the WTO's Committee on Trade and Environment (CTE) might do well to start by examining the past and potential of NAFTA's side accord on the environment.

Conclusion

This analysis leads to three conclusions. First, it is vital for the side accords to work and to be elaborated in North America. Second, for all of its Canadian 'habits of mind' and idiosyncrasies, if trade sanctions were removed, the NAAEC would remain a possible model for other bilateral or multilateral potential trade agreements.

Third, there is considerable research that suggests how it can be improved (Kirton and Maclaren 2002; Rugman, Kirton, and Soloway 1999). Certainly, it is too early to apply or append both the side accords to the WTO. But it is time to ask if and how the original NAFTA environmental undertakings could be added to the WTO. As for the NAAEC side accord, one could examine the value of adding the increased transparency obligations at a minimum as the price of continued membership for developed countries in the WTO. The positioning of the environment with trade in the original NAFTA trade agreement before the side accords is worthy. Such an incorporation merits exploring and perhaps pursuing by the WTO. A commitment to do so might not only speak to the constructive protestors but also add purpose to the WTO agenda. Adding some insights from the NAAEC side accords could do so as well.

The year before the Doha WTO ministerial meeting in Qatar opened in November 2001, Sylvia Ostry elegantly outlined its predicament. Citing the uncertainty principle of Werner Heisenberg, who won the Nobel Prize in 1932, Ostry offered the following paraphrase: 'You know where you are but not where you are going, or you know where you are going but not where you are.' That predicament remained as the Doha meeting opened. And even well after its successful conclusion, it still describes very well the path the world trade system faces in the years ahead.

Note

1. Although the author was not directly involved in the environmental part of NAFTA until 1993, he had observed what had gone before in his work in Uruguay and Montreal, working on the agriculture and financial services part of the Uruguay Round and NAFTA negotiations. The side accords, however, engaged him, and the Canadian Ministry of International Trade, in the environmental and labour files in a central way.

References

Alanis Ortega, Gustavo (2002). 'Public Participation within NAFTA's Environmental Agreement: The Mexican Experience'. In J. J. Kirton and V. W. Maclaren, eds., *Linking Trade, Environment, and Social Cohesion: NAFTA Experiences, Global Challenges*, pp. 183–186. Ashgate, Aldershot.

Carlisle, Charles (1996). 'Is the World Ready for Free Trade?' *Foreign Affairs* vol. 75, no. 6, pp. 113–127.

GATT Panel (1987). 'United States – Taxes on Petroleum and Certain Imported Substances'. Document symbol L/6175.

Hockin, Thomas A. (1971). 'The Voluntarist Tradition in Canadian Foreign Policy'. In S. Clarkson, ed., *An Independent Foreign Policy for Canada*. McClelland and Stewart, Toronto.

Hockin, Thomas A. (2001). *The American Nightmare: Trade Politics after Seattle*. Lexington Books, Lanham.

Kirton, John J. and Virginia W. Maclaren, eds. (2002). *Linking Trade, Environment, and Social Cohesion: NAFTA Experiences, Global Challenges*. Ashgate, Aldershot.

North American Commission for Environmental Cooperation (1993). 'North American Agreement on Environmental Cooperation'. <www.cec.org/pubs_info_resources/law_treat_agree/naaec/index.cfm> (April 2004).

North American Free Trade Agreement (1994). 'North American Free Trade Agreement'. <www.nafta-sec-alena.org> (April 2004).

Reid, Escott (1992). 'Forming the North Atlantic Alliance, 1949'. In D. Munton and J. J. Kirton, eds., *Canadian Foreign Policy: Selected Cases*, pp. 27–41. Prentice Hall, Scarborough.

Rugman, Alan M., John J. Kirton, and Julie A. Soloway (1999). *Environmental Regulations and Corporate Strategy: A NAFTA Perspective*. Oxford University Press, Oxford.

Thomas, Christopher (1998). 'NAFTA Speaking Notes on Trade and the Environment'.

Tollefson, Chris (2002). 'Stormy Weather: The Recent History of the Citizen Submission Process of the North American Agreement on Environmental Cooperation'. In J. J. Kirton and V. W. Maclaren, eds., *Linking Trade, Environment, and Social Cohesion: NAFTA Experiences, Global Challenges*, pp. 153–182. Ashgate, Aldershot.

Wilson, Serena (2002). 'Article 14-15 of the North American Agreement on Environmental Cooperation: Intent of the Founders'. In J. J. Kirton and V. W. Maclaren, eds., *Linking Trade, Environment, and Social Cohesion: NAFTA Experiences, Global Challenges*, pp. 187–193. Ashgate, Aldershot.

Wolfers, Arnold (1956). *The Anglo-American Tradition in Foreign Affairs*. Yale University Press, New Haven.

World Trade Organization (2001). 'Ministerial Declaration'. Doha, 14 November. <www.wto.org> (April 2004).

Chapter 14

Integrating Environment and Labour into the World Trade Organization

Roy MacLaren

The first time that trade was linked overtly with labour and the environment was in the 1994 side agreements to the North American Free Trade Agreement (NAFTA). Those two side agreements were introduced late in the negotiations, when U.S. president Bill Clinton came into office. In part in response to Democratic congressmen and women and, more precisely, at the behest of the American Federation of Labor and Congress of Industrial Organizations (AFL-CIO), Clinton had already promised during the presidential campaign that the two subjects would be explicitly injected into the trade agreement, as a condition of support. This was in spite of the fact that, at least in the era of the General Agreement on Tariffs and Trade (GATT), labour and the environment had been considered extraneous to the formulation of international trade policy.

As the price of hard-won congressional support for NAFTA as a whole, the Clinton administration advocated the two side agreements as preventing 'eco-dumping' or, put another way, as preventing the creation of pollution havens for footloose U.S. corporations with no sense of social responsibility. Equally, corporations seeking to exploit low-wage Mexican workers further — and this was very much in the mind of the AFL-CIO — would be discouraged from doing so. The Clinton administration made it abundantly clear to Canada and Mexico that it required the side agreements if it was to win congressional approval for NAFTA — despite Republic Party opposition. In the event, they presented no practical challenge either to Canada or to Mexico, the former already applying generally higher standards and the latter being able to point to its own regulations — not, alas, consistently applied — as similar or identical to the proposed NAFTA regulations. The rub was, however, at least for trade policy purists, the acceptance in principle that environmental and labour rules had any place in a trade agreement. That principle was ultimately accepted even by developing country Mexico, fully aware that it had become the American price — or at least the price of the Democratic Party and the AFL-CIO — for the coveted NAFTA itself. For Canada, once Mexico had acceded, the choice had become a stark one: either accept the side agreements or reject NAFTA. Canada took them.

The acceptance of the principle that environmental and labour standards have a place in international trade agreements — at least among contiguous states — did not mean that Mexico and Canada, and indeed the U.S. Republican Party, had accepted that they had any place in global trade policy or in any pending multilateral agreements involving non-contiguous states. This dichotomy in policy did not arise from observing the NAFTA side agreements in practice — an interesting subject in itself but beyond the scope covered here. It derived from genuine misgivings on the part of Canada and Mexico, in time shared by most developing countries, that the introduction into multilateral trade agreements of labour and environmental standards, including sanctions, would supply a new weapon to the latent, and not so latent, protectionists in developed countries.

Developing country opponents of hard rules certainly smelled a rat in all this. If a developed country wanted to prevent a low-cost import from a developing country from entering its domestic market, especially now that more traditional trade barriers and quotas were being reduced or even eliminated, a ready new tool would be at hand: 'We do not want your low-cost textiles or garments, so we shall claim that they were not produced in accordance with international environmental or labour standards. Your lower wage rates threaten jobs at home, but under the World Trade Organization (WTO) rules we cannot exclude your garments simply because you can pay your workers a dollar a day. But if we set international environmental and labour standards, enforceable by sanctions, we may have an additional way of keeping your products out of our markets.' Developing countries, as a whole, responded with a flat no to all such thinking, adding that any such rules would not only be exploitable by protectionists, but would also be a major invasion of national sovereignty. Trade policy is increasingly intrusive of sovereignty, given that trade is no longer regulated at the border but is increasingly 'within borders', involving domestic policies. For an international organisation, such as the WTO, to pass judgement on environmental and labour practices would be too much. And to propose such measures at the same time as proclaiming the WTO to be more sensitive to the special needs of the least developed countries would be wonderfully incongruous, if not ironic.

The major differences, largely although not exclusively between developed and developing countries, over trade and labour and the environment remain far from resolution. The major players may from time to time modify their positions — in Washington a Republican administration is not a Democratic administration, and in Berlin there is a coalition government with Greens — but since its inception less than two years after the Rio Earth Summit, the WTO has become a cockpit for continuing contention. Before the WTO ministerial meeting in Singapore in 1996, the issues of both labour and environmental standards — which many feared would break up the meeting — were resoundingly rejected by developing countries as subjects that belonged elsewhere in the galaxy of international organisations. So pronounced was their opposition that the meeting concluded with nothing more than a vague consensus to relegate the question of labour standards to the International Labour Organization (ILO). There was, however, no specialised agency of the United Nations at hand to which environmental questions could be quietly assigned. Despite a myriad of international environmental agreements, neither the United Nations Environment

Programme (UNEP) nor the WTO's newly created Committee on Trade and Environment (CTE) were regarded by the environmental advocates as any sort of answer to their concerns. As was widely expected, the Clinton administration went to the 'WTO ministerial meeting in Seattle in 1999 determined to continue to push the issues of labour and the environment, but the fiasco that the meeting soon became prevented any discussion, although there was certainly ample on the streets outside.

Given the *opera buffa* of Seattle and the uncertain handling of the two issues by the Director General of the WTO, the two issues will not simply fade away. The basic question abides: Can a multilateral trade organisation serve as a mechanism to impose sanctions for the development, monitoring, and enforcement of internationally agreed standards? The advocates, almost entirely in developed countries, continue to answer an awkward yes. They welcome the proliferation of international organisations active in some aspect or other of the environment, but continue to deplore the absence of a centre of gravity able to offer a coherent and hard law response to the challenge of fostering sustainable development. The United Nations Conference on Environment and Development (UNCED) did not produce environmental governance at Rio, and Johannesburg was not promising. UNEP never had any bite, remaining an entirely voluntary organisation. The advocates continue to see WTO as a *rara avis* among international organisations that could be shaped so as to be capable of monitoring judging and, as necessary, endorsing sanctions. No specialised agency has, in their view, that capacity: the World Health Organization cannot, even in the area of the impact on health of environmental degradation; the ILO cannot, even in the area of the impact on the workplace; the World Meteorological Organization cannot, even despite the deleterious effects of global warming; the International Maritime Organization cannot, even in spite of pollution at sea. But the WTO, with its unique dispute settlement mechanism, might just do it.

The genuine advocates of sustainable development on an international plane — not those who seek a new protectionist tool — will have to look elsewhere for more progress despite the growing ecological deficit from population pressure. Possibly the World Summit on Sustainable Development at Johannesburg in 2002 will have proven to be a step in an incremental approach, paralleling the incremental approach to trade liberalisation itself. But no big bang is likely, either on the environmental front or on the trade liberalisation front. Hostility to the idea of the WTO engaging in environmental policing will, in fact, likely grow, as the role of the developing countries in the organisation grows, especially if China makes itself the champion for those who increasingly contend that the WTO (and the GATT before it) does little enough for the poorer members, now easily the majority of its 140-plus members.

There remain wide divergences on the treatment of environmental issues in relation to market access and rules-related initiatives. There is no consensus to clarify the relationship between economic measures, however vague, in various multilateral environmental agreements and WTO rules. Nor is there any consensus that product labelling or the non-protectionist application of precaution must be defined in the

WTO. Discussions in the context of the Doha round of the WTO negotiations sidestep all such issues. All that was agreed upon in the Doha mandate was that the CTE should study specified issues and report to the next ministerial meeting on whether to attempt to negotiate environmental rules. The language used is significant. The draft did not call upon the CTE to negotiate rules, but simply to recommend to the next ministerial conference whether environmental rules should be discussed in the WTO.

In short, the effort to impose on the WTO the means to resolve problems with the environment, labour, and even human rights must be seen as impracticable, best left to be dealt with by other international organisations, however inadequate they currently are, through policy measures other than trade, and by the gradual evolution of soft law. Consensus remains the cornerstone of the WTO architecture; sanctions for non-trade purposes would weaken its non-discriminatory provisions and could contravene co-operative efforts. The answer here, as elsewhere, is greater development assistance and technology transfer to the poor countries so as to enable them to participate more in the benefits of rules-based trade. The less poverty stricken one is, the more room there will likely be to apply and enforce domestic environmental and labour regulations, frequently lofty on paper but, lamentably, seldom enforced.

Chapter 15

The Future of the World Trading System: Beyond Doha

Sylvia Ostry

The future of the world trading system is shrouded in fog even though a 'round' was launched at the ministerial meeting of the World Trade Organization (WTO) in Doha in November 2001. (The term 'round' was not used because it has become politically incorrect in some WTO circles.) In order to understand why this is so, it is essential to understand the current context of the trade policy process and the implication of the transformative changes of the 1990s. Three aspects of the new and evolving trade policy template are discussed in this chapter: the political economy, the ambience, and the rise of new actors in the policy-making process.

The Political Economy of Trade Policy

This section could have an alternative title: 'The Unintended Consequences of the Uruguay Round'. The Uruguay Round was the eighth negotiation under the auspices of the General Agreement on Tariffs and Trade (GATT), created in 1948 as part of the post-war international economic architecture. The GATT's primary mission was to reduce or eliminate the border barriers that had been erected in the 1930s and had contributed to the Great Depression and its disastrous consequences. The GATT reflected its origins in the post-war world in that it provided rules to buffer or serve as interface between the international objective of sustained liberalisation and the objectives of domestic policy — primarily full employment and the creation of the welfare state.

Before the Uruguay Round, the GATT worked very well. Tariffs and nontariff barriers were significantly reduced, and trade grew faster than output as each fed the other. Most recent rounds were essentially managed by the European Community and the U.S., although smaller countries were able to play a useful mediating role. The developing countries were largely ignored as players. Agriculture was virtually excluded from negotiations, so the transatlantic alliance, helped by the cold war's constraint on trade friction, was the effective manager of the international trading system.

The Uruguay Round was a watershed in the evolution of that system. For the first time, agriculture was at the centre of the negotiations. The European effort to block the launch of the negotiations in order to avoid coming to grips with Europe's heavily subsidised and protected Common Agricultural Policy (CAP) went on for half a decade. This foot-dragging also spawned a new single-interest coalition — the Australian-led Cairns Group, which included Southern countries from Latin America and Asia determined to ensure that liberalisation of agricultural trade would not be relegated to the periphery by the Americans and the Europeans, as it always had in the past.

But the role of a group of developing countries, tagged the 'G10 hardliners' and led by Brazil and India, was in many ways even more important in the Uruguay Round's transformation of the system. The G10 members were bitterly opposed to the inclusion of the so-called 'new issues' — trade in services, intellectual property, and investment — central to the U.S. negotiating agenda.

Although those new issues are not identical — obviously negotiations on telecommunications or financial services differ from intellectual property rights — they do have one common or generic characteristic. They involve not the border barriers of the original GATT but domestic regulatory and legal systems embedded in the institutional infrastructure of the economy, as well as the flows of factors — investment and labour. The degree of intrusiveness bears little resemblance to the shallow integration of the GATT, with its focus on border barriers and its buffers to safeguard domestic policy space.

The inclusion of the new issues in the Uruguay Round was a U.S. initiative, and this policy agenda was driven largely by U.S. multinational corporations (MNCs) that were market leaders in the services and high tech sectors. These corporations made it clear that without a fundamental rebalancing of the GATT they would not continue to support a multilateral policy, and preferred a bilateral or regional track. But they did not just talk the talk; they also walked the walk, organising business coalitions in support of services and intellectual property in Europe and Japan as well as some smaller members of the Organisation for Economic Co-operation and Development (OECD). The activism paid off. It is fair to say that U.S. MNCs played a key — perhaps even *the* key — role in establishing the new global trading system.

By the onset of the 1990s, a major change in economic policy was underway. The debt crisis of the 1980s, and thus the role of the International Monetary Fund (IMF) and the World Bank, plus the fall of the Berlin Wall — a confluence of two unrelated events — ushered in a major transformation in the economic policy paradigm. Economic reforms, namely deregulation, privatisation, and liberalisation, were seen as essential elements for launching and sustaining growth. Economic regulatory reform is at the heart of the concept of trade in services. Even without the thrust from the Uruguay Round, many developing countries began to see reform of key service sectors such as telecommunications as essential building blocks in the soft infrastructure underpinning growth and the General Agreement on Trade in Services (GATS) as a means to furthering domestic reform.

Thus, well before the end of the Uruguay Round the hard-line coalition had disappeared and coalitions of developing countries concentrated on liberalisation of agriculture and textiles and clothing. Many undertook unilateral liberalisation of tariffs and other trade barriers, and by the round's conclusion in December 1993 were among the strongest supporters of the negotiations they had so adamantly opposed in the 1980s. What might be called a North-South Grand Bargain was completed and was quite different from old-time GATT reciprocity — I'll open my market if you'll open yours. It was essentially an implicit deal: the opening of OECD markets to agriculture and labour-intensive manufactured goods, especially textiles and clothing, for the inclusion into the trading system of trade in services, intellectual property, and (albeit to a lesser extent than originally demanded) investment. Also, there was a last-minute piece of the deal: the creation of a new institution, the WTO, with almost no executive or legislative powers (aside from negotiating rounds), but with the strongest dispute settlement mechanism in the history of international law. Because the WTO consisted of a 'single undertaking' (in WTO legal-ese), the deal was pretty much take-it-or-leave-it for the Southern countries. They took it, but, it is safe to say, without a full comprehension of the profoundly transformative implication of this new trading system (an incomprehension shared by their Northern negotiators).

The Northern side of the bargain consisted of some limited progress in agriculture, with a commitment to go further in new negotiations in 2000, limited progress in textiles and clothing with most of the restrictions to be eliminated later rather than sooner, and a rather significant reduction in tariffs in goods in exchange for deeper cuts by developing countries. On the whole, this was not bad when compared with previous rounds based on traditional GATT-type negotiations on market access. But this was not a GATT negotiation, as the Southern side of the deal so amply demonstrates.

The essence of the South side of the deal — the inclusion of the new issues and the creation of the new institution — requires a major upgrade and change in the institutional infrastructure of many or most Southern countries, namely in their governance. These changes will take time and cost lots of money. The transition periods for implementation for developing countries were arbitrary and not based on any analysis or, indeed, on any awareness of this systemic problem. The technical assistance promised by the North was not followed up. And the new increasingly litigious and evidentiary-intensive dispute settlement system required a level of legal expertise rare in non-OECD countries, or else it required pots of dollars to purchase Northern legal services.

There were three significant unintended consequences to this Grand Bargain (or bum deal). One is a wide and serious North-South divide in the WTO. While the South is hardly homogeneous, there was a broad consensus among all developing countries that the Uruguay Round should be rebalanced before any new negotiations were launched. Rebalancing was defined in the implementation demands before the 1999 Seattle and 2001 Doha ministerial meetings. However, the Seattle meeting ended with the walkout of almost all the developing countries. What is more striking is the proactive role of the Southern countries both before and after Seattle. They became far better organised and informed, in part because of the rise of democracy and the growing awareness of trade policy issues in the general public and political institutions and the business community. But this development was also because of the role of a number of nongovernmental organisations (NGOs), established in developing countries during the 1990s, to provide information ranging from technical research to policy strategy papers. And since the mid 1990s, the internet has accelerated the links between Southern NGOs with a number of Northern partners in both Europe and the U.S. Together, these NGOs act, in effect, as a 'virtual secretariat' and often collaborate with a newly revitalised United Nations Conference on Trade and Development (UNCTAD). Their role both before and during Doha was striking, as they briefed the African countries and sought to co-ordinate a Southern position to block the round. Although their attempts failed (only India stood firm until the last moment), the virtual secretariat continues to play a proactive role during the negotiations and will certainly affect the outcome — if there is one.

Another unintended consequence also exacerbated the North-South divide. Because of the focus of attention on economic regulation, the negotiations on social regulation concerning product standards, health and safety measures, and the environment received little publicity and little attention from the senior policy ranks. In the OECD countries, social regulation started in the late 1960s and has been accelerating ever since. The OECD has called the phenomenon 'regulatory inflation'. One could — with perhaps a bit of a stretch — say that the post-war economic regulatory state of the advanced countries is withering away, while the social regulatory state is alive, well, and growing: a curious mix of *laissez-faire* and *laissez-régler*. Thus social regulation, covering the environment, labour, food safety, product labelling, and so on, has grown by as much as 400 percent in the industrialised countries since 1970 (OECD 1997, 191–248). This is decidedly not the case in the developing countries, nor are they likely to embrace the social regulatory state unilaterally. Moreover, although the positive regulatory approach in social regulation — technical barriers to trade (TBTs) and sanitary and phytosanitary measures (SPS) — is more procedural than substantive, the implications for infrastructure are equally demanding, involving sophisticated administrative procedures law as well as highly trained scientific human resources, which are sadly lacking in many if not most developing countries.

Moreover, since the establishment of the WTO, the most high-profile and contentious disputes have concerned social regulatory issues (food safety and environment), which are very sensitive in terms of public opinion in the OECD countries. Inevitably, the dispute panels and, especially, the Appellate Body have been forced to interpret the relevant WTO rules. And equally inevitably, as is the case with all courts and all legal rules (but even more so because some WTO rules are deliberately imprecise or 'creatively ambiguous'), the interpretation has essentially involved these internationally appointed judges making law, that is, defining the boundary for domestic policy space and reinforcing the perception — and reality — of the intrusive nature of the new system. And it is this issue of sovereignty that has enraged and emboldened a new group of policy actors — the NGOs.

In case the new actors needed more fuel for their fire, the third, and perhaps most pervasive and persistent, unintended consequence of the Uruguay Round has been the rise in profile of the MNCs, in large part due to their role in the round. For the more paranoid, the round was simply a conspiratorial collusion between corporations and rich country governments — especially the U.S. government. In any case, the global current of deepening integration, accelerated by the Uruguay Round, has evoked a counter-current focussed both on the MNCs and the WTO. It would be too simplistic, however, to attribute the rise of antiglobalisation (or anti-corporate globalisation, as it is termed by many protestors) only to trade developments. Other counter-currents were also flowing in the 1990s.

The Ambience of Trade Policy

There had been demonstrations against the GATT in Geneva at the outset of the decade because of a ruling concerning a U.S. ban on imports of tuna caught by a method that harmed dolphins. Posters of a monstrous 'GATTZILLA' were plastered on walls, and environmentalists were alerted to the 'dangers' of the negotiations. During the final stages of the negotiations, environmental NGOs covered Swiss highway bridges with graffiti admonishing 'GATT: No Patents on Life'. There were demonstrations at the meetings of the World Bank and IMF later on in the 1990s. But the scale and complexity of the choreography and drama at Seattle in 1999 were unprecedented. A new trend seemed to have been established. Something more than aversion to trade negotiations was now involved.

The antiglobalisation movement reflects a more pervasive secular change — an alienation from the elite. V. O. Key Jr. (1964, 27–53) wrote about the 'permissive consensus' of the earlier post-war decades. Although the broad public had little detailed knowledge of international policy, opinion polls demonstrated consistent support for governmental foreign policy, including trade policy. The change since the mid 1970s in all OECD countries has been fundamental — a clear, marked decline in confidence in government and all political institutions, that is, in the political underpinnings of the entire democratic system (Pharr, Putnam, and Dalton 2000). (After the events of 11 September 2001, however, support for the U.S. government's war on terrorism shot up to unprecedented levels but declined after the 2003 Iraq war.)

There are many differing views on the reasons for this striking phenomenon, and no doubt different factors are operative in different countries, given that the alienation trend is longer and steeper in the U.S. than in Europe. So, although there is no agreed or simple explanation, among the factors most often cited are the greater importance of information, the role of the media, and, by the end of the 1990s, the rise of the internet. As will be discussed shortly, Seattle would not have been possible otherwise.

In writings on the antiglobalisation two words appear with striking frequency: alienation and anomie (Ostry 2003). The first term is borrowed from Marx and is essentially a moral critique of capitalism. Marx argued that because labour becomes a commodity, the worker loses all power to control the processes by which decisions are made that affect his or her life. Anomie, first used by the French sociologist Emile

Durkheim at the beginning of the twentieth century, stresses the individual's sense of powerlessness and the loss of social cohesion. When anomie is confined to a small number, they can form groups — alienated from the rest of society but unified among themselves. The anarchists and the neo-fascists provide good examples. But if widely dispersed groups can unite around one theme — say, feminism or environmentalism — that is a qualitatively different phenomenon. And if divergent groups with divergent objectives can, by means of the internet, unite around one theme, say, anti-corporate globalisation, that could produce a quantum change.

Alienation and anomie are useful concepts in explaining the rise of the antiglobalisation movement. The disjuncture between the goals of free trade — rising living standards for all and the distributional impact, especially between rich and poor countries — is taken as evidence (not accurate, to be sure) that the WTO does not deliver. Unfortunately, this view was reinforced by overzealous proponents of free trade that unleashed unrealistic expectations. If Nirvana is promised but not delivered, it is bound to evoke cries of Armageddon from the disapproving and the disappointed. In many developing countries, this anger and disappointment are threatening already fragile democracies. Moreover, increasing inequality both within and, especially, among countries has been highlighted by the media and used effectively by many NGOs. The figures are stunning, and many are quoted. And it is not just poverty that marks the increasing marginalisation of the poorest. There is also disease — and not only AIDS but also ebola, malaria, and drug-resistant bacterial infections. (The NGOs were quick to seize on the AIDS epidemic, and their campaign yielded the one high-profile item on the agenda at Doha — see below.)

But, of course, freer trade did not create the growing inequality in economic and social conditions, nor will blocking trade cure these problems. *Au contraire.* Something far more complex would be needed in terms of policy co-ordination (including compensating the losers), and since that seems improbable at present, it is easier for politicians to yield to the temptation of promising simple solutions. And a vicious circle emerges. Governments lacking credibility promise to deliver, are attacked for promises that were undeliverable, and watch their credibility decline further. But the situation (as usual) is rather more complex. What about the credibility of the new global players in the trade policy process — the NGOs?

The Process of Trade Policy: The Nongovernmental Organisations

Before an examination of the concerted attack on the WTO, it must first be made clear that there is no homogeneous set of institutions called NGOs. Even if the development groups in poor countries are separated from the advocacy NGOs, whose main objective is to shape policy, the latter group must be divided into several categories. For example, as already noted, there is a new virtual secretariat for Southern countries, there has been a remarkable proliferation of groups centred on establishing business codes of conduct, and there are also groups rich in technical and legal expertise that usually consult 'inside' the system; and all of these are rather different from what can be termed the 'mobilisation networks', for which a major objective is to rally support for dissent at a specific event, such as a WTO ministerial meeting, the Summit of the Americas, a meeting of the World Bank and the IMF, the G8 Summit, and so on.

Other main objectives of the mobilisation networks are to heighten public awareness of the target international institution's role in globalisation and, by doing so, to change its agenda and mode of operation — or, in the case of the more extreme members, to shut it down. While these networks are loosely knit coalitions of very disparate groups, an analysis of the networks at Seattle (in 1999), Washington, Bangkok, and Prague (in 2000), and Quebec City (in 2001) show that a significant proportion are NGOs working on the environment, human rights, and gender issues. The 2001 G8 Summit in Genoa included these groups as well as a number of left-wing political parties from Europe and an unknown number of anarchist groups known as the Black Bloc. However, one must be wary of the view (often stressed by the NGOs themselves) that these loose and diverse coalitions represent a new form of globalised participatory democracy on the internet or a movement that has no leader, no centre, or even a common name. This resembles what P. J. O'Rourke has termed (in another context) 'a clarion call to whatever'. Participatory demonstrations as a form of participatory democracy? Fine, but the most significant development facilitated by the internet — and vividly demonstrated at Seattle — has been the emergence of a new service industry, namely the business of dissent. And there is a centre of business — call it dissent.com — that is very effectively operated by a core group of NGOs headed by a new breed of policy entrepreneurs. It is important to stress that the dissent industry is largely a product of the internet. Inexpensive, borderless, real-time networking provides advocacy NGOs with economies of scale and also of scope by linking widely disparate groups with one common theme. As is the case for all innovations, there are also important positive feedback loops. An NGO network established at the Rio Earth Summit in 1992 was used by U.S., Canadian, and Mexican groups advocating against the North American Free Trade Agreement (NAFTA) and this experience was vital to mobilising the fight against the Multilateral Agreement on Investment (MAI). The lessons from the MAI were put to use in preparing for Seattle, and the Seattle experience was helpful for planning to Washington and Prague and Porto Alegre and Davos and Quebec City and Genoa and Kananaskis and Johannesburg and Evian and Mumbai, and so on — and on.

The key assets of dissent.com are the ability to use the media to deliver the message. Since the networks are so diverse both in mission and location, the message must carry a simple, common theme: anti-corporate globalisation and pro-democracy. The charge is that the WTO (or the IMF or the World Bank) is dominated by the interests of the MNCs while harming the environment and increasing inequality and that their rules and procedures are undemocratic and secretive.[1] Even when the message and the media are combined with money (for those NGOs successful at mass mailings and securing funds from foundations — mainly in the U.S.), the viability of the new business will depend not only on the 3Ms of the message, the media, and money, but also ultimately on the saleability of its product — anti-corporate globalisation (see below).

In tracking the NGOs on the internet preparing for Seattle and subsequent meetings, a core group managed the protests, crafted the message, and organised the circulation of sign-up lists or manifestos. That group included Ralph Nader's Public Citizen, Global Trade Watch, and the International Forum on Globalization (IFG) in the U.S.; Corporate Watch and ATTAC in France, Corporate Europe Observatory in Amsterdam, Oxfam in the United Kingdom and Belgium, Friends of the Earth (FOE) in the UK and elsewhere, and the Third World Network of Malaysia, which was the most prominent transnational Southern NGO. A number of other NGOs were involved in facilitating the demonstrations in Seattle, Washington, and Prague. Thus, the Direct Action Network (DAN), with offices in California and New York, began as a coalition of activist groups dedicated to shutting down the WTO ministerial meetings. ('Direct action' is a term derived from the anarchist movement in the nineteenth and twentieth centuries — indeed, as is discussed later, it was the central core of the movement's ideology.) DAN was inspired by Peoples' Global Action (PGA), which was formed in Geneva for the first WTO ministerial meeting in 1998. PGA organised the first Carnival against Capitalism in London in June 1999, which almost shut down the City of London's financial district. The anarchist Black Bloc was central to the London meeting. PGA was also involved in the Quebec demonstrations through its Canadian arm, the Anti-Capitalist Convergence of Montreal (CLAC).

There are a number of interesting developments apparent in the evolution of the network strategy since Seattle, although it is still far too early to cite these as trends. One concerns the co-ordination of global and regional protests. This was clearest in the planning for the April 2001 Quebec Summit of the Americas. While some of the 'Seattle plus' core groups were actively involved, the umbrella network for Quebec City — the Hemispheric Social Alliance — was established in Brazil in May 1997 and included many of the anti-NAFTA networks extended to Latin America and the Caribbean. In Asia, protests organised during the May 2001 meeting of the Asian Development Bank in Honolulu were much smaller but included NGOs from across the Asia-Pacific region as well as a local coalition of indigenous, human rights, and environmental groups. The protests were also co-ordinated with the IFG in California, which was very active in planning for Seattle. The hope was that the mobilisation networks formed for the Honolulu meeting could become the basis of an expansion across Asia — as the NAFTA networks did in Latin America. The speed with which the dissent industry has been adopted in Europe has been impressively rapid. In Prague, the Czech group INPEG was formed to carry out a week of protest (18 to 28 September 2000), based on Seattle, and was regarded as an impressive model. ATTAC, a French group established in 1998 after the 1997 financial crisis, has since expanded considerably; it organised the anti-Davos 2001 demonstrations in Porto Alegre, which, in turn, fed into the planning for Quebec City. ATTAC was pre-eminent at Genoa 2001, as it has expanded throughout Europe as well as Latin America. The demonstrations and discussions were co-ordinated by the Genoa Social Forum, but it was not able to plan for or control the violence of either the extremist groups or the police — not a new aspect of this form of advocacy, but one now very much heightened.

Because the main objective of mobilisation networks is to influence public opinion and, through that route, initiate change in the policy processes of the international institutions, has the dissent industry been successful? In the case of the WTO, as explained further below, the answer is probably yes. The claim that the protests shut down the Seattle meeting does not stand up to closer scrutiny. The meetings failed because the North-South divide and Clinton's remarks about using sanctions to enforce labour standards caused the Southern countries to walk out.

The NGOs also claim that Seattle was the 'big bang' of a new social movement. There are certainly signs of spreading activism in Europe and the western hemisphere, albeit less so (thus far) in Asia. But there are also signs that the dissent.com brand may have been seriously tarnished by the violence at Genoa. This is worth a brief exposition.

There was serious destruction of property at the J18 Carnival against Capitalism in London and at Seattle and other meetings, but nothing was comparable to Genoa, where one protestor was killed. A number of mainline NGOs stayed away because they were fearful of being associated with the violence planned by the Black Bloc and others, although no one could have foreseen what occurred, including the extraordinary brutality of the police. Allegations of neo-fascist *agents provocteurs* were repeated endlessly on the internet. While the full story may never be known, the presence of extreme right and extreme left coalitions has become a feature of demonstrations since Seattle. As the documented by the Southern Poverty Law Center (an American NGO), neo-fascist groups in Britain, Europe, and the U.S. have embraced the antiglobalisation credo with enthusiasm. In Europe, the development of the so-called International Third Position (ITP), created by a breakaway faction of the UK's neo-fascist National Front and Italian Fascists, has increasingly gained a following among the international hard right. The ITP represents a fusion of left and right (including environmentalism, wealth redistribution, and workers' co-operatives), but its basic objective is to identify the prime enemies (Jews, blacks, and immigrants). The Southern Poverty Law Center (2000) underlines that the 'left and right did not exactly march arm in arm' in Seattle — nor did they in Prague or Genoa. Yet there is a probably inevitable tendency for all demonstrations to attract extremists; a free ride is hard to decline. Escalating violence generates the need for more police security, which likewise generates more violence and attracts more extremists. And, of course, Genoa was hardly the end of a story.

The terrorist attacks of 11 September, while unrelated to the antiglobalisation movement *per se,* greatly added to the pressure for dissent.com to adopt a new strategy. Indeed, some believe that the movement has been decimated. There appears to be a widening split among the NGOs that fear being tarnished by the events of Genoa and, *a fortiori*, 11 September and those more strident and extreme coalitions that regard global capitalism as bereft of all values other than greed and are active in the peace movement. Even before the annual meetings of the IMF and the World Bank were cancelled in the fall of 2001 because of the terrorist attack on Washington, a number of organisations (the AFL-CI0, FOE, the Sierra Club, and Oxfam, for example) announced that they would not join in any protests and were scathingly denounced by others (especially anarchists) as cowardly traitors to a just cause. The need to transform strategy seems clear enough. And some innovation is already evident.

Thus, to demonstrate or not to demonstrate is not the only question facing the mobilisation networks. It may be well that the international institutions are increasingly apprehensive about holding meetings for fear of violence and cost. Certainly, security measures have been vastly increased. One indication of a possible alternative could be to conference 'online', as the World Bank did after cancelling a planned meeting in Barcelona in May 2001. It was well aware of another new global movement — 'hacktivism', or the use of computers to organise 'electronic civil disobedience' by defacing websites, organising virtual sit-ins by generating massive traffic to block other users, or creating denial of service. But the World Bank hired an outside firm to handle the conference and provide expert anti-hacking protection.[2] The conference went ahead with no glitches (despite threats to 'crack the whole event'), but the spread of hacktivism has begun and cannot be dismissed. Some firms have set up special units for threat analysis — IBM being a leader, since it provides service for corporate computer systems around the world.[3] But new security arrangements instituted as a result of 11 September have certainly made hacktivism more difficult. It is still too early to judge the potential for what one anarchist group has termed 'cyber trench warfare'. Nonetheless, whatever emerges from this new phenomenon, hacktivism is unlikely to provide an alternative to the major objectives of the mobilisation networks, that is, to build a global social movement against corporate globalisation.

But to forge a truly global social movement uniting diverse single-interest NGOs — a coalition of 'hedgehogs', in Isaiah Berlin's terminology — would require more than agreement on who is the enemy. It would require some agreement on an alternative. And, indeed, since Seattle, there is a growing and lively debate on the issue of defining a systemic alternative to the neoliberal paradigm. Not surprisingly, that goal seems to be a long way off. One idea gaining currency is localisation. While not exactly a yearning for some mythical past, a close reading of the popular *Localization: A Global Manifesto* (Hines 2000) raises many more questions than answers. For example, just how would does one define local — Colin Hines has many different suggestions — and how would could one get countries to agree to stop all trade except that which involves essential products that are not produced locally?

One is reminded of the anarchist writings of the nineteenth and early twentieth centuries and the failed effort to establish an anarchist regime that was destroyed by the Spanish Civil War.[4] The anarchist movement did not have a single, powerful spokesman as did their most hated enemies — the communists — so their ideas are diffuse and somewhat amorphous. But a unifying theme is confederal arrangements of small communities with no central or hierarchical authority. The term 'anarchy' is today associated with violence, yet the anarchist ethic was libertarian (as extreme as today's right wing) and violence was an individual right. The ideas of 'direct action' and 'diversity of tactics' are stressed by many of the NGOs today, so that the echoes are clear. Be that as it may, an Edenic confederal bottom-up model — a model of deglobalisation — is reminiscent of the definition of a romantic: one who prefers to travel hopefully rather than arrive at a destination.

It must be stressed, however, that many of the groups currently engaged in the discussion are debating very specific proposals to shrink (rather than sink) the WTO. Among a number of NGOs, there is a discernible shift from demonstrating to debating. The shift to the South is also a very important trend.

Conclusion: The Future of the World Trading System

On 15 November 2001 in Qatar, the fourth ministerial meeting of the WTO concluded with a declaration outlining a work agenda — termed the Doha Development Agenda by Mike Moore, the head of the WTO (2001). The words are significant: the term 'round' appears nowhere in the declaration, despite references throughout to negotiations and an emphasis on the term 'development' in an effort to assuage the anger and disappointment of many Southern countries about the poor harvest on their implementation demands.

Nonetheless, it could be argued that the great success of Doha was the absence of failure. The negotiations included agriculture and services — which were, in any case, mandated as part of the built-in agenda of the final act of the Uruguay Round — but also industrial products, some aspects of government procurement, and trade facilitation (specifically reducing red tape at borders). This market access agenda was a top U.S. priority and was presented to Congress as a major achievement. The European Union managed to include a reference to environment — the first in the WTO — as well as a commitment to current 'negotiating modalities' of the so-called Singapore issues of investment and competition policy, transparency in government procurement, and trade facilitation, carried over to the fifth ministerial meeting in 2003, which would consider how to proceed. This was considered a major achievement by the European trade commissioner. Although the outcome at Cancun in 2003 seemed rather uncertain, to put it mildly, this factor even so was claimed as a victory by the Southern countries, which had firmly opposed these 'new issues'.

At Doha, the Southern countries made few gains on implementation and lost out most seriously on liberalisation in textiles and clothing. On agriculture, a breakthrough of sorts seemed to be the reference to the elimination of export subsidies, although at the last minute the EU was able to insert some wording to provide wiggle room. The agenda also included a reference to antidumping and subsidies — a high priority for the South — although careful parsing suggests that there might be less than meets the eye. As with all of the above, the devil is the details, and the details will dominate the negotiations. There were numerous references to technical assistance and capacity building in the declaration and a reference to beginning discussions on funding before the end of the year, but there were no commitments to increase the budget.

Many newspapers ran headlines about Doha that presented the outcome as a triumph for the South because of the declaration on the Agreement on Trade-Related Aspects of Intellectual Property Rights (TRIPS) in relation to public health emergencies such as HIV/AIDS. This remarkable development was undoubtedly the result of a highly co-ordinated and skillful campaign on the part a number of NGOs, especially Oxfam and Médecins Sans Frontières. The declaration (the details remained to be negotiated) would allow governments in conditions of national emergency concerning health to license the production of drugs by generic firms and would be a boon to the generic pharmaceutical industries in Brazil, India, and China. It would also allow the import of generic drugs into countries where the drugs are produced by the patent holder (known as parallel imports), but not in countries with no production — most of the poorest countries. This latter issue was a priority for the NGOs but remained to be negotiated in the TRIPS council.

What Doha represents remains to be seen. It is one thing to launch a negotiation — a symbolic triumph — and quite another to conclude it. The North-South divide remains, as do the transatlantic differences on the environment. The WTO remains a 'Mercedes Benz without gas' (a description coined by this author some years ago) and reform is essential to strengthen it. But the second symbolic triumph — the TRIPs declaration — suggests that the NGOs may have learned that in selecting issues it is better to concentrate on Southern concerns because 80 percent of the WTO members are not rich, industrialised countries.

Postscript: Cancun

At first, the news from the fifth ministerial meeting in Cancun in September 2003 invoked a strong sense of *déjà vu* all over again. Cancun was a mid-term ministerial meeting, as was Montreal in 1988. On the last morning of the Montreal meeting, at around 6 a.m., the bleary-eyed negotiators waited for the arrival of the European Community and United States warriors, who had been up all night dealing with agriculture. When they arrived, they announced that it was too bad but they had not reached an agreement, so all that was left was to tidy up the other agenda items and finish the communiqué. A group of Latin American countries headed by Brazil said no: if there was no agriculture, there was no agreement on anything. It was a moment of shock (and maybe awe for some), but the officials handled it with great finesse by announcing that the meeting was adjourned and would be reconvened shortly in Geneva. There were no big headlines.

The feeling of *déjà vu* soon dissipated. As noted earlier, among the many unintended consequences of the Uruguay Round was a sharp cleavage between the North and the South over the Grand Bargain, which sealed the negotiations: the inclusion of the new issues of services and intellectual property by the South, and the reform of agriculture and improved access for labour-intensive goods, especially textiles and clothing by the North. The Grand Bargain turned out to be a bum deal, and Southern hostility was evident at every subsequent ministerial meeting. And so were the NGOs present at all previous meetings. Cancun, however, was very different. At Cancun, there appeared to be an axial shift in the political economy of policy making that would require a fundamental reorientation of the players and the game. Granted, it may be too early to tell, but it is worth reviewing Cancun in that light.

Cancun and the Nongovernmental Organisations

The *Wall Street Journal* editorial headline after 11 September was entitled 'Adieu Seattle'. Now maybe it should be 'Bonjour Cancun'.

It is not yet possible to assess the role played by the NGOs in Cancun. However, what is evident is that there has been a marked increase in the numbers and activities of NGOs in the South. The virtual secretariat has expanded and deepened and is of great assistance to developing countries in the political and technical aspects of trade policy. Instead of screaming outside the tent, many NGOs are now quietly at work inside a growing number of tents. There were no violent demonstrations at Cancun, and there was respect and grief over the suicide of the Korean farmer.

The precise role of NGOs at Cancun will, of course, be a subject of continuing debate. It is claimed that African civil society played a crucial role in the ministerial rejection of compromise on the Singapore issues, which, in effect, shut down the meeting. There were many other factors (including the rather odd decision of the Mexican chair not to resume the agriculture discussions), but that is not really the point. The role of NGOs at Cancun was important on substantive as well as process grounds. And that implies a significant change in the dynamics of the policy process of the WTO. The loud cheers of joy from many NGOs at the collapse of talks are not exactly a hopeful sign.

Although it was the Singapore issues that triggered the Cancun collapse, in fact there was a much more serious conflict related to agriculture. The draft proposal of the U.S. and the EU polarised the debate into a North-South struggle, with a new Southern coalition led by Brazil. This coalition, the G20 (also sometimes referred to as the G21 or the G20+, and not to be confused with the G20 finance ministers and central bank governors), included Brazil, India, and China (the 'Big Three') and a group of Southern countries with varying views on economic policies. It seemed an unlikely grouping, but it did not collapse under pressure at Cancun and, if it persists, it will represent a very significant challenge to the 'Big Two' of the U.S. and EU. But that is a very large 'if'. In his closing remarks Brazilian ambassador Celso Amorin thanked civil society for its support for the G20. Was that significant? Perhaps, but perhaps not. Since Cancun, the U.S. has actively put pressure on a number of countries and has been successful (so far) in 'persuading' Colombia, Peru, Ecuador, Guatemala, Thailand, and Costa Rica to drop out of the G20. But even if the G20 dwindles, it is interesting that Brazil, India, China, and South Africa are talking about a new trade agreement — a Southern region agreement. The U.S. is well aware that a shift of power to the South buttressed by a growing NGO infrastructure would have major consequences for the multilateral system and for the Free Trade Agreement of the Americas (FTAA).

Implications for the Multilateral System

There is no doubt that the collapse at Cancun will damage the smaller, poorer countries the most. They have no other options, and the expected delay in completing the Doha Development Agenda will mean that there will be no agriculture reform for some time. Nor will there be any movement on other issues of concern, including special treatment for the least developed or, more importantly, any reform of special treatment to strengthen the co-ordination between trade technical assistance and financial assistance for development aid. That is crucially important if the poor developing countries are to increase their capability to export and thus benefit from liberalisation.

Signs of a new foreign policy were evident at Cancun and in Washington. The anger of Robert Zoellick (2003), the U.S. Trade Representative, was hardly surprising. Nor were the threats to countries that were 'pontificating' and engaging in 'inflammatory rhetoric' that the U.S. would actively pursue bilateralism and regionalism, but not with what Zoellick christened the 'won't do' countries (such as the G20, but especially Brazil) but with the 'can do'. Trade bilaterals are like birthday gifts for the well behaved. The views of Senator Charles Grassley, chair of the Senate Finance Committee, were very clear: 'We will continue to pursue broad market liberalization on all fronts ... The United States evaluates potential partners on an ongoing basis [and] I'll take note of those nations that played a constructive role in Cancun, and those nations that didn't' ('Support for WTO in Congress, Private Sector Wanes after Collapse of Cancun Trade Talks' 2003, 1).

This view was echoed by some business executives at Cancun. The Vice-President for International Economic Affairs at the National Association of Manufacturers was very explicit: 'If we can't get a level playing field in the WTO, we are going to take our poker chips out of the game and start dealing in free trade agreements where we know we can get a fair deal' ('Support for WTO in Congress, Private Sector Wanes after Collapse of Cancun Trade Talks' 2003, 2). One U.S. official stated that the G20 is just a 'third-world chest thumping festival led by Brazil and India' ('Commission Likely to Face Tough Questions from Member States on Cancun' 2003, 7). And this added a danger to the most important of the American regional negotiations — the FTAA. The impact of Cancun on the FTAA is already apparent. Brazil is trying to narrow the agenda and the U.S. is trying to isolate Brazil. And, indeed, criticism of Brazil's position is mounting in a number of Latin American countries that are desperately keen on securing access to the U.S. market. But it would be unwise to ignore the growing importance of the Chinese market for Brazilian natural resource products as a counterweight to the U.S. The game may be chess, not poker.

Unlike the U.S., the EU is not interested in favouring bilateralism as a policy option because it is already overloaded in that domain. The only new bilateral on the EU agenda at present is with the African, Caribbean, and Pacific group of states (ACP) — the signatories to the Cotonou Agreement. That could change, of course, since this form of competitive liberalisation by the U.S. — or domino game — can evoke a response by others (as is evident from the activity of China, Japan, and India in Asia). But the EU is facing a changing of the guard in 2004 and the loss of its current trade commissioner, as well as the challenge of integrating new members. It will be much more difficult for the EU to play a positive role in global trade policy, a fact already clearly evident as the EU has declined to become engaged in WTO efforts to resuscitate the negotiations. Moreover, the transatlantic alliance was fragile even with the unusual rapport among the negotiators. Therefore, the growing European-U.S. hostility over trade and non-trade issues is unlikely to abate in the near future.

One major consequence of Cancun could be the marginalisation of the WTO as a talking shop and forum for disputes. The fragmentation of the global system into bilateral and regional blocs and the political fallout from this messy bowl of spaghetti could be very serious. The increasing emphasis on reducing inequality and poverty has been an important achievement of the NGOs, and a seriously weakened WTO is unlikely to be able to play an effective role. As already stated, the poorer countries will suffer the most. The plea to the U.S. to cease cotton subsidies from the small African countries dependent on cotton exports was highly symbolic. It has been compared with the drugs issue at Doha, which, however difficult the negotiations, did result in greater welfare for the poorest countries. At Cancun, however, the demand about cotton subsidies was summarily dismissed in rather complex technical language.

Of course, the withering away of the WTO may be too gloomy an assessment. Perhaps the negotiations, albeit delayed, will produce a reasonable agreement. Estimates by various economists suggest a significant improvement in growth and a reduction in poverty, especially if agriculture is reformed and Southern countries reduce barriers to increase South–South trade. But that would require significant political change in most countries, especially the most powerful, as well as some reform of the WTO as an institution. Pascal Lamy, the European Trade Commissioner, was wrong in calling it medieval — that is, in fact, a compliment.

WTO Reform?

Niccolo Machiavelli wrote:

> There is nothing more difficult to plan, more doubtful of success, nor more dangerous to manage than the creation of a new system. For the initiation has the enmity of all who would profit by the preservation of the old institutions and merely lukewarm defenders in those who would gain by the new ones.

There is no doubt that the Uruguay Round in effect created a 'new system' — largely by accident and not intent. Among its many features was the establishment of a new institution, the WTO. This was a last-minute development, and the Canadian proposal was supported by the European Community in large part to constrain American unilateralism. The WTO is a peculiar institution: it has virtually no executive or legislative powers (aside from negotiations of rounds) but has the strongest dispute settlement system in the history of international law. The absence of institutional infrastructure worked when the membership was small and the GATT, a club of a few powerful traders with a common set of norms and standards and a shared view of their objectives. By the end of the Uruguay Round, members numbered 111; today there are 147 and there will likely be 170 or more in the not-too-distant future. The members do not share a common vision. *Au contraire*.

The consistent and insistent demand for greater transparency — WTO-ese for democracy — has been a feature of the NGO movement since Seattle. The concept of internal transparency broadly concerns the management of the institution. External transparency involves more access to WTO documents, more participation in WTO activities, and the right to observe and present *amicus curiae* briefs in dispute settlements. Although the WTO has made considerable progress in providing information and has engaged civil society groups in symposiums and informal briefings, these actions have hardly satisfied the *démandeurs*. And no progress has been made in even discussing the question of improving the management of the institution. After the debacle at Seattle, there was some desultory discussion about internal and external transparency. But after strong opposition from many countries — especially Southern — to discussing the issues, the subject was dropped. After Doha, the General Council (having deliberated for four years) agreed on a set of procedures for de-restricting documents and discussed some proposals for ministerial preparatory processes. Nothing was agreed. A group of NGOs just before Doha sent an 'Open Letter on Institutional Reforms in the WTO' (World Wildlife Foundation et al. 2001). There was no response at Doha. Just before Cancun, a group of NGOs launched the 'Campaign for Democracy at the WTO' dealing with the decision-making process for ministerial meetings, which essentially proposes that 146 (now 147) countries fully participate in all meetings and discussions. It was not discussed in Cancun. But perhaps it helped provoke Lamy's outburst.

As a member of an WTO advisory committee established by Michael Moore when he was Director General, the author wrote a chapter for a yet-to-be published book on the need for reform (Ostry 2002). It included proposals for a management committee to take care of day-to-day business (which was also proposed by three former heads of GATT/WTO in 2001), a policy forum to provide the locus for discussion of basic policy issues (such as the Consultative Group of 18 established in 1975 and disbanded at the end of the Uruguay Round), and a plan for external transparency to be housed in the Trade Policy Review Mechanism in order to encourage a participatory policy process in member countries. There has been no response to these proposals.

Now there is a new advisory committee, chaired by former Secretary General Peter Sutherland. After Cancun, several newspapers reported that the committee would propose how to reform the WTO. Ideas circulating include opting out (as in the Tokyo Round), which would create a variable geography institution, or going back to basics — in other words, a WTO centred on eliminating trade barriers on manufactured goods and agricultural products and so on. The drive for reform is primarily an EU objective. The U.S. remains rather cool to any change. As one U.S. official recently remarked, 'pounding our heads against the wall trying to reform the WTO' is not a 'good way to spend our time' ('WTO Director General Cancels All Special Negotiating Sessions' 2003, 3). Given the sharp North-South disputes over reform and the continuing NGO attack on the need for democracy and the worsening paralysis by consensus of the decision-making process, optimism is hard to rally.

Yet some reform is essential. The grand visions proposed by experts, committees, eminent persons, and so on and so forth are wonderful, but might take some time to come to fruition (*pace* Machiavelli). The Management Committee and Policy Forum and the voluntary use of the Trade Policy Review Mechanism would be of considerable assistance in the mean time. Furthermore, it is essential to be frank and straightforward. The concept of democracy is really not relevant to an international institution, as the eminent legal scholar Robert Hudec (2001, 295–300) has so brilliantly explained.

> The World Trade Organization is a member-driven organisation and the focus of its orders is the product of a process in which the governments agree to participate and ultimately control.

The demand for democracy and legitimacy should begin at home. The NGOs can help by exerting pressure on their home governments to establish ownership of the trade policy process at home. It may just be a first step, but at least the journey would begin.

Notes

1. Examples at Seattle were 'fix it or nix it'; after Seattle, it was 'shrink it or sink it'. At Washington, it was 'De-fund the Fund! Break the Bank! Dump the Debt!'.
2. Communication to author from Matts Karlsson, World Bank, 8 September 2001.
3. The IBM unit is called the Threat Analysis Group. For a useful review of hacktivism, see Denning (1999) and references cited therein.
4. International Anarchism (2002) lists recent anarchist writings on the Spanish Civil War. For a useful summary of the history and philosophy of anarchism, see Sonn (1992); see also the classic study by Thomas (1986).

References

'Cancun's Charming Outcome' (2003). *Economist*. 18 September.

'Commission Likely to Face Tough Questions from Member States on Cancun' (2003). *Inside U.S. Trade*. 3 October, pp. 6–7.

Denning, Dorothy E. (1999). 'Activism, Hacktivism, and Cyberterrorism: The Internet as a Tool for Influencing Foreign Policy'. Paper presented at 'The Internet and International Systems: Information and Technology and American Foreign Policy Decisionmaking',10 December. San Francisco. <www.nautilus.org/info-policy/workshop/papers/denning.html> (April 2004).

Hines, Colin (2000). *Localization: A Global Manifesto*. Earthscan, London.

Hudec, Robert E. (2001). 'Comment'. In R. B. Porter, ed., *Efficiency, Equity, and Legitimacy: The Multilateral Trading System at the Millennium*, pp. 295–300. Brookings Institution Press, Washington DC.

Key, Valdimer O. (1964). *Public Opinion and American Democracy*. Knopf, New York.

Organisation for Economic Co-operation and Development (1997). *Report on Regulatory Reform. Volume II: Thematic Studies*. Organisation for Economic Co-operation and Development, Paris.

Ostry, Sylvia (2002). 'External Transparency: The Policy Process at the National Level of the Two-Level Game'. WTO Advisory Group. <www.utoronto.ca/cis/WTO_Transparency.doc> (April 2004).

Ostry, Sylvia (2003). 'Global Integration: Currents and Counter-Currents'. In P. MacRory and A. Appleton, eds., *The Kluwer Companion to the World Trade Organization*. Kluwer Academic Publishers, Amsterdam.

Pharr, Susan J., Robert D. Putnam, and Russell J. Dalton (2000). 'A Quarter-Century of Declining Confidence'. *Journal of Democracy* vol. 11, no. 2, p. 5.

Sonn, Richard D. (1992). *Anarchism*. Twayne Publishers, New York.

Southern Poverty Law Center (2000). 'Neither Left nor Right'. *Intelligence Report*. 28 September.

'Support for WTO in Congress, Private Sector Wanes after Collapse of Cancun Trade Talks'. (2003). *International Trade Daily*, 16 September.

Thomas, Hugh (1986). *The Spanish Civil War*. Harper & Row, New York.

World Trade Organization (2001). 'Ministerial Declaration'. Doha, 14 November. <www.wto.org> (April 2004).

World Wildlife Foundation, Center for International Environmental Law, Oxfam International, et al. (2001). 'Open Letter on Institutional Reforms in the WTO'. October. <ciel.org/Publications/Reform.pdf> (April 2004).

'WTO Director General Cancels All Special Negotiating Sessions' (2003). *Inside U.S. Trade*. 3 October, pp. 1–3.

Zoellick, Robert B. (2003). 'America Will Not Wait for the Won't Do Countries'. *Financial Times*, 22 September.

Chapter 16

Enhancing Global Governance: Corporate Social Responsibility and the International Trade and Investment Framework

Christopher Wilkie[1]

This chapter straddles two issue areas, both of which have already contributed — and have the potential to contribute further — to processes associated with global governance.[2] The first of these issues is corporate social responsibility (CSR), which addresses the relationship and responsibilities of corporations not only to shareholders but to other stakeholders as well, particularly with respect to non-economic issues.[3] The second is the issue of trade and investment rules, particularly those of the World Trade Organization (WTO) but also including rules in regional and other forums — rules that are based on the rights and obligations accorded by states to each other, largely with respect to the activities of private actors buying and selling goods and services internationally, or engaging in activities facilitating this.

Despite possible synergies, specialists in both policy domains tend to be wary of advocating greater interaction and real co-operation between them. At first glance, links between these two domains seem evident. After all, to some degree, both ultimately concern the framework for private entities underpinned by nation-states in the domestic and international arena. However, different subject matter, intellectual traditions, and sources of political legitimacy have made for uneasy and infrequent bedfellows.

Unlike trade rules, which are usually accompanied by dispute settlement procedures, the recent acceptance of norms and standards related to CSR has not necessarily meant commensurate observation or enforcement. In response, some proponents of stronger enforcement have sought to varying degrees to link CSR norms with rules governing trade and investment (Kelley 2001; Aaronson 2002; Picciotto 2003). However, others in the CSR community are wary of the issue becoming dependent on enhanced prospects for enforcement through the trade and investment system (Graymore 2003). Complicating matters in this regard is the fact that under trade rules with developed dispute settlement procedures, rights of retaliation remain limited and could be a counterproductive instrument through which to advance CSR norms *vis-à-vis* vulnerable social or environmental interests.

From the perspective of the trading system, as the scope and possible implications of the post–Uruguay Round WTO have begun to be realised, calls for attention to other than purely trade issues in its deliberations have increased. Human rights issues, the environment, labour, and broader concepts such as democratic legitimacy and justice are no longer strangers to a discussion of WTO-related issues (Charnovitz 2002; Beviglia Zampetti 2003; Petersmann 2003). Nonetheless, there are relatively few practical approaches to bridging these different perspectives.

Even if CSR could be accommodated on the trade agenda, it remains unclear whether using trade rules is an effective means of upholding CSR norms, particularly in the short term. After all, trade and investment policy, which is backed by a relatively strong means of enforcement, still essentially amounts to legally sanctioned rights of limited economic retaliation by governments in specified circumstances. WTO and many other trade and investment agreements are usually predicated on the belief that non-discrimination is simply in the interests of the efficient allocation of resources and ultimately economic growth. Only incidentally, one can note a strong correlation between economic growth and relatively high environmental and social standards. Furthermore, even if mechanisms associated with current trading arrangements could be used in this regard, sanctions may not only seem a poor mechanism for, say, helping workers in low-income countries, but could even hurt those who are the intended beneficiaries the most (Brown 2001, 107, 109–110).

Following from this, there is another factor in explaining why CSR issues have been only tangentially or obliquely addressed in the trading system, and are the subject of relatively soft multilateral disciplines even within their own policy domains (such as the labour conventions of the International Labour Organization [ILO] and multilateral environmental agreements). It is the sensitivity of these issues among stakeholders in civil society, among governments, and, of course, between North and South. In turn, one of the main reasons for this is the fear of protectionist interests using labour, environmental, and other issues as a smokescreen for their own ends. This remains the overriding fear of developing countries (and many economists). Perhaps the best example of this concern appeared immediately prior to the 1999 WTO ministerial meeting at Seattle, when Jagdish Bhagwati and more than 75 of his colleagues published a manifesto outlining their strong opposition to linking labour and environmental issues to the WTO and trade treaties (Bhagwati et al. 1999).[4] How to begin to square this seemingly intractable circle is the underlying theme of this chapter.

This chapter addresses a recent institutional success in the CSR domain, namely the 2000 review of the OECD Guidelines for Multinational Enterprises published by the Organisation for Economic Co-operation and Development ([OECD] 2000b). Reasons for this success are discussed, as well as how they relate to the governance and trade agendas. The second part of the chapter focusses on the relationship between CSR and the trade and investment framework, and offers suggestions for incorporating aspects of the CSR agenda into the trading system. A number of examples have already been acted upon, by virtue of the OECD declaration, of which the guidelines form a part. In addition, a number of others will be noted, in particular one suggestion that has received hardly any attention: the WTO's Trade Policy Review Mechanism (TPRM).

Political considerations as well as different traditions and concerns suggest that progress in this vein will be incremental. However, following the guidelines review, and in the wake of the failure of the WTO ministerial meeting at Cancun in September 2003, it is perhaps an appropriate juncture at which to examine the potential for more constructive approaches to issues of relevance to these two domains, tempered by realistic expectations from each. Notably, narrowing the impasse between CSR and trade and investment also has considerable potential to enhance global governance.

Corporate Social Responsibility and the Organisation for Economic Co-operation and Development's Guidelines for Multinational Enterprises

The Trade and Investment Context

While at first glance it could be argued that the trading system is ill suited to address issues only tangentially related to it, such as labour and environment, it is increasingly asked to respond to them. As 'behind the border' trade irritants are now often addressed in trade agreements, subjects such as environment, labour, and other issues such as corruption and corporate governance that are more closely linked to the domestic polity similarly arise. In fact, they have longstanding antecedents in trade policy — Dani Rodrik (1997, 33) notes that Article 7 of the ill-fated Havana Charter contained an article on fair labour standards, and Peter Muchlinski (1995) documents elements with a longer history. And these concerns already register on the international trade and investment agenda: in the preamble to its trade agreements, the European Community, for instance, defers to human rights, and WTO members recognised the objective of sustainable development in the preamble to the Agreement Establishing the World Trade Organization in 1994 (WTO 1994b).[5] Thus it is argued that these issues should be addressed in their own right, and now with a view to how their interaction with trading arrangements institutionalised in the WTO, and elsewhere, can be managed.

In this context, it is important to outline the chief characteristics of the OECD guidelines, which address many of the issues of other CSR initiatives in both the domestic and international domains. It is also evident that processes associated with the review contributed to its ultimate endorsement at the OECD ministerial meeting in 2000, which may also have lessons for their applicability to other trade and investment agreements, including the WTO. Ultimately, this should help to begin to address how soft law approaches to labour and the environment (and perhaps others such as corruption and human rights) can interact constructively with a trade and investment framework.

At the outset, it is fundamentally important to realise that the guidelines are one of four integral elements in the OECD Declaration on International Investment and Multinational

Enterprises (OECD 2000e).[6] This declaration is a loose collection of instruments that allows different elements to be reviewed at different times. Thus, the guidelines were reviewed in 2000, whereas the National Treatment Instrument — the other main part of the OECD declaration — was last reviewed in the early 1990s. This flexibility continues to be an important factor in fostering the legitimacy of the declaration, as well as component instruments such as the guidelines, including with respect to their chief client group to which the guidelines recommendations are directed, namely OECD-based business organisations. By anchoring the guidelines within a framework of investment relations, business could more easily support a review that resulted in stronger and more focussed recommendations to it.[7] This in itself may be viewed as an important precedent for the trade agenda: notwithstanding differences between an investment-focussed, non-binding OECD declaration and the WTO trading system, a successful CSR initiative such as the guidelines should be viewed as simply one element within a broader framework for international investment relations.[8] Indeed, in the case of the guidelines, this relationship is symbiotic — it is difficult to envision the success or relevancy of either the declaration or the guidelines without the existence of the other.[9]

The Guidelines: Substantive Issues

The OECD's guidelines are recommendations from governments to business firms (OECD 2000b; Acconci 2001). These recommendations are substantive (contained in ten main chapters), as well as procedural (implementation procedures largely addressed to government representatives known as National Contact Points, or NCPs). In addition, 'commentaries' on both the chapters and implementation procedures supply a wealth of background and other contextual information for businesses, governments, trade unions, nongovernmental organisations (NGOs), and other interested observers.

Given the scope of the recent review, as well as the relevance to the nexus of CSR and trade and investment policy, it is useful to highlight a number of provisions, chapter by chapter. It should be noted that these provisions capture most of the concerns on the CSR agenda. They also include recommendations incorporating principles that have hitherto been the subject of discussion and agreement at the multilateral level. Perhaps the most important in this regard are those involving human rights, labour issues, and environmental standards, which not coincidentally also underlie the appeal of the United Nations Global Compact to firms through multilateral instruments on these issues that have been already negotiated (see Table 16-1).

Importantly, the guidelines also include principles more often associated with private or self-regulatory initiatives on the part of industry and industry associations responding to social pressure and changing market conditions. There is indeed a growing relationship between multilateral and self-regulatory initiatives, epitomised by the guidelines themselves. The inclusion of recommendations based on progress in self-regulation was a key accomplishment of the review and has unquestionably contributed to their broader relevance to governance issues, as discussed below.

A number of CSR issues may also be considered related to trade. This is not to say that they relate directly to the trade rules of the WTO or other institutions, but rather to international trade relations in a broader sense, whether in respect to the cost of inputs or terms of trade with respect to specific tradable goods or services, the attractiveness of a specific location for investment and trade, or, more broadly still, to the political and economic environment in which decisions on trade and investment are made. Viewed more narrowly, CSR issues may be considered trade related through the way in which they affect the trade and investment decisions of individual companies.

Chapter I (Concepts and Principles) of the guidelines outlines their scope, and is the sole chapter addressed chiefly to governments. Among the concepts of direct relevance to the trade and investment domain is an implicit recognition of the possibility of misuse in this regard: 'Governments adhering to the *Guidelines* should not use them for protectionist purposes nor use them in a way that calls into question the

Table 16-1 The Nine Principles of the United Nations Global Compact

The Global Compact's nine principles in the areas of human rights, labour, and the environment enjoy universal consensus being derived from:

- The Universal Declaration of Human Rights
- The International Labour Organization's Declaration on Fundamental Principles and Rights at Work
- The Rio Declaration on Environment and Development

The nine principles are:

Human Rights

- Principle 1: Businesses should support and respect the protection of internationally proclaimed human rights within their sphere of influence; and
- Principle 2: make sure that they are not complicit in human rights abuses.

Labour Standards

- Principle 3: Businesses should uphold the freedom of association and the effective recognition of the right to collective bargaining;
- Principle 4: the elimination of all forms of forced and compulsory labour;
- Principle 5: the effective abolition of child labour; and
- Principle 6: eliminate discrimination in respect of employment and occupation.

Environment

- Principle 7: Businesses should support a precautionary approach to environmental challenges;
- Principle 8: undertake initiatives to promote greater environmental responsibility; and
- Principle 9: encourage the development and diffusion of environmentally friendly technologies.

Source: United Nations (1999).

comparative advantage of any country where multinational enterprises invest' (OECD 2000b, 18). At the outset, then, the guidelines refer to this thorny issue — the ostensible competitive advantages that accrue to firms in countries with poor (and declining) labour and environmental standards. Although comprehensive literature surveys point to scanty evidence for this phenomenon (for example, Brown 2000; Nordström and Vaughan 1999), it is a powerful political assertion, particularly in developed countries undergoing structural changes emphasising the growth of 'cleaner' and capital-intensive industries. Similarly, it can sound ominous to developing countries seeking to secure enhanced market access in wealthy developing country markets. Following the Bhagwati et al. (1999) manifesto, this debate did not figure prominently in debates of the WTO after the Doha declaration, and it has played a relatively minor although nonetheless important role in WTO-related work since then. But this reflects a stalemate over these issues, rather than any resolution or consensus amongst the membership.

Chapter II of the guidelines is titled 'General Policies' and combines initiatives emanating largely from multilateral sources. It sets the stage for the rest of the instrument by pointing to the legitimate role of 'other stakeholders' in the countries and societies in which firms operate (OECD 2000b, 19). It also introduces the concept of 'sustainable development' (19) as introduced in the World Commission on Environment and Development's report in 1987. Two other recommendations are important. The first is the responsibility of firms to encourage 'corporate conduct compatible with the *Guidelines*' among suppliers and sub-contractors (OECD 2000b, 19) — an increasing challenge for firms outsourcing with perhaps hundreds of suppliers and locations with widely divergent social and environmental standards to meet competitive pressures. Second, a recommendation on human rights, responsibility for which is among the primary duties of nation-states, was also introduced: 'Enterprises should ... respect the human rights of those affected by their activities consistent with the host government's international obligations and commitments' (19).[10] As noted later, business has become a partner in the debate over human rights and trade, but there are limitations to how much can be expected without the ultimate support of the state in upholding these rights.

Chapter III (Disclosure) includes elements from both multilateral and self-regulatory sources, and embodies an important link between the guidelines and the corporate governance agenda (including an OECD instrument on the subject). It also encourages the provision of additional information elements on a company's 'social, ethical and environmental policies' (OECD 2000b, 20). Although pressure from government and securities authorities for improved corporate governance and reform has increased recently, it has also been a longstanding market-driven concept, extending at times to embrace triple bottom-line corporate reporting, as well as large institutional investor support for elements of the CSR agenda, including human rights (see below).

Chapter IV (Employment and Industrial Relations) was the most highly developed chapter in previous versions of the instrument. Indeed, the 1977 Tripartite Declaration of Principles Concerning Multinational Enterprises and Social Policy was influenced by the first version of the guidelines one year earlier — with the early participation of business representatives (ILO 1977). In 2000, the chapter was updated to mirror the 1998 ILO Declaration on Fundamental Principles and Rights at Work, so that it covers recommendations on all internationally recognised core labour standards (OECD 2000b, 21).[11] This was an important reflection of what had happened among ILO signatories (which includes all OECD members). In 1998, ILO members had committed to promote the declaration, even if they had not ratified relevant conventions — and now firms based in countries adhering to the guidelines were being encouraged to do the same (ILO 1998).[12]

Along with provisions for employment and industrial relations, recommendations on the environment are among the most important CSR provisions in the guidelines. The chapter on the environment (Chapter V) implicitly defers to the principles of a number of recent multilateral and regional agreements, including the Rio Declaration on Environment and Development (and Agenda 21 within it) and the Convention on Access to Information, Public Participation in Decision Making, and Access to Justice in Environmental Matters (United Nations Conference on Environment and Development [UNCED] 1992; United Nations Economic Commission for Europe 1998), all noted in Chapter V's commentary. It also does the same with respect to a precautionary approach to business activity where there are serious threats of environmental damage (OECD 2000b, 23). The chapter also contains a recommendation for establishing and maintaining an appropriate environmental management system (22), whose antecedents emanate largely from self-regulatory initiatives. Management systems that internalise environmental objectives are powerful tools in enhancing the profile of environmental issues in the policies and conduct of enterprises. There are also internationally recognised standards for doing so, the most successful of which are administered through the International Organization for Standardization's (ISO) 14000 standards.

The ISO's experience with environmental management standards represents a precedent for the consideration of social and labour standards in the self-regulatory domain, which as noted exist already in the multilateral domain through the ILO's Declaration on Fundamental Principles and Rights at Work. ISO committees have been conducting studies on this issue for a number of years. The introduction of such standards will likely have a major influence on the inclusion of more labour-related issues in firm management and ultimately conduct.[13] The fact that they have not been agreed to yet again reflects in part the sensitivity attached to two main issues, the first of which relates very much to trade: the fear of protectionism by developed countries against imports from low-wage economies (not strictly an issue of labour standards), as well as the fact that relevant ILO conventions are often ignored by ILO member states.

ISO activity is significant in these two issue areas of interest both to firms and to CSR. The ISO is, of course, a respected and longstanding part of the self-regulatory movement, with roots in the development first of scientific standards and then of quality standards. Links between the ISO and the trading system on these issues already formalised, principally through WTO decisions related to the Agreement on Technical

Barriers to Trade (TBT).[14] If the ISO is successful in developing social standards (and so far the experience with ISO 14000 standards has exceeded expectations), those standards will likely prove highly influential with respect to CSR on the corporate agenda, and ultimately possibly will be considered under international trading arrangements.

Chapter VI on Combating Bribery implicitly defers to the OECD Anti-Bribery Convention, although it addresses demand as well as supply — an important precedent for both the OECD and the Draft UN Convention against Corruption.[15] Pressure to combat bribery has been strongly backed by OECD members (particularly the U.S.) as well as by some firms and NGOs such as Transparency International. Finally, Chapter VII on Consumer Interests and Chapter VIII on Science and Technology defer to issues that have been of more interest in the past, namely consumer interests by NGOs and others interested in market framework policy, and science and technology — or more specifically technology transfer — to developing country states in particular (United Nations Conference on Trade and Development [UNCTAD] 2001). However, these issues have a lower profile than they did during the first wave of interest in CSR a generation ago. Chapters IX and X on Competition and Taxation, respectively, cover framework issues that are primarily the concern of governments. Although important, they are also not high on the contemporary governance agenda.

The 2000 review of the guidelines thus succeeded in incorporating a number of recommendations to enterprises based on progress at both the multilateral and particularly self-regulatory levels. It was the most comprehensive review of the guidelines by far, and succeeded in placing the instrument at the heart of debates over CSR in the international arena. Some of the provisions also were related to issues on the trade and investment system, not least among them the fear of using them for protectionist purposes. It also echoed and intensified work on the nine principles underlying the more universal UN Global Compact, as well as continuing work on these and other issues in the international business community. Importantly, some of this success was also due to the identity of the guidelines as part of a wider package of instruments in the OECD declaration, thereby confirming this important initiative for CSR within the international trade and investment framework.

The Guidelines: Implementation Procedures and the Review Process

Implementation procedures of the guidelines were also significantly revised during the 2000 review. Indeed, the guidelines were already unique in that they enjoyed a mechanism to aid in their implementation. Unlike the substantive provisions, which are recommendations directed to business, implementation procedures are directed primarily to governments.

Through a network of NCPs, governments are responsible for promoting the guidelines, handling inquiries, and helping to resolve issues that arise in specific instances. The review includes a section called 'Procedural Guidance' to help those NCPs in carrying out their duties, and includes suggestions for improving institutional arrangements, information, and promotion, and particularly for implementation in specific instances, as well as for reporting.[16] This procedural guidance was added in order to assist the NCPs in each country in becoming a key vehicle for encouraging implementation of the guidelines. What amounted to a decentralisation of implementation procedures freed them to respond to national and local concerns, subject to shared notions of what the OECD termed 'functional equivalence'. However, the OECD Committee on International Investment and Multinational Enterprises (CIME) remained the responsible body for clarifying the meaning of the guidelines and overseeing their effectiveness. In addition, social partners, particularly business and labour groups, were called upon to contribute to the smooth functioning of these procedures and, subject to individual government practices, may participate in the activities of the NCPs.

The guidelines review process also was an important element of the review's success. The CIME conducted extensive consultations with the business community, labour representatives, NGOs, and nonmember countries. Nonmember countries that did not adhere to the OECD instruments were consulted on related issues at two meetings at the OECD during the review (in September 1999 and February 2000). Opportunities for public comment were offered via the internet. This increased transparency and openness provided essential inputs and reflected an important evolution in the way OECD reviewed the instrument. In the wake of the failure of the Multilateral Agreement on Investment (MAI), this new commitment to transparency also contributed to the confidence OECD members had in endorsing by far the most comprehensive review of the instrument since its inception.

Once the review was complete, many spoke highly of the guidelines' potential contribution not only to the world of CSR, but also to their possible influence more broadly. OECD ministers began by noting in their June 2000 statement entitled 'Shaping Globalization' that 'the Guidelines provide a robust set of recommendations for responsible corporate behaviour worldwide consistent with existing legislation' (OECD 2000c). The revised guidelines were also welcomed by a relatively broad constituency, including representatives of business, labour, NGOs, and others, and the confirmation that they applied to the business operations of enterprises based in signatory countries world-wide also enhanced their relevance to a wider audience.[17] Notably, this development enhanced the guidelines' relevance to the trade and investment framework that facilitates them.

Admittedly, the fact that no NCPs existed in non-adhering countries was a complicating factor in the implementation of the guidelines. But the guidelines nonetheless represent expectations of enterprise behaviour wherever they operate. Clearly, further dialogue on this issue is necessary with developing countries, and has already begun.[18] Implementation procedures will differ by country in any case, whether or not there is an NCP directly involved.[19] Among countries adhering to the guidelines, how NCPs carry out their work is not important, as long as the results can be deemed to be positive by stakeholders. As long as 'functional equivalence' is achieved, flexibility can be seen as a key attribute of the implementation procedures. This is another advantage of soft law instruments, in that they can make it easier to accommodate different mechanisms for the observance of agreed-to norms and standards.

The contents of the guidelines versus other CSR initiatives have been documented elsewhere (OECD 2001b; Global Reporting Initiative 2000; World Bank Group–Consumer Protection in the Global Market Working Group 2002, 19). Although some overlap is evident, the relevance of individual recommendations in all instruments will vary from situation to situation. Reflecting this, the precise content of the recommendations also varies from instrument to instrument. Nonetheless, also apparent are recommendations consistent with evolving norms and standards affecting corporate conduct on social and environmental issues as well as on financial issues — and sometimes including attention to the means to achieve them. In this context, it can be noted that some CSR initiatives go further toward accommodating the interests of outside stakeholders in their implementation. In the draft operational guidelines of the Compliance Advisor Ombudsman, for instance, mechanisms have been included for handling complaints made by any community or group likely to be affected by a project of the International Finance Corporation or the Multilateral Investment Guarantee Agency, both parts of the World Bank Group (World Bank Group–Compliance Advisor Ombudsman 2003). On the other hand, the Global Compact initiative of the United Nations, despite working closely with private partners, makes it clear that, although it will defend itself from abuse of its programme, it will not formally assist in the observance of its recommendations.[20] Notably, however, a growing consensus on core provisions associated with human rights, environment, and labour characterising comprehensive initiatives as diverse as the OECD guidelines, the UN Global Compact, and the Global Reporting Initiative, supplemented by other issues including corruption and corporate governance, points to the contribution that CSR can make to international governance. This potential can be further advanced as trade-related aspects of these issues are considered in international trading arrangements.

Business Support for Corporate Social Responsibility

As has been the case with all recent successful CSR initiatives, business — through the OECD's Business and Industry Advisory Committee (BIAC) — was a willing participant during the guidelines review. This involvement reflected growing comfort on the part of international business, both individually and collectively, to participate in initiatives emanating from governments at the multilateral level that can aid in the management of thorny social issues.

The support and involvement of the business community are crucial for managing the linkages between CSR and the trading system. Without active business support, it is difficult to see how recommendations to business on CSR can serve as a useful reference point for dealing with these issues in the context of trade policy. How businesses respond to public and shareholder concerns about their conduct, both with respect to the self-regulatory movement as well as in co-operation with other stakeholders, is evident. And business is not immune to the potential market advantages of such tools (or any activities associated with them). As in the guidelines, should attention to CSR in the context of trade bear fruit for both domains, it will be partly because of the acceptance of assurances that short-term protectionist interests will not be indulged, but rather linked to an overall framework fostering openness, free trade, and transparency with respect to both CSR and trade.

As noted, self-regulation has provided considerable impetus to items on the CSR agenda. This contribution has been both substantial and relatively ignored (Haufler 2001). Businesses have been responding to public concerns over their activities in a variety of ways. Sectoral initiatives abound.[21] OECD research on well over 200 private initiatives in this regard indicate commitments covering a vast array of issues, including core labour standards, environmental stewardship, human rights, disclosure, corporate governance, public safety, protection of privacy, and consumer protection. Fair employment and labour rights as well as environmental stewardship are the most common broad issue areas covered in the codes, but they are by no means universal (OECD 1998).

Why have companies and industry associations increasingly adopted codes, many of which include mechanisms for aiding internal compliance? There are a number of reasons, including those of public concern and pressure from NGOs. But these are by no means the only influences on the growth of self-regulatory responses. Two of the most important have already been noted, and have largely evolved in what may be termed as corporate self-interest — but, in any case, with minimal public sector involvement. The first of these is ISO 14000 on environmental management, which enjoys widespread business, NGO, and, lately, government support (ISO 2002). The Global Reporting Initiative is another example, and encourages an international, multistakeholder effort to create a common framework for voluntary reporting of economic, environmental, and social impact activities.[22]

Another influence, partly in response to changing market preferences from investors (and consumers) is the growing ethical investment movement. Ethical investors, whether individual or institutional, still seek to make returns for their investments, but factor in social and environmental concerns as well as broad financial concerns (including corporate governance). This form of investment has spawned a new services industry in advice advocating the benefits of this perspective. For instance, when setting up their investment service in this vein in 1999, Dow Jones Sustainability Asset Management argued that 'companies with an eye on their "triple bottom line" — economic, environmental, and social sustainability [tend to] outperform their less fastidious peers on the stock market'.[23]

Another related reason for the growth in self-regulation is a growing tendency for large institutional investors to include corporate social responsibility in considering their investment strategies. In 1999, the board of one of the world's largest institutional investors, the California Public Employees Retirement System (CalPERS, with assets around US$146 billion), announced the endorsement of the Global Sullivan Principles on human rights. According to *Global Proxy Watch*, a newsletter on international corporate governance and shareholder value, if a firm breaches the Global Sullivan Principles, CalPERS 'would consider writing to and meeting with company executives, or sponsoring dissident shareholder resolutions'.[24] In Canada, one of the largest institutional investors (the Ontario Municipal Employees Retirement System, or OMERS), with assets of around CA$34 billion, also includes provisions for 'Socially Responsible Investing' in its statement of investment policies and procedures (OMERS 2003). In the UK, the Universities Superannuation Scheme (USS), with assets of more than £22 billion, writes of the fund's management committee 'monitoring expectations' of social, ethical, and environmental issues.[25] These are just a few examples.

Whatever the influences, self-regulation has had a profound influence on independent corporate acceptance and indeed on the internalisation of many of the concepts that underlie provisions in the OECD guidelines, the Global Compact, and other initiatives. This has been a gradual but nonetheless significant accomplishment. It should be recalled that business was often wary of unrealistic expectations with respect to CSR, particularly those emanating from multilateral sources.[26] To a greater or lesser degree, business plays a social role wherever it operates, but it should not be expected to usurp functions that are ultimately those of the state. Decrying the power of corporations while calling on them to exercise powers normally the prerogative of the state is neither fair nor realistic. While it may be legitimate to expect corporations to refrain from violating basic civil and political rights (that is, to uphold 'negative' rights as enshrined in the International Covenant on Civil and Political Rights), it is another to expect them to enshrine 'positive' rights to economic development and social rights such as those contained in the International Covenant on Economic, Social, and Cultural Rights (Kelley 2001, 532). It is the state that is primarily responsible for providing an environment conducive to upholding the negative human rights embodied in the International Covenant on Civil and Political Rights, as well as defending the positive human rights of the International Covenant on Economic, Social, and Cultural Rights. These are also vitally important issues rightly discussed, defended, and enforced in the political and legal arenas, and are ultimately a *sine qua non* of the state.

Nonetheless the growing relationship between self-regulatory and multilateral initiatives captured by the OECD guidelines in its recent review is an excellent example of soft law on all these issues taken in partnership between business and government, as well as with the co-operation of other actors. In this context, it may also be noted that business groups' increasing embrace of CSR in their own self-interest has also allowed them to support exercises such as the review of the guidelines — in particular recalling its characteristics as part of a broader package of instruments in the OECD declaration.

Summary

Given the widely differing interest groups involved, as well as the rapidly evolving norms and standards underlying new provisions, the guidelines' review in 2000 was highly successful. This is particularly true when measured against the very limited character of previous reviews. In turn, a number of additional serendipitous factors played a role. The first of these was timing. With the review coming on the heels of the failure of the MAI, OECD members were eager to promote a success and to show that they were heeding some of the public criticism associated with the processes surrounding the MAI. Second, the review incorporated elements of the self-regulatory movement as well as multilateral initiatives. It encompassed key CSR accomplishments whether emanating from business, NGOs, trade unions, governments, or international organisations. The OECD also made draft texts available on the internet for public comment and consulted with non-OECD members in the developing world. By focussing on the instrument as a means of helping to resolve specific problems in particular circumstances, it helped demonstrate the instrument as offering practical assistance to firms in difficulty.

As also demonstrated by the UN Global Compact, the guidelines review illustrated that multilateral initiatives could take account of and complement corporate self-regulatory efforts, thereby increasing the comfort level many businesses had with CSR-related instruments. Furthermore, and also reflecting the latter dynamic, the revised guidelines remained part of broader and revitalised investment instrument — the OECD declaration — which confirmed their place in the broader trade and investment framework in which firms operate and made them more relevant to broader questions of governance related to the nexus of CSR and international trade and investment. This was important to the international trade and investment framework itself, including to the way in which these issues could be addressed in trading arrangements at the WTO.

Corporate Social Responsibility and International Trading Arrangements

Corporate Social Responsibility, the World Trade Organization, and Other Trade Agreements

In the wake of the Bhagwati et al. (1999) statement and other concerns expressed by or on behalf of developing countries, Peter Sutherland, the first director general of the WTO, was among those who noted that an impasse on issues related to CSR affected prospects for future trade liberalisation under the WTO. As part of a volume exploring the role of the WTO in global governance, he and his co-authors also argued that a reluctance to address labour and environmental issues compromised the WTO's effectiveness.[27] Despite few concrete results on these issues since the Doha declaration, progress has been made, both at the WTO and with respect to the increasing attention they receive in relevant literature, including by trade specialists. Implications for further work in the WTO is already evident.

As has been noted, the complex relationship between the self-regulatory movements and multilateral initiatives has played a part in the success of CSR-related instruments.

In the same way that the success of the self-regulatory movement has been relatively under-appreciated in the growth in influence of CSR, the increasing role of private actors has also received relatively little attention, including in the trade domain. Although the subject matter of the WTO is agreements between governments, a nascent change in the profile of private entities is evident. In an investigation into human rights and the law at the WTO, Ernst-Ulrich Petersmann (2003, 274) noted that the agreements on preshipment inspection, government procurement, General Agreement on Trade in Services (GATS), the Agreement on Trade-Related Aspects of Intellectual Property Rights (TRIPS), and the Agreement on Trade-Related Investment Measures (TRIMs) already indirectly protect private parties. These entities have been implicitly rendered subject to the WTO, with references to exporters in the Agreement on Preshipment Inspection, service suppliers in GATS Article XXVIII, established and new enterprises in TRIMs, and service suppliers as well as users in the GATS Telecoms Reference Paper. Although the implications of this status are just beginning to be addressed, it is clear that one aspect of the evolution of the WTO is consideration of actors and interests other than nation-states in its agreements.[28] And indeed this evolution has been echoed in proceedings under the Dispute Settlement Understanding, not least by the Appellate Body when it disagreed in the shrimp-turtle case with the decision to exclude unsolicited *amicus curiae* briefs from two NGOs *a priori* (although it did not object to a panel's right to disregard them).

Under current WTO agreements, environmental and labour issues are implicitly included under the General Exceptions (Articles XX of the GATT and XIV) of the GATS, as well as the preamble to the Marrakesh Agreement establishing the WTO.[29] These have also been the subject of several decisions under the Dispute Settlement Understanding, at least with respect to environmental matters (Nordström and Vaughan 1999; Trebilcock and Howse 1999). Although Article XX exceptions continue to be interpreted relatively narrowly, the implications continue to be debated, not only for the environment, but also for labour issues, human rights, and other issues (Biermann 2001; Lim 2001; Marceau 2001; Petersmann 2003). In addition to exploring, for instance, the potential scope of Article XX(b) and perhaps (d) with respect to labour and human rights issues, or XX(b), (g), and perhaps (d) with respect to the environment, broader questions are now being asked. Indeed, with respect to the environment this has led two respected trade observers to ask whether Article XX amounts to a GATT environmental charter (Trebilcock and Howse 1999, 397).

Many efforts to expand the reach of WTO agreements in this vein are advocated under existing rules. For instance, a 1996 OECD study on trade, employment, and labour standards wondered whether trade in goods facilitated through neglected labour standards could constitute dumping under GATT Article VI or be considered an illegal subsidy under GATT Article XVI (OECD 1996). Others have gone further to suggest minor amendments to existing agreements. Kyle Bagwell and Robert Staiger (2001) have suggested that labour standards be dealt with under GATT Article XXIII on nullification and impairment.[30]

Environment and trade constitute a vast and more mature subject area than do CSR and trade. Therefore, environment and trade serve as a useful proxy for how aspects of CSR issues might be addressed in the broader debate. Above all, the complexity of the subject area is evident, with many subsidiary questions, including how to approach assessing the environmental effects of trade agreements, the subject of continued debate (OECD 2000a). Partly in recognition of these complexities, a number of suggestions for dealing with these issues under existing WTO arrangements have been made. There is the implication that negotiations particularly on labour issues, but also with respect to the Committee on Trade and Environment (CTE) and related work, would not only be unproductive, but are perhaps superfluous as well.[31] Indeed, this has been implicitly recognised by the WTO membership since the Singapore Declaration (WTO 1996).

Critics of the shrimp-turtle case were quick to argue that environmental causes were not served, since what can be may be deemed 'unjustifiable discrimination' or a 'disguised restriction in international trade' in the *chapeau* to Article XX is interpreted rather broadly. However, there was an apparent recognition of the contextual issues, including with respect to *amicus* briefs, the scope of Article XX(g), as well as the importance of the consistency between WTO agreements and the objective of sustainable development.[32] This, plus a recognition of the importance of environmental issues in other WTO agreements, as well as the institution of the CTE, point to an organisation already keenly aware of the sensitive tasks of reconciling free trade with other legitimate priorities. Nonetheless, given its adversarial nature, a key problem with cases under the Dispute Settlement Understanding is that, ultimately, the result of a case — even if successful — would be a right of retaliation that would not necessarily help the immediate environmental, human, labour, or other rights under threat. In addition, it would take a large number of challenges even to begin to determine acceptable practices under existing rules.

Apart from within the Dispute Settlement Understanding, it also appears that there is little appetite to address these issues more directly under existing rules, and still less to negotiate new ones. This is particularly true of core labour standards, although as Michael Trebilcock and Robert Howse (1999, 462) point out, developed countries can and do take them into account when they impose their own sanctions through the withdrawal of Generalised System of Preferences (GSP).[33] The picture is slightly more complicated with respect to environmental considerations — as noted, they are already taken into account in a number of WTO agreements, including exceptions to GATT (Article XX), GATS (Article XIV), as well as in the agreements on TBT, sanitary and phytosanitary measures, agriculture, and TRIPS, and in the CTE itself (see WTO 2003c).

Nonetheless, it can be noted that despite the reluctance of developing countries in the WTO to examine the links between many CSR initiatives and the trade and investment system, as the review of the OECD guidelines revealed in 2000, these sorts of links were acceptable to Argentina, Brazil, and Chile, all of which participated in the review. In much the same way as the business community found the guidelines useful as an instrument within the OECD declaration, this demonstrates that developing countries, too, can recognise the advantages of anchoring CSR within a broader trade and investment context. Admittedly, this is not yet universally shared, but there are signs that this may be a plausible *modus operandi*.

Interestingly, at the WTO, it is also when investment has been discussed at the Working Group on the Relationship between Trade and Investment that CSR issues have been raised. Two documents, one published by the European Community in 1999 and one from developing countries China, Cuba, India, Kenya, Pakistan, and Zimbabwe, and released in 2002, raised the issue of links between investor behaviour and the prospect of investment rules (WTO 1999; 2002). Although the developing country submission did not highlight labour issues or human rights in this context, it nonetheless noted a number of other subjects on the CSR agenda, among them disclosure and accounting, consumer protection and environmental protection, and technology transfer. It also raised the prospect of 'legally enforceable norms of investors' or corporate conduct' in the context of an international investment agreement. It may be briefly noted that while international investment, which occurs 'behind the border' and therefore has historically been considered part of the purview of domestic policy, increasing complementarities between investment and trade — as well as the decline in traditional trade barriers — mean that investment is now seen as a relevant issue for governments whenever international trade agreements are discussed. The same is beginning to be true of CSR, particularly *vis-à-vis* investment. Indeed, in its 2003 *World Investment Report*, UNCTAD (2003, 167–170) also notes the importance of 'good corporate citizenship' in helping developing countries to maximize the benefits of foreign direct investment, and sets out a number of approaches for addressing these issues in international investment agreements.[34]

In addition, there are also indications that developing country NGOs as well as their governments would be willing to support such as discussion. In its consideration of investment issues at the WTO in 2001, for instance, the Consumer Unity and Trust Society (2001), an influential Indian NGO, raised the issue of a code of conduct for 'regulating investors' and advocated the inclusion of provisions on corruption, accounting and reporting, transfer pricing, technology transfer, information disclosure, environmental and labour standards, dispute settlement, etc.'

There are indeed many examples of developing country acceptance of CSR-related issues that have already been incorporated in international trade and investment agreements, mostly with the U.S. (The tradeoff for market access to the world's largest market is, of course, an important element in developing country considerations.) One approach exists in NAFTA, whereby provisions in the agreement state that environmental or labour standards will not be lowered in order to attract investment are coupled with commitments to enforce (and work to enhance) domestic laws in these areas.[35] A newer version of this approach, which has enjoyed support from the labour community, is to elevate these sorts of commitments to the treaty itself, as has been done in Articles 5 (on the environment) and 6 (on labour) of the United States–Jordan Free Trade Agreement

(United States Trade Representative [USTR] 2001). This has been done in the U.S.–Singapore and U.S.–Chile agreements as well.

Notably, the U.S. has recently reasserted the links between environmental and labour issues and the trading system through the publication of a 'toolbox' that includes an illustrative list of actions that could be taken 'in combination with trade negotiations to promote the protection of children, adherence to core labor standards, and mutually supportive trade and environmental protection policies' (USTR 2002).[36] As noted earlier, the EU includes human rights in the preambles to its trade agreements. And in the negotiations for the Free Trade Agreement of the Americas (FTAA), the accommodation of civil society interests at the negotiating stage suggests that CSR issues will also be considered in this — and other — free trade agreements in the future.[37]

Corporate Social Responsibility and the World Trade Organization's Trade Policy Review Mechanism

There have been several calls from environment, labour, and human rights specialists that their concerns warrant greater attention at the WTO, whether under existing trade agreements or under amended agreements. These calls have been echoed by some observers more closely associated with trading arrangements *per se* (Charnovitz 2002; Trebilcock and Howse 1999). More directly, Susan Aaronson and James Reeves (2002) have called for the USTR to 'link trade and global CSR' explicitly.[38] Glen Kelley (2001) has argued for greater 'balance' with respect to multinational corporations and investment treaties.

As observed at the outset, however, little practical attention has been devoted to the question of how to link these two domains, particularly in the absence of further negotiations on these issues or through the Dispute Settlement Understanding. One suggestion worth additional consideration is attention paid to CSR through the third of the three pillars of WTO institutions: the Trade Policy Review Mechanism.[39]

To date, principal advocates of this approach have been noted with respect to environment issues through the USTR's Labor and Environment Toolbox, and with respect to labour issues by Bill Jordan, General Secretary of the International Confederation of Free Trade Unions (ICFTU), and in passing by two other labour specialists (Wolffgang and Feuerhake 2002).[40] Little consideration has been given to practicalities in this regard, however.

The possibility of including pertinent CSR issues among those considered through the TPRM may not, in fact, require any further amendment of existing rules. The formal text governing WTO trade reviews is reproduced in Appendix 16.1. While it focusses on trade-related issues pertaining to the WTO agreements, the introductory paragraph and arguably other provisions give scope for including broader issues related to the trading environment.

It is notable that until 1996, the TPRM contained a section entitled 'Foreign Direct Investment and Trade' in its country reports (Keesing 1998). Since then, it is usually contained in the introductory section entitled 'The Economic Environment' (sometimes 'Recent Economic Developments'), or in a second section called 'Trade Policy Regime: Framework and Objectives' (or a variation). Implicitly deferring to the links between trade and investment, and in contrast to the immediate prospects for comprehensive investment rules at the WTO, the domestic investment regime is nonetheless accepted as a legitimate and important area of inquiry in trade policy review, in both government and secretariat reports, in both developed and developing countries.

The way in which the current mechanism works has been discussed elsewhere (for example, Curzon Price 1991; Keesing 1998; Laird 1999; WTO 2003d). Although the terms of reference for writing country reports are relatively tight, there is considerable latitude for government contributions to indicate issues that governments consider important. Following from this, there is also a relationship between the issues highlighted by the government and coverage in the secretariat's report. These issues can include trade-related aspects of issues on the CSR agenda. There are several recent examples of this occurring: in the recent review of Honduras, for instance, domestic efforts against corruption were highlighted by both the government and WTO secretariat (WTO 2003b).[41] Notably, the same was true with respect to environment and labour issues in the U.S. reports, when the U.S. repeated that it recognised a connection between these issues and trade issues (WTO 2001b; 2001c).[42] In addition, at subsequent meetings of the Trade Policy Review Body, there is nothing stopping members from addressing other trade-related issues they consider important when questioning the member under review.

There is no obvious window for the incorporation of CSR issues in trade policy reviews as such. However, as some reviews have already indicated, and in light of the underlying importance of transparency to WTO agreements, including the 'inherent value' of domestic transparency referred to in the text of the TPRM, CSR-related issues are not excluded from consideration. Indeed, they may be argued to be encompassed by the text of the TPRM, which states that in the interests of the 'smoother functioning of the multilateral trading system ... the review mechanism enables the regular collective appreciation and evaluation of the full range of individual Members' trade policies and practices and their impact on the functioning of the multilateral trading system" (WTO 1994c, sec. A[i]).

In another manifestation of the flexibility inherent in the TPRM, Article A(i) separates the reviews from dispute settlement procedures. Although not formally noted in the TPRM (which was an 'early harvest' of the Uruguay Round in 1989), processes associated with it may be characterised the world community's attempt to institutionalise 'peer pressure' as a means of furthering the goals of the WTO, set up to complement dispute settlement procedures as well as other rules (Curzon Price 1991, 238).[43] This non-adversarial aspect is a key attribute in any discussion of how to approach evolving CSR-related norms and emerging standards in the WTO.

Again, the TPRM could fulfil this function under existing mandates. The format of the reports is based on one established by a 1989 decision (BISD 36S/406-409) 'amended as necessary to extend the coverage of reports to all aspects of trade policies covered by Multilateral Trade Agreements' (WTO 1994c, sec. D); the wording goes on to say that 'this format may be revised by the [Trade Policy Review Body] in the light of experience' (sec. G). The review board, in turn, is a body of official WTO representatives of member countries that choose to participate in a country's review (Keesing 1998). The review board is also responsible for issuing an 'overview of developments in the international environment which are having an impact on the multilateral trading system', assisted by an annual report from the Director General 'setting out major activities affecting the trading system' (sec. G). To date, the Director General overview reports have varied considerably from year to year.[44] The lengthiest report to date, released in 2001, sounded a note of caution with respect to regional trade agreements for the WTO system (a theme common to other reports in this series), and included a section on controversies to do with health issues and the trading system, in addition to addressing issues related to WTO agreements. It would therefore not be unprecedented for future reports to address additional issues, including those associated with corporate responsibility (see WTO 2001a, ch. 3).[45] The same is true of annual reports of the Trade Policy Review Body, which are also mandated as part of the TPRM under Article G, although these reports have thus far all been very brief and limited to a factual account of reviews undertaken in the previous year.

In any case, if uniform reporting on trade-related CSR issues proved particularly problematic with respect to country reports, the annual reports of the Trade Policy Review Body under Section G of the TRPM — either by the review board or in particular through the Director General's overview — would appear to facilitate investigations of more systemic aspects of links between CSR and trade. With respect to individual countries, recent experience has already indicated that countries can volunteer information on these issues in the government reports. A voluntary but co-ordinated approach to doing this among like-minded countries would contribute additional momentum in this regard and add to the legitimacy of this approach under the existing mechanism.

It should also be recalled that in the absence of such work at the WTO, it is proceeding apace outside the organisation. As noted, dialogue is beginning to take place as part of the self-regulatory movement at the ISO, the Global Reporting Initiative, within sectors, and by individual businesses, and also within states, multilaterally (including at the OECD), as well as at the UN through the Global Compact initiative. Nor can public interest in these questions, particularly through NGOs, be expected to abate.

In addition, it is worth noting that relatively independent of dialogue with the trading system, work is also continuing at a rapid pace. This has considerable potential to affect trade and investment flows. For example, the UN's Office of the High Commissioner for Human Rights ([OHCHR] 2003a) was mandated to report to the WTO at the 2003 Cancun ministerial meeting on 'human rights, trade and investment'.[46] It had already presented a series of reports on human rights and trade to the United Nations Economic and Social Council (ECOSOC), which itself continues work on a code of responsibilities for international firms (see OHCHR 2003b, since updated; 2003c). Very shortly after the formal adoption of the Rome Statute of the International Criminal Court in 1998, corporate liability for breaches of international criminal law under ICC rules was being discussed, noting its roots in domestic law (Schabas 2001).

The interests of the trading system, whether articulated by governments or other interested parties, warrant consideration in this debate. The TPRM could implicitly complement all this activity, and would offer a means of devoting attention to this work that emanates from other areas in the academic arena and within member states. (The work of other international organisations can already be accommodated by virtue of their status as observers at meetings of the Trade Policy Review Body [Keesing 1998].) It should be understood that the TPRM should continue not to impose new policy commitments on members. However, reports that incorporate findings based on standard TPRM processes — as well as those that address issues related to, for example, environmental or labour standards issues that have influenced debates over trade-related implications issues in other countries — would help act as an early warning system for WTO members under review. Any resulting reports would still be expected to enjoy influence and contribute to debate among decision makers and policy makers domestically. The debates of the review board over country reports would also be expected to have an influence on other more conventional trade-related debates elsewhere in the WTO, including under the Dispute Settlement Understanding.

Flexibility is a necessary attribute that would be encompassed if CSR-related issues were to be among those addressed in TPRM country and annual reports. Even among CSR initiatives such as the OECD's guidelines, this was a feature through the institution of functional equivalence for NCPs. It is necessary because social concerns with respect to corporate conduct are not static. Work on CSR-related issues in the 1970s and 1980s, for instance, revealed an emphasis on disclosure and accounting, as well as bribery and corruption, rather than labour and the environment.[47] Again, the TPRM is not 'intended to serve as a basis for the enforcement of specific obligations under the Agreements or for dispute settlement procedures, or to impose new policy commitments on Members' (WTO 1994c, sec. A[1]) — a vitally important provision in the context of CSR, where standards simply do not exist in all areas. In order to be acceptable to all members, not to mention in the interests of global governance, attention to CSR issues in TPRMs must not be seen as an overt threat to non-discriminatory trading practices or to prospects for the continued evolution of CSR outside the WTO.

Further work is necessary on exactly how to accommodate relevant issues in the context of a TPRM examination in country or annual reports. Nonetheless, as this chapter has illustrated, there is a growing consensus among a rather disparate group of observers that the detailed recommendations of the OECD guidelines or the broader nine principles of the UN Global Compact embody an excellent starting point. There would also be nothing to stop consideration of, for example, traded goods and human rights abuses in circumstances relevant to an economy under review. There is a growing convergence over the appropriateness of discussion of human rights, labour, environmental, and other issues when trade and investment issues are discussed, and it is important for the trade policy community to engage productively in this debate. The TPRM offers a means of doing so, and facilitates it at a sufficiently concrete level of detail and relevance.

The TPRM does not necessarily have to be used or seen to be used as a mechanism for the enforcement of CSR, which is evolving rapidly in its own right, including through self-regulation in the private sector. Rather, the emphasis would be on exploring the trade-related aspects of these issues in a nonlegalistic, non-adversarial manner that is nonetheless at the heart of a better understanding of economic and other relations between states. As this work developed, the WTO would become an important repository internationally, the importance of which will increase as the field of inquiry matures.

This change of emphasis in the way in which the Trade Policy Review Division of the WTO performs on behalf of WTO members would imply greater resources to be devoted to the TPRM, which is already overstretched (Keesing 1998). Recalling commitments to greater domestic transparency in Article B and assessment under Article C(iii), it will also be important to reinforce other understandings and assuage concerns in the wider world. The developing world must not have cause to fear protectionism in a new guise, but rather needs a means of assisting with the furtherance of trade in its broader context.

Conclusion

The OECD guidelines and other CSR initiatives exist and influence (and in turn are influenced by) the trade and investment framework. Institutionalising this relationship — without at this stage formalising it more than necessary — would benefit the understanding of the interaction between labour, environmental, and other goals that are espoused and defended in most codes and in the self-regulatory movement and a trade and investment system based on non-discriminatory treatment of goods and services traded between states.

The formal institutionalisation of CSR initiatives in WTO agreements with dispute settlement procedures is not a practicable way to establish the groundwork for this relationship. But insulating the WTO from this evolving debate will not serve it well either, given that commitments to enforce existing standards in agreements already in place (such as NAFTA) within the OECD declaration and given recent experience with the OECD guidelines, the evolution of corporate practice, and the acceptance of aspects of the debate in the WTO (in the CTE and elsewhere). And it should also be recalled that the CSR norms and standards, as well as reporting and relevant management systems — not to mention the international legal environment for these issues — are all rapidly evolving.

The informal institutionalisation of CSR initiatives in the TPRM, however, is one possibility that could facilitate a rational and productive dialogue within the WTO. In addition to being explicitly non-adversarial, such a facility would be less of an administrative burden on its developing and, particularly, least-developed members. Reporting by most of these members is done at six-year intervals, as opposed to every two years for the members considered most important in terms of their share in world trade, and every four years for the next 16 most important members in these terms. Technical assistance for reporting for developing countries and the least-developed countries is also facilitated in the TRPM.

More work is necessary into how to incorporate CSR-related issues such as those included in the OECD guidelines or Global Compact in the context of trade, or indeed on whether it should be precisely specified. But there appears to be scope for doing so in the WTO, especially through the TPRM, and in particular by member governments in their reports. To this end, the following premise should underlie this consideration: paying greater attention to the implications of CSR in the international trading system and intensifying work on non-discriminatory trading arrangements are not mutually exclusive concepts. Shifting balances of rights and obligations accorded by states to each other, as well as to corporations and other private entities — including with respect to human and economic rights in a world economy committed to sustainable development — make progress on this understanding fundamentally important. The review of the OECD guidelines conducted in 2000 offers reasonable grounds for optimism in this regard, as do the Global Compact and the activities of many other actors, including trading and investment entities within the self-regulatory movement.

The use of the TPRM risks further burdening a WTO secretariat and membership already overstretched. However, the non-adversarial and relatively informal nature of the mechanism offers a promising means of making progress on the effectiveness of environmental, labour, and other CSR-related norms and standards within the context of improving prospects for a rules-based, non-discriminatory trading system in the WTO. This would also be a constructive contribution to the not incidental goal of enhancing global governance.

Notes

1. Parts of this chapter are based on presentations made on behalf of the Organisation for Economic Co-operation and Development to the Evangelische Akademie/Brot für die Welt conference on 'The Challenge of Better Practice in the Oil and Gas Industry', in Mülheim an der Ruhr, Germany, in November 2000 and to the World Trade Organization's Working Group on Trade and Investment in Geneva in March 2001. The opportunity to work with all these groups, as well as the support of colleagues at Industry Canada in Ottawa, is gratefully acknowledged, as are comments on an earlier draft of this text from John Hancock, Vern MacKay, Peter Muchlinski, and Rico Zampetti. The views expressed herein are solely the author's.
2. When the term 'global governance' received its first main fillip in the early 1990s (under an eponymous commission established under the chairmanship of Ingvar Carlsson and Shridath Ramphal), international trade issues received little attention on its agenda. The introduction

to the Commission on Global Governance's report stated: 'Governance is the sum of the many ways individuals and institutions, public and private, manage their common affairs … At the global level, governance has been viewed primarily as intergovernmental relationships, but it must now be understood as also involving non-governmental organizations, citizens' movements, multinational corporations, and the global capital market' (Carlsson and Ramphal 1994). A few years later, the Centre for the Study of Global Governance at the London School of Economics and Political Science defined global governance as a minimum framework of rules necessary to tackle global problems guaranteed by a diverse set of institutions including both international organisations and national governments. Although vague, the term enjoys increasing currency: the academic journal *Global Governance* was founded in 1995, it was one of the themes of the 2003 World Economic Forum meeting at Davos, and the Brazil's Universidade Federal do Rio Grande do Sul also manages the Research Centre for Global Governance, which maintains links with the World Social Forum. Governance and development had long been a key concern of the World Bank, encompassing a focus on 'public sector development' (see World Bank 1992) and corruption. The latter is also one of the issues that features on the corporate governance agenda (along with issues such as disclosure and transparency), which in turn are also among the concerns on the corporate social responsibility and — ultimately — global governance agendas. Following the birth of the World Trade Organization, multilateral trade and investment issues also became core issues for global governance (see, for example, Sampson 2001; Kirton and von Furstenberg 2001).

3. The term 'corporate social responsibility' has broad connotations, implicitly including environment issues, human rights, labour issues, and others typically covered by comprehensive instruments such as the Organisation for Economic Co-operation and Development's guidelines and the United Nations Global Compact. Calls by some to increase attention to monitoring and verification issues recently have led some to prefer the term 'corporate accountability', but given the continued evolution of underlying norms and standards this may be argued to be premature. The term 'corporate social responsibility' is used throughout this chapter.
4. The Group of 77 (G77) of developing countries, in its 'Declaration of the South Summit' in Havana 2000, strongly echoed the Bhagwati et al. position: 'while recognizing the value of environmental protection, labour standards, intellectual property protection, indigenous innovation and local community, sound macroeconomic management and promotion and protection of all universally recognized human rights and fundamental freedoms, including the right to development, and the treatment of each issue in its competent international organization, we reject all attempts to use these issues as conditionalities for restricting market access or aid and technology flows to developing countries' (G77 2000).
5. See, for example, European Community (2000; 1999). With respect to the WTO, the reference to sustainable development was cited in an important Appellate Body report in the shrimp-turtle case, which, as the report noted, also underpinned the WTO decision to establish the Committee on Trade and Environment (see WTO 1998).
6. The other principal instrument of the declaration is the 'National Treatment Instrument', which contains commitments by governments adhering to the declaration to treat foreign enterprises based in countries of other adhering governments as domestic. Two other elements of the declaration include commitments to consult on 'Conflicting Requirements' imposed on enterprises from different jurisdictions, as well as on 'Investment Incentives and Disincentives' (see OECD 2000e). There are two other OECD investment instruments, known as the OECD Codes of Liberalisation (the Code of Liberalisation of Capital Movements, and the Code of Liberalisation of Current Invisible Operations; see OECD 2003a; 2003b).
7. The link between the guidelines and the rest of the package of instruments in the OECD declaration is continuously noted by the OECD's Business and Industry Advisory Committee (BIAC). See, for example, BIAC (2000; 2001).

8. There are examples of bilateral and multilateral comprehensive trade agreements that include provisions for labour and environmental issues. The first is the North American Free Trade Agreement (NAFTA), which in 1993 included side agreements on co-operation and the setting up of institutions to further the observance of labour and environmental standards in all three NAFTA countries (see the North American Commission for Labor Cooperation <www.naalc.org> and the North American Commission for Environmental Cooperation at <www.cec.org>). These commitments were elevated to the status of provisions within the agreement in the 2000 U.S.–Jordan Free Trade Agreement (United States Trade Representative 2001), as well as in the 2003 U.S.–Singapore and U.S.–Chile free trade agreements (see, for example, United States Trade Representative 2003).
9. A somewhat oblique but nonetheless important relationship between investment and trade also exists under current WTO rules. Despite the lack of a comprehensive investment agreement *per se* in the WTO, by virtue of the third mode (that is, the commercial presence) of four modes of service delivery, the General Agreement on Trade in Services (GATS) is an investment agreement with respect to investment in services industries. In addition, the Agreement on Trade-Related Investment Measures (TRIMs), the Agreement on Trade-Related Aspects of Intellectual Property Rights (TRIPS), and the Agreement on Subsidies and Countervailing Measures (ASCM) all pertain to investment behaviour to a greater or lesser degree.
10. The Universal Declaration of Human Rights is noted in the commentary to this chapter.
11. Following the inclusion of ILO Convention No. 182 on Prohibition and Immediate Action for the Elimination of the Worst Forms of Child Labour (including child prostitution and pornography, the use of children to traffic in drugs, and work that is likely to harm the health, safety, or morals of children) in 2000, the ILO declaration includes the following: Convention No. 138, Minimum Age for Admission to Employment (1973); Convention No. 87, Freedom of Association and Protection of the Right to Organize (1950) and No. 98, Application of the Principles of the Right to Organize and to Bargain Collectively (1951); Convention No. 29, Forced or Compulsory Labour (1932) and No. 105, Abolition of Forced Labour (1959); and Convention No. 111, Discrimination in Respect of Employment and Occupation (1960) and No. 100, Equal Remuneration of Men and Women Workers for Work of Equal Value (1953).
12. Article II of the ILO Declaration includes the following: 'all Members, even if they have not ratified the Conventions in question, have an obligation arising from the very fact of membership in the Organization to respect, to promote and to realize, in good faith and in accordance with the Constitution, the principles concerning the fundamental rights which are the subject of those Conventions' (ILO 1998, para. 2). Not one OECD member had formally ratified all eight conventions underlying the 1998 ILO declaration at the time of the guidelines review in 2000. Even by late 2003, no underlying Convention of the ILO Declaration enjoyed universal ratification among OECD countries (ILO 2003).
13. Health and safety issues already figure prominently among corporate codes (see Global Reporting Initiative 2000; OECD 1998). For recent work on ISO social standards, see in particular the report of the Consumer Protection in the Global Market Working Group of the ISO Consumer Policy Committee (2002b); also the International Institute for Sustainable Development website (2003) on 'Standards and Trade'. Another system of workplace standards (SA8000) and verification is run by Social Accountability International (2003).
14. See WTO (1994a). Both this decision and the annexes to the TBT agreement implicitly recognise the development of ISO standards in particular in technical domains consistent with non-discriminatory trade practices.
15. At time of writing, the UN Convention against Corruption was still under negotiation (UN 2003).
16. See especially the section titled 'National Contact Points' (OECD 2000b, 35–37), as well as further elucidation in the relevant commentary (58–62).

17. In addition to statements of support from business, labour, and NGOs associated with the review at the time of the OECD ministerial meeting in June 2000, representatives from these groups continue to draw attention to the instrument. See, for example, the issue of the *OECD Observer* entitled 'Spotlight on the OECD Guidelines for Multinational Enterprises' (OECD 2000d), which included short articles from Bruno Lamborghini, Chairman of Olivetti Lexikon, S.p.A ('Leading from the Front'), John Sweeney, President of the AFL-CIO ('Putting the Guidelines to Work'), and Patricia Feeney, Human Rights Advisor, Oxfam UK ('The Trust Business'). For reaction in the mainstream media, see Dale (2000).
18. Initial reaction from non-OECD countries was relatively circumspect, including with respect to the broader implications for the guidelines and international trade (see OECD 2001a). Among developing countries that participated in the review (such as Brazil), however, reaction was more positive. See in particular Lauro Eduardo Soutello-Alves' article, 'A View from the South', in the *OECD Observer* (OECD 2000d).
19. Operational procedures for NCPs are summarised by country in the OECD's 2001 Annual Report on the guidelines (OECD 2001b, Annex 1).
20. UN secretary general Kofi Annan, in unveiling the Global Compact at Davos in 1999, 'challenged world business leaders to "embrace and enact" the Global Compact, both in their individual corporate practices and by supporting appropriate public policies. These principles cover topics in human rights, labour and environment' (UN 2001).
21. Examples include the European Chemistry Industry Council's Responsible Care programme <www.cefic.be>; the Global Mining Initiative <www.globalmining.org>, which concluded in November 2002, and is succeeded by the International Council on Mining and Metals <www.icmm.org>; the Rugmark Foundation <www.rugmark.org>, overseeing trade in carpets; and Fairtrade Labelling Organizations International <www.fairtrade.net>, a programme concentrating on agricultural foodstuffs from the developing world. See also Appendix C for mandatory reporting requirements on social and environmental standards, as well as Appendix D for a summary of Standards and Guidelines in 19 countries surveyed in the KPMG International Survey of Corporate Sustainability Reporting (2002).
22. See the Global Reporting Initiative's website at <www.globalreporting.org>. In the same vein, the Canadian government sponsored the development of the Sustainability Reporting Toolkit, which facilitates online reporting against sustainability standards consistent with the Global Reporting Initiative (Government of Canada 2003).
23. *Economist*, 11 September 1999, p. 24. The study, released as the 'Dow Jones Sustainability Group Index', maintained that the world's top 200 'sustainable' firms tended to outperform their peers in almost all sectors, including basic materials, energy, and technology.
24. *Global Proxy Watch*, 17 September 1999. A senior CalPERS lawyer confirmed that the organisation would also apply its social responsibility criteria to non-U.S. corporations, the newsletter also noted. The human rights, environmental, and labour principles embodied in the Global Sullivan Principles of Social Responsibility (2003) are an offshoot of the Sullivan principles developed for corporations investing in apartheid South Africa.
25. See 'Statement on Socially Responsible Investment' in the Responsible Investment section and the 'Statement of Investment Principles' at the USS website <www.usshq.co.uk>.
26. A more market-oriented school of thought warns firms of too much distraction from their core business, and that market signals are adequate, including with respect to social issues. 'The social responsibility of business is to increase its profits', wrote Milton Friedman (1970). See a more recent version of this argument in Henderson (2001).
27. Peter Sutherland, John Sewell, and David Weiner (2001, 104) wrote that: 'pressure for the inclusion in trade agreements of provisions on labour standards and the environment has been firmly opposed by developing country governments. The resulting tensions have damaged the WTO's effectiveness and played a significant role in stalling further liberalization efforts. These pressures have been brought to bear on the WTO not only

because the institutions that might be expected to deal with labour and environment issues either do not exist or are weak.'

28. In the investment domain, this is even more evident with the growth of investor-state dispute settlement, as well as the setting up of mechanisms to deal with labour and environmental issues.
29. Exceptions relating to the import or export of gold or silver, products of prison labour, and others, make the GATT list in Article XX longer (but not necessarily more comprehensive) than that of GATS Article XIV.
30. Under this suggestion, countries that weaken labour (or presumably environmental) standards are viewed as implicitly altering the balance of rights attained by virtue of the GATT, and would have to offer foreign suppliers a tariff reduction to restore the volume of trade to the previously agreed upon level. It is suggested that this would weaken tendencies to race to the bottom, or the chill over environmental regulations.
31. With respect to labour issues, the WTO ministers renewed their commitment to the observance of internationally recognised core labour standards, which is the prerogative of the ILO (WTO 1996, para. 4). They also stated that: 'We believe that economic growth and development fostered by increased trade and further trade liberalization contribute to the promotion of these standards. We reject the use of labour standards for protectionist purposes, and agree that the comparative advantage of countries, particularly low-wage developing countries, must in no way be put into question. In this regard, we note that the WTO and ILO Secretariats will continue their existing collaboration.' Environmental issues have a much higher profile at the WTO. After accommodating the possibility of discussing trade and environment issues in the GATT, the CTE was established in Marrakesh at the birth of the WTO (see WTO 2003a).
32. The Appellate Body in the shrimp-turtle case noted twice that the preambular language to the WTO Agreement demonstrated 'a recognition by WTO negotiators that optimal use of the world's resources should be made in accordance with the objective of sustainable development'.
33. Recent examples include the withdrawal of GSP for trade with Burma by Canada, the European Union, and the U.S., as well as with Belarus by the U.S. It should be noted that as tariff rates decline over time, this option loses its potency, which also suggests that the issue could be productively addressed in the WTO.
34. Americo Beviglia Zampetti and Torbjörn Fredriksson (2003) also point to the fact that international agreements addressing these issues can assist in enhancing internal governance processes in developing countries.
35. Article 1114.2 of the NAFTA text on 'not lowering standards' pertains only with respect to the environment, which is coupled with a side agreement that includes institutional mechanisms for follow-up on environmental law enforcement through the North American Agreement on Environmental Cooperation (NAAEC). Labour issues are covered by a side agreement to which signatories pledge domestic enforcement of laws and other regulations in this area, facilitated through the North American Agreement on Labor Cooperation (NAALC).
36. The toolbox also highlights the WTO in this regard.
37. In addition to soliciting civil society contributions, FTAA negotiating parties have instituted the Committee of Government Representatives on the Participation of Civil Society (Free Trade Area of the Americas 2003).
38. Aaronson and Reeves (2002, 99) call on the USTR to 'examine whether and to what extent trade and investment agreements and codes of conduct should be linked as a way to build greater public support for international trade and investment agreements. Corporate conduct is still almost exclusively regulated at the national level. Trade and investment agreements delineate the rights of international corporate actors, but say little about corporate responsibilities or the rights of citizens. Linking trade agreements to codes of conduct could formally delineate the rights and responsibilities of international actors.'

39. Article IV of the Marrakesh Agreement established the General Council responsible for carrying out the functions of the WTO. The council also convenes as appropriate to discharge the responsibilities of the 'Dispute Settlement Body' (Article IV.3) and of the 'Trade Policy Review Body' (Article IV.4) (WTO 1994b).
40. Regarding environmental issues, the U.S. proposed the inclusion of the 'discussion of ways in which a country's or a member economy's trade and environment policies mutually reinforce each other ... in WTO Trade Policy Review' (USTR 2002). Concerning labour issues, Bill Jordan, General Secretary of the ICFTU suggested that 'the WTO's Trade Policy Review Mechanism (TPRM) is potentially a powerful tool to ensure that member states follow WTO rules, and it is gradually developing its methods of exercising pressure on WTO members. Its scope should be expanded to include trade-related environmental, social, and gender concerns, including core labour standards. The UN's specialized agencies should provide input to the TPRM process as part of the increased collaboration of the WTO with the broader family of UN institutions' (quoted in Sampson 2001, 253).
41. There are many other examples in this vein and, in the case of developing or transition economies, domestic reform efforts are often commended, for example, the Czech Republic, Malaysia, and India. The secretariat reports are available from the WTO's website at <www.wto.org> under the WT/TPR/S series.
42. The U.S. government also stated that 'the subject of implementation of core labor standards is relevant for TPRM reviews' (WTO 2001b, para. 78). Although few members were similarly explicit, trade-related aspects of labour and particularly environmental issues were noted by Norway, the EU, and Canada.
43. Peer review and peer pressure are the chief mechanisms through which the OECD and other international organisations (including at the UN) exercise influence on their members and other groups (Pagani 2003).
44. These reports are available at the WTO website <www.wto.org> under the WT/TPR/OV series.
45. Just prior to his appointment as Director General in 2002, Supachai Panitchpakdi expressed an interest in a code of conduct framework in the WTO (Fabian Global Forum 2002).
46. This report addressed CSR 'as an integral part of trade and investment liberalisation' and noted that 'as traders and investors benefit from freer trade, it is important to ensure that free trade is also fair and that business enterprises respect human rights, labour standards and environmental standards when trading and investing' (OHCHR 2003a, 5). For a theoretical perspective on these and similar initiatives, see Muchlinski (2003).
47. These were reflected in the relatively few corporate codes developed at the time as well, for example, Caterpillar's Code of Worldwide Business Conduct, IBM's Business Conduct Guidelines, or Xerox's An Understanding (see Behrman 1981; Walton 1977; Wallace 1982).

References

Aaronson, Susan (2002). 'Global Corporate Social Responsibility Pressures and the Failure to Develop Universal Rules to Govern Investors and States'. *Journal of World Investment* vol. 3, no. 3, pp. 487–506.

Aaronson, Susan and James T. Reeves (2002). *Corporate Responsibility in the Global Village: The Role of Public Policy*. National Policy Association, Washington DC.

Acconci, Pia (2001). 'The Promotion of Responsible Business Conduct and the New Text of the OECD Guidelines for Multinational Enterprises'. *Journal of World Investment* vol. 2, no. 1, pp. 123–150.

Bagwell, Kyle and Robert W. Staiger (2001). 'The WTO as a Mechanism for Securing Market Access Property Rights: Implications for Global Labor and Environmental Issues'. *Journal of Economic Perspectives* vol. 15, no. 3, pp. 69–88.

Behrman, Jack N. (1981). *Discourses on Ethics and Business*. Oelgeschlager, Ginn, and Hain, Cambridge, MA.

Beviglia Zampetti, Americo (2003). 'Democratic Legitimacy in the World Trade Organization: The Justice Dimension'. *Journal of World Trade* vol. 37, no. 1, pp. 105–126.

Beviglia Zampetti, Americo and Torbjörn Fredriksson (2003). 'The Development Dimension of Investment Negotiations in the WTO: Challenges and Opportunities'. *Journal of World Investment* vol. 4, no. 3, pp. 399–450.

Bhagwati, Jagdish, Muchkund Dubey, Arvind Panagariya, et al. (1999). 'Third World Intellectuals and NGOs Statement against Linkage (TWIN-SAL)'. <www.columbia.edu/~jb38/twin-sal12.pdf> (April 2004).

Biermann, Frank (2001). 'The Rising Tide of Green Unilateralism in World Trade Law: Options for Reconciling the Emerging North-South Conflict'. *Journal of World Trade* vol. 35, no. 3, pp. 421–448.

Brown, Drusilla (2000). 'International Trade and Core Labour Standards: A Survey of the Recent Literature'. Labour Market and Social Policy Occasional Paper No. 43. Organisation for Economic Co-operation and Development.

Brown, Drusilla (2001). 'Labor Standards: Where Do They Belong on the International Trade Agenda?' *Journal of Economic Perspectives* vol. 15, no. 3, pp. 89–122.

Business and Industry Advisory Committee to the OECD (2000). 'BIAC Chairman's Remarks to the Ministerial Special Session on the Revision of the OECD Guidelines'. 27 June. <www.biac.org> (April 2004).

Business and Industry Advisory Committee to the OECD (2001). 'BIAC Secretary General Letter to Chairman of OECD CIME on Positive Promotion'. 22 November. <www.biac.org> (April 2004).

Carlsson, Ingvar and Shridath Ramphal, eds. (1994). *Our Global Neighbourhood: The Report of the Commission on Global Governance*. Oxford University Press, Oxford.

Charnovitz, Steve (2002). *Trade Law and Global Governance*. Cameron May, London.

Consumer Protection in the Global Market Working Group of the ISO Consumer Policy Committee (2002). 'The Desirability and Feasibility of ISO Corporate Social Responsibility Standards'. May. <europa.eu.int/comm/employment_social/soc-dial/csr/isoreport.pdf> (April 2004).

Consumer Unity and Trust Society (2001). 'Investment Issues at the WTO: Viewpoint'. <www.cuts.org/viewpnt-invstmnt.htm> (April 2004).

Curzon Price, Victoria (1991). 'GATT's New Trade Policy Review Mechanism'. *World Economy* vol. 14, no. 2, pp. 227–238.

Dale, Reginald (2000). 'Rules Even Business Should Love'. *International Herald Tribune*, 12 December.

European Community (1999). 'Agreement on Trade, Development, and Cooperation between the European Community and Its Member States, of the One Part, and the Republic of South Africa, of the Other Part'. *Official Journal of the European Communities* pp. L311/313–297. 4 December.

European Community (2000). 'Economic Partnership, Political Coordination, and Cooperation Agreement between the European Community and its Member States, of the one part, and the United Mexican States, of the other part'. *Official Journal of the European Communities* pp. L276/245–261. 28 October.

Fabian Global Forum (2002). 'Change from Within?' Edited transcript of a debate between Dr. Supachai Panitchpakdi, Martin Khor, and Mohau Pheko, 8 June. <www.fabianglobalforum.net/forum/article006.html> (April 2004).

Free Trade Area of the Americas (2003). 'Participation of Civil Society'. <www.ftaa-alca.org/SPCOMM/COMMCS_E.ASP> (April 2004).

Friedman, Milton (1970). 'The Social Responsibility of Business Is to Increase Its Profits'. *New York Times Magazine*. 13 September, pp. 32–33.

Global Reporting Initiative (2000). 'Sustainability Reporting Guidelines on Economic, Environmental, and Social Performance'. <www.getf.org/file/toolmanager/O16F22722.pdf> (April 2004).

Global Sullivan Principles of Social Responsibility (2003). 'Principles'. <globalsullivanprinciples.org/principles.htm> (April 2004).

Government of Canada (2003). 'Sustainability Reporting Toolkit'. <www.sustainabilityreporting.ca> (April 2004).

Group of 77 (2000). 'Declarations and Programmes of Action — Declaration of the South Summit'. Havana. <www.g77.org/main/main.htm> (April 2004).

Graymore, Daniel (2003). 'Corporate Social Reponsibility and FDI'. In 'The Development Dimensions of FDI: Policy and Rule-Making Perspectives', Expert Meeting, 6–8 November. Report UNCTAD/ITE/IIA/2003/4.

Haufler, Virginia (2001). *A Public Role for the Private Sector: Industry Self-Regulation in a Global Economy*. Carnegie Endowment for International Peace, Washington DC.

Henderson, David (2001). *Misguided Virtue: False Notions of Corporate Social Responsibility*. Institute of Economic Affairs, London.

International Institute for Sustainable Development (2003). 'Standards and Trade — CSR Briefings and Documents'. <www.iisd.org/standards/csr_documents.asp> (April 2004).

International Labour Organization (1977). 'Tripartite Declaration of Principles concerning Multinational Enterprises and Social Policy'. <www.ilo.org/public/english/standards/norm/sources/mne.htm> (April 2004).

International Labour Organization (1998). 'ILO Declaration on Fundamental Principles and Rights at Work'. <www.ilo.org/public/english/standards/decl/declaration/text> (April 2004).

International Labour Organization (2003). 'Number of Ratifications of the ILO Conventions'. <webfusion.ilo.org/public/db/standards/normes/appl/index.cfm> (April 2004).

International Organization for Standardization (2002). 'ISO 14000'. <www.iso.ch/iso/en/iso9000-14000/iso14000/iso14000index.html> (April 2004).

Keesing, Donald B. (1998). *Improving Trade Policy Reviews in the World Trade Organization*. Institute for International Economics, Washington DC.

Kelley, Glen (2001). 'Multilateral Investment Treaties: A Balanced Approach to Multinational Corporations'. *Columbia Journal of Transnational Law* vol. 39, no. 2, pp. 483–532.

Kirton, John J. and George M. von Furstenberg, eds. (2001). *New Directions in Global Economic Governance: Managing Globalisation in the Twenty-First Century*. Ashgate, Aldershot.

KPMG (2002). 'KPMG International Survey of Corporate Sustainability Reporting (2002)'. <www.kpmg.co.uk/services/ras/pubs.cfm> (April 2004).

Laird, Sam (1999). 'The WTO's Trade Policy Review Mechanism: From Through the Looking Glass'. *World Economy* vol. 22, no. 6, pp. 741–764.

Lim, Hoe (2001). 'Trade and Human Rights'. *Journal of World Trade* vol. 35, no. 2, pp. 275–300.

Marceau, Gabrielle (2001). 'Conflicts of Norms and Conflicts of Jurisdictions: The Relationship between the WTO Agreement and MEAs and Other Treaties'. *Journal of World Trade* vol. 35, no. 6, pp. 1081–1132.

Muchlinski, Peter (1995). *Multnational Enterprises and the Law*. Blackwell, Oxford.

Muchlinksi, Peter (2003). 'Human Rights, Social Responsibility, and the Regulation of International Business: The Development of International Standards by Intergovernmental Organisations'. *Non-State Actors and International Law* vol. 3, no. 1, pp. 123–152.

Nordström, Håkan and Scott Vaughan (1999). *Trade and Environment*. World Trade Organization, Geneva.

Office of the High Commissioner for Human Rights (2003a). 'Human Rights and Trade'. Report to the 5th WTO Ministerial Conference, 10–14 September. Cancun. <www.unhchr.ch/html/menu2/trade/cancunfinal.doc> (April 2004).

Office of the High Commissioner for Human Rights (2003b). 'Human Rights, Trade, and Investment'. Report no. E/CN.4/Sub.2/2003/9. 2 July.

Office of the High Commissioner for Human Rights (2003c). 'Draft Norms on the Responsibilities of Transnational Corporations and Other Business Enterprises with Regard to Human Rights'. Document No. E/CN.4/Sub.2/2003/L.8.

Ontario Municipal Employees Retirement System (2003). 'Statement of Investment Policies and Procedures for the Ontario Municipal Employees Retirement System'. <www.omers.com/investments/statementofinvpol.html> (April 2004).

Organisation for Economic Co-operation and Development (1996). *Trade, Employment, and Labour Standards: A Study of Core Workers' Rights and International Trade*. Organisation for Economic Co-operation and Development, Paris.

Organisation for Economic Co-operation and Development (1998). 'Codes of Corporate Conduct: An Inventory'. Report No. TD/TC/WP(98)74/FINAL. 13 October.

Organisation for Economic Co-operation and Development (2000a). *Assessing the Environmental Effects of Trade Liberalisation Agreements*. Organisation for Economic Co-operation and Development, Paris.

Organisation for Economic Co-operation and Development (2000b). 'The OECD Guidelines for Multinational Enterprises'. Organisation for Economic Co-operation and Development, Paris. <www.oecd.org/dataoecd/56/36/1922428.pdf> (April 2004).

Organisation for Economic Co-operation and Development (2000c). 'Shaping Globalisation'. Press release, 27 June. Paris. <www1.oecd.org/media/release/nw00-70a.htm> (April 2004).

Organisation for Economic Co-operation and Development (2000d). 'Spotlight on the OECD Guidelines for Multinational Enterprises'. *OECD Observer*. December. <www1.oecd.org/publications/pol_brief/2000/2000_09.pdf> (April 2004).

Organisation for Economic Co-operation and Development (2000e). 'Text of the OECD Declaration on International Investment and Multinational Enterprises'. 27 June.

Organisation for Economic Co-operation and Development (2001a). 'Non-Member Economies and the OECD Guidelines for Multinational Enterprises'. 12 December. <www.oecd.org/dataoecd/55/51/2542956.pdf> (April 2004).

Organisation for Economic Co-operation and Development (2001b). 'OECD Guidelines for Multinational Enterprises: Annual Report 2001'. <www1.oecd.org/publications/e-book/2101171e.pdf> (April 2004).

Organisation for Economic Co-operation and Development (2003). 'OECD Codes of Liberalisation of Capital Movements and Current Invisible Operations: Users' Guides'. <www.oecd.org/dataoecd/14/13/1935919.pdf> (April 2004).

Pagani, Fabrizio (2003). 'Peer Review: A Tool for Cooperation and Change'. Document no. SG/LEG(2002)1. Organisation for Economic Co-operation and Development, <www.oecd.org/dataoecd/33/16/1955285.pdf> (April 2004).

Petersmann, Ernst-Ulrich (2003). 'Human Rights and the Law of the World Trade Organization'. *Journal of World Trade* vol. 37, no. 2, pp. 241–282.

Picciotto, Sol (2003). 'Corporate Social Responsibility for International Business'. Expert Meeting, 6–8 November. Report UNCTAD/ITE/IIA/2003/4.

Rodrik, Dani (1997). *Has Globalization Gone Too Far?* Institute for International Economics, Washington DC.

Sampson, Gary P., ed. (2001). *The Role of the World Trade Organization in Global Governance*. United Nations University Press, Tokyo.

Schabas, William A. (2001). 'Enforcing International Humanitarian Law: Catching the Accomplices'. *International Review of the Red Cross* no. 42, pp. 439–459.

Social Accountability International (2003). 'SA8000'. <www.sa8000.org/SA8000/SA8000.htm> (April 2004).

Sutherland, Peter, John Sewell, and David Weiner (2001). 'Challenges Facing the WTO and Policies to Address Global Governance'. In G. P. Sampson, ed., *The Role of the World Trade Organization in Global Governance*. United Nations University Press, Tokyo.

Trebilcock, Michael J. and Robert Howse (1999). *The Regulation of International Trade*. 2nd ed. Routledge, New York.

United Nations (2001). 'Global Compact'. 18 April. <www.un.org/Depts/ptd/global.htm> (April 2004).

United Nations (2003). 'Seventh Session of the Ad Hoc Committee on the Negotiation of a Convention against Corruption'. Vienna, 29 September–1 October. <www.unodc.org/unodc/crime_cicp_committee_corruption_session_7.html> (April 2004).

United Nations Conference on Environment and Development (1992). 'Rio Declaration on Environment and Development'. <www.unep.org> (April 2004).

United Nations Conference on Trade and Development (2001). 'Social Responsibility'. Report no. UNCTAD/ITE/ITT/22. <www.unctad.org> (April 2004).

United Nations Conference on Trade and Development (2003). 'World Investment Report 2003 — FDI Policies for Development: National and International Perspectives'.

United Nations Economic Commission for Europe (1998). 'Convention on Access to Information, Public Participation in Decision-Making, and Access to Justice in Environmental Matters'. Aarhus, Denmark, 25 June. <www.unece.org/env/pp/documents/cep43e.pdf> (April 2004).

United States Trade Representative (2001). 'The U.S.-Jordan Free Trade Agreement'. 17 December. <www.ustr.gov/regions/eu-med/middleeast/US-JordanFTA.shtml> (April 2004).

United States Trade Representative (2002). 'Labor and Environment "Toolbox"'. <www.ustr.gov/toolbox.pdf> (April 2004).

United States Trade Representative (2003). 'United States-Free Trade Agreement'. 6 May. <www.ustr.gov/new/fta/Singapore/final.htm> (April 2004).

Wallace, Cynthia D. (1982). *Legal Controls of Multinational Enterprise*. Nijhoff, The Hague.

Walton, Clarence C., ed. (1977). *The Ethics of Corporate Conduct*. Prentice Hall, Englewood Cliffs, NJ.

Wolffgang, Hans-Michael and Wolfram Feuerhake (2002). 'Core Labour Standards in World Trade Law: The Necessity for Incorporation of Core Labour Standards in the World Trade Organization'. *Journal of World Trade* vol. 36, no. 5, pp. 883–902.

World Bank (1992). *Governance and Development*. World Bank, Washington DC.

World Bank Group–Compliance Advisor Ombudsman (2003). 'CAO General Information'. <www.cao-ombudsman.org> (April 2004).

World Commission on Environment and Development (1987). *Our Common Future*. Oxford University Press, Oxford.

World Trade Organization (1994a). 'Decision on Proposed Understanding on WTO-ISO Standards Information System'. <www.wto.org/english/docs_e/legal_e/37-dtbt1_e.htm> (April 2004).

World Trade Organization (1994b). 'Marrakesh Agreement Establishing the World Trade Organization'. <www.wto.org/english/docs_e/legal_e/04-wto_e.htm> (April 2004).

World Trade Organization (1994c). 'Trade Policy Review Mechanism'. <www.wto.org/english/docs_e/legal_e/29-tprm_e.htm> (April 2004).

World Trade Organization (1996). 'Singapore Ministerial Declaration'. WT/MIN(96)/DEC/W. <www.wto.org/english/thewto_e/minist_e/min96_e/wtodec_e.htm> (April 2004).

World Trade Organization (1998). 'United States — Import Prohibition of Certain Shrimp and Shrimp Products. AB-1998-4. Report of the Appellate Body'. Document no. 98-3899.

World Trade Organization (1999). 'Where Are We on Foreign Investors' Behaviour?' Working Group on the Relationship between Trade and Investment, 20 September, Document symbol WT/WGTI/W/81. Geneva.

World Trade Organization (2001a). 'Annual Report 2001'. Geneva. <www.wto.org/english/res_e/booksp_e/anrep_e/wto_anrep01_e.pdf> (April 2004).
World Trade Organization (2001b). 'United States: Report by the Government'. Trade Policy Review Body, Document symbol WT/TPR/G/88, 20 August.
World Trade Organization (2001c). 'United States: Report by the Secretariat'. Trade Policy Review Body, Document symbol WT/TPR/S/88, 15 August.
World Trade Organization (2002). 'Investors' and Home Governments' Obligations'. Working Group on the Relationship between Trade and Investment, Communication from China, Cuba, India, Kenya, Pakistan, and Zimbabwe, 19 November, Document symbol WT/WGTI/W/152. Geneva.
World Trade Organization (2003a). 'Early Years: Emerging Environment Debate in GATT/WTO'. <www.wto.org/english/tratop_e/envir_e/hist1_e.htm> (April 2004).
World Trade Organization (2003b). 'Honduras: Report by the Government'. Trade Policy Review Body, Document symbol WT/TPR/G/120, 3 September.
World Trade Organization (2003c). 'Relevant WTO Provisions: Texts'. <www.wto.org/english/tratop_e/envir_e/issu4_e.htm> (April 2004).
World Trade Organization (2003d). 'Trade Policy Reviews'. <www.wto.org/english/tratop_e/tpr_e/tpr_e.htm> (April 2004).

Appendix 16.1
Trade Policy Review Mechanism
(Annex 3 of Marrakesh Agreement
Establishing the World Trade Organization)

Members hereby agree as follows:

A. Objectives

i) The purpose of the Trade Policy Review Mechanism ('TPRM') is to contribute to improved adherence by all Members to rules, disciplines and commitments made under the Multilateral Trade Agreements and, where applicable, the Plurilateral Trade Agreements, and hence to the smoother functioning of the multilateral trading system, by achieving greater transparency in, and understanding of, the trade policies and practices of Members. Accordingly, the review mechanism enables the regular collective appreciation and evaluation of the full range of individual Members' trade policies and practices and their impact on the functioning of the multilateral trading system. It is not, however, intended to serve as a basis for the enforcement of specific obligations under the Agreements or for dispute settlement procedures, or to impose new policy commitments on Members.
ii) The assessment carried out under the review mechanism takes place, to the extent relevant, against the background of the wider economic and developmental needs, policies and objectives of the Member concerned, as well as of its external environment. However, the function of the review mechanism is to examine the impact of a Member's trade policies and practices on the multilateral trading system.

B. Domestic Transparency

Members recognize the inherent value of domestic transparency of government decision-making on trade policy matters for both Members' economies and the multilateral trading system, and agree to encourage and promote greater transparency within their own systems, acknowledging that the implementation of domestic transparency must be on a voluntary basis and take account of each Member's legal and political systems.

C. Procedures for Review

i) The Trade Policy Review Body (referred to herein as the 'TPRB') is hereby established to carry out trade policy reviews.
ii) The trade policies and practices of all Members shall be subject to periodic review. The impact of individual Members on the functioning of the multilateral trading system, defined in terms of their share of world trade in a recent

representative period, will be the determining factor in deciding on the frequency of reviews. The first four trading entities so identified (counting the European Communities as one) shall be subject to review every two years. The next 16 shall be reviewed every four years. Other Members shall be reviewed every six years, except that a longer period may be fixed for least-developed country Members. It is understood that the review of entities having a common external policy covering more than one Member shall cover all components of policy affecting trade including relevant policies and practices of the individual Members. Exceptionally, in the event of changes in a Member's trade policies or practices that may have a significant impact on its trading partners, the Member concerned may be requested by the TPRB, after consultation, to bring forward its next review.

iii) Discussions in the meetings of the TPRB shall be governed by the objectives set forth in paragraph A. The focus of these discussions shall be on the Member's trade policies and practices, which are the subject of the assessment under the review mechanism.

iv) The TPRB shall establish a basic plan for the conduct of the reviews. It may also discuss and take note of update reports from Members. The TPRB shall establish a programme of reviews for each year in consultation with the Members directly concerned. In consultation with the Member or Members under review, the Chairman may choose discussants who, acting in their personal capacity, shall introduce the discussions in the TPRB.

v) The TPRB shall base its work on the following documentation:
   a) a full report, referred to in paragraph D, supplied by the Member or Members under review;
   b) a report, to be drawn up by the Secretariat on its own responsibility, based on the information available to it and that provided by the Member or Members concerned. The Secretariat should seek clarification from the Member or Members concerned of their trade policies and practices.

vi) The reports by the Member under review and by the Secretariat, together with the minutes of the respective meeting of the TPRB, shall be published promptly after the review.

vii) These documents will be forwarded to the Ministerial Conference, which shall take note of them.

D. Reporting

In order to achieve the fullest possible degree of transparency, each Member shall report regularly to the TPRB. Full reports shall describe the trade policies and practices pursued by the Member or Members concerned, based on an agreed format to be decided upon by the TPRB. This format shall initially be based on the Outline Format for Country Reports established by the Decision of 19 July 1989 (BISD 36S/406-409), amended as necessary to extend the coverage of reports to all aspects of trade policies covered by the Multilateral

Trade Agreements in Annex 1 and, where applicable, the Plurilateral Trade Agreements. This format may be revised by the TPRB in the light of experience. Between reviews, Members shall provide brief reports when there are any significant changes in their trade policies; an annual update of statistical information will be provided according to the agreed format. Particular account shall be taken of difficulties presented to least-developed country Members in compiling their reports. The Secretariat shall make available technical assistance on request to developing country Members, and in particular to the least-developed country Members. Information contained in reports should to the greatest extent possible be coordinated with notifications made under provisions of the Multilateral Trade Agreements and, where applicable, the Plurilateral Trade Agreements.

E. Relationship with the Balance-of-Payments Provisions of GATT 1994 and GATS

Members recognize the need to minimize the burden for governments also subject to full consultations under the balance-of-payments provisions of GATT 1994 or GATS. To this end, the Chairman of the TPRB shall, in consultation with the Member or Members concerned, and with the Chairman of the Committee on Balance-of-Payments Restrictions, devise administrative arrangements that harmonize the normal rhythm of the trade policy reviews with the timetable for balance-of-payments consultations but do not postpone the trade policy review by more than 12 months.

F. Appraisal of the Mechanism

The TPRB shall undertake an appraisal of the operation of the TPRM not more than five years after the entry into force of the Agreement Establishing the WTO. The results of the appraisal will be presented to the Ministerial Conference. It may subsequently undertake appraisals of the TPRM at intervals to be determined by it or as requested by the Ministerial Conference.

G. Overview of Developments in the International Trading Environment

An annual overview of developments in the international trading environment which are having an impact on the multilateral trading system shall also be undertaken by the TPRB. The overview is to be assisted by an annual report by the Director-General setting out major activities of the WTO and highlighting significant policy issucs affecting the trading system.

Source: World Trade Organization (1994).

Chapter 17

A Corporate Perspective on Globalisation, Sustainable Development, and Soft Law

Michael E. Cloghesy

How Business Seeks Sustainable Development

The way in which business seeks to promote sustainable development, through the use of hard law as well as soft law, can be seen in the work of one major Canadian business association, the Conseil patronal de l'environnement du Québec (CPEQ). This organisation has more than 150 corporate and association members, including most of the major industry sectors and corporations with operations in Quebec: pulp and paper, mining, manufacturing, and service firms. The 2001 chair, Jean Piette, was the principal author of Quebec's environmental law when he worked for the province's government. CPEQ is thus well equipped to forward its mission of sustainable development. To do so, it works co-operatively with all levels of government — municipal, provincial, and federal — and with international organisations on major environmental issues. It seeks consensus among its members and works with government to develop realistic, enforceable legislation and regulation and to find alternative means of achieving the desired results, such as voluntary measures and covenants.

In many instances, reducing or eliminating pollution or waste results in an economic benefit to the firm by lowering costs related to raw material usage, energy usage, or pollution abatement processes. It makes good business sense to manage environmental issues not separately, but as part of a total business plan.

The CPEQ's representatives have participated in a number of multistakeholder processes that have developed the basis for major pieces of federal environmental legislation, in particular the *Environmental Protection Act* and the *Environmental Assessment Act*. They have also been involved in the processes that led to the Accelerated Reduction or Elimination of Toxics (ARET) programme, which was a voluntary approach to reducing emissions of pollutants. There was much negative criticism of the programme by environmental groups, mainly because its results were not verified by a third party.

The Issue

The globalisation of the marketplace and sustainability are not necessarily incompatible concepts. On the contrary, in many instances, particularly in developing countries, foreign investments in the local economy may provide the only source of financing for projects that improve the quality of the environment. In addition, multinational corporations (MNCs) may provide the needed incentives for developing countries to strive toward good governance. Investments will not occur unless there is a stable government infrastructure that minimises corruption.

In the context of the environment, hard law and soft law are not mutually exclusive but, in fact, should complement each other. Hard law is defined as the legislative and regulatory context put in place by the government to meet certain objectives, whereas soft law may be defined as other measures such as voluntary programmes, covenants, or bilateral agreements between a government and a company. Nonregulatory measures, in such cases, may be as effective — or more so — as regulatory instruments in obtaining the desired results. Market forces are another powerful means of influencing the way firms do business domestically and internationally. As Thomas d'Aquino (2003) said:

> The new expectations of corporate governance sweeping across the world are based on a simple reality — you can run, but you cannot hide … In short, anything a company does anywhere in the world can affect its reputation everywhere in the world. Reputation has become the cornerstone of both investor confidence and public trust. What happens to a company's reputation therefore has an increasingly direct impact on its relationships with everyone from governments and regulators to customers, employees and investors.

It is a well-known principle of physics that for every action, there is a reaction. Globalisation and the drive to create an international marketplace have met with considerable opposition by certain segments of civil society. One of the main reasons for this disenchantment is the view that as the power of a few mega-corporations becomes greater and there is increasing focus on trade, some individuals feel their voices are being ignored and their values and concerns appear to be disregarded by government. This phenomenon can be related to a certain loss of sovereignty incurred by a nation when it signs a bilateral or multilateral agreement, whether on trade or otherwise. When one joins a club, one does so to gain benefits, but at the same time one must abide by the club's rules and be subject to some loss of self-determination.

The European Union is an excellent example of this type of tradeoff. Its task was to remove trade barriers and create a single currency and open borders, but with a common set of rules that, granted, can be adjusted to some degree by individual states.

The International Challenge

In the broader international context, the major reality is a world split between developed and developing nations. The priorities on both sides of the great divide differ considerably. Where the 'have' nations talk about sustainable development, the 'have not' nations seek to provide the basic needs of life. It is important, if not essential, to discuss sustainability in the global context, because as global population increases, the stresses on the planet become more acute. Pollution knows no borders, whether in the air, in the water, or as waste. The actions of one nation can affect others. All nations are in the same boat together, and must agree on how to steer the right course and manage the global environment.

Currently, the United Nations stands as the sole unifying force that deal with global environmental and other key societal issues. It has had some relative successes in tackling specific issues, such as stratospheric ozone depletion via the Montreal Protocol (United Nations Environmental Programme [UNEP] 1999b). However, it is not clear whether the Basel Convention on hazardous waste management could be classified as a success, for it is extremely bureaucratic and has perverse effects on the trading of recyclable materials (UNEP 1999a).

Also questionable is the United Nations Framework Convention on Climate Change (UNFCCC) and in particular the Kyoto Protocol (UNFCCC 1992). Those who attended as members of the Canadian delegation at the meetings of the Conference of the Parties, including the sessions in the Hague and Bonn, can easily question the practicality of attempting to attain consensus among the 184 member countries, each having its own agenda and belonging to one of several factions. The Kyoto Protocol may appear to focus on the environment, but, in reality, it is more about politics, foreign aid, wealth redistribution, the creation of winners and losers, and the fostering of new industries involved in emissions trading and technological development. For all this effort, what are the real environmental gains, especially after the United States backed out? They are next to nil, and mainly symbolic.

UNEP is also of questionable value. It is highly bureaucratic and quite inefficient. It is hardly surprising that the U.S. hesitates to pay its dues to the UN.

Discounting its many inadequacies, however, the UN nevertheless remains a valuable forum that contributes to the resolution of many complex issues that the world faces. One must recognise that its success is limited by the fact that, in many instances, the UN is powerless in enforcing its various conventions and protocols. There is no world police force and sanctions are of limited effectiveness. Thus, other means of achieving results must be considered — including those referred to as soft law and market forces, which have the potential for exercising considerable influence, particularly on MNCs.

A Vision for an International Approach

In developed countries, the general trend is toward smaller governments. Budgets for environmental departments are continuously being reduced, while the complexity of dealing with environmental issues is increasing. As knowledge increases about the potential toxicity of the 30 000 or so substances used in commerce worldwide, the need for more control and better management becomes apparent. Yet it is unlikely that governments can regulate everything that may pose a risk. The onus for establishing the level of risk and for reducing or eliminating the risk is thus shifting to the private sector, because it alone has the resources to do the job.

With the concept of sustainable development, the previous command-and-control regime and the confrontational relationship between government and industry have largely given way to a more constructive approach. The dominant attitude is currently to collaborate in developing legislation and regulations that work and that achieve environmental gains as well as ensure economic growth and job creation.

Market forces have been very effective in producing major corporate behavioural changes to this end. Several such changes stand out, such as the following.

- When the client specifies a certain standard, the supplier must comply.
- When a banker or insurer will not deal with an irresponsible polluter, the company must conform.
- When shareholders want to dump their stock, the company must improve itself.
- When employees complain or denounce a company to the media, the company must adjust.
- When a local community wants a company to move out or shut down, that company had better change.

The reputation of companies can also suffer if they adopt double standards — one for their operations in developed countries and another, less stringent, for their activities in developing countries. The Nike case well illustrates the point (Klein 2000).

Many of the world's largest MNCs now have only one code of ethics for all their operations. In addition, they pressure local governments to eliminate corruption and provide assistance to develop a sustainable infrastructure, because business needs to operate in a climate of security and predictability as much as possible. Thus, the private sector can become a positive force capable of improving the quality of life, particularly in developing countries.

International Standards

MNCs also see the need to adopt international standards, such as ISO 14000, for several reasons. One fundamental reason is that it makes good business sense to manage environmental issues. Another reason is that it may be a requirement to obtain a contract. ISO 14000 has the advantage of having a multiplier effect as it may reach out to a firm's suppliers and clients, but it does not necessarily guarantee compliance with regulations (International Organization for Standardization [ISO] 2002).

Another soft law means of achieving results is through sector-based programmes, such as the chemical industry's Responsible Care programme. This programme, developed in Canada by the Canadian Chemical Producers Association (2003), has spread worldwide. The concept has been adopted by other sectors such as mining, pulp and paper, petroleum products, and paints. One of the big advantages is that ISO 14000 and Responsible Care imply continuous improvement.

As the world heads toward a context in which the marketplace is the world, there is a need to develop global standards or face of the prospects of 'pollution leakage' — industries moving from one country to another because of varying regulatory regimes. For example, it would make more sense to develop environmental emission levels based on eco-regions and by type of industry on a North American basis, and similarly for a European context, than on a national level.

Domestic Action

Although all agree on the need for a regulatory framework that is simple, realistic, and enforceable, it is clear that, in some cases, regulations cannot do the job. At a sub-federal level, how do jurisdictions set one emissions limit, for example volatile organic compounds, when the products, technologies, and capabilities vary from industry to industry? Moreover, governments do not have the resources to regulate further, and it is bad regulatory policy to produce regulations that government neither can enforce because of a lack of resources nor wants to enforce for political reasons, such as loss of jobs. A government's credibility can suffer even more if citizens realise that the law is not imposed.

Thus, there is a need for a toolkit of various instruments, regulatory and nonregulatory, to meet objectives. This toolkit should include:

- voluntary measures, such as ARET;
- covenants that are legally binding, such as the Dutch government's agreement with the petroleum industry on environmental issues;
- sector-based programmes of a voluntary nature, such as Responsible Care;
- sector-based approaches with a legal basis, such as Quebec's sector-based pollution reduction programme;
- agreements between companies and governments on pollutant reductions; and
- standards such as ISO 14000 on environmental management systems.

There is also a need for voluntary measures to be credible, transparent, and verifiable by third parties in order to be politically acceptable.

Conclusion

There cannot be put too much faith placed in international conventions or other means of hard law at the international level to bring about needed changes to ensure global sustainability. This is primarily because of the ponderous and inefficient process of achieving agreement from the many members of the UN and because of the considerable difficulty in enforcing these instruments at the global level.

Rather, both internationally and nationally, results and gains toward sustainability can, in many cases, be more easily achieved through international standards and voluntary programmes such as Responsible Care. In addition, there is a need for more positive, free-flowing dialogue on sustainable development among representatives of government, the private sector, and civil society, and a recognition that the regulatory framework or hard law approach has its limitations. The best way to achieve results is to develop a more flexible and open approach to the use of various tools, both hard and soft. Although globalisation may at first lessen national sovereignty and create certain instability in areas such as the job market and investments, it may also provide the sole means to improve the economic status of a number of developing countries, thus allowing for a movement toward sustainability objectives by these nations.

Furthermore, if globalisation is to go forward, changes may need to occur at the level of the World Trade Organization (WTO). Rather than creating separate bodies to deal with environmental and other social issues, the WTO should become more of a sustainability organisation that will be transparent and open to civil society. To be sure, it will be a challenge to move trade issues ahead in this fashion, but the potential gains may make the change worthwhile.

References

Canadian Chemical Producers' Association (2003). 'Reponsible Care'. <www.ccpa.ca/english/who/rc/index.html> (April 2004).

d'Aquino, Thomas (2003). 'You Can Run, But You Cannot Hide: Transparency and Disclosure in Corporate Governance'. Notes for an address to 'Intention at the Top: 2003 Corporate Governance Conference', 30 January, Toronto. <www.ceocouncil.ca/English/Media/speeches/jan30-03.htm> (April 2004).

International Organization for Standardization (2002). 'ISO 14000'. <www.iso.ch/iso/en/iso9000-14000/iso14000/iso14000index.html> (April 2004).

Klein, Naomi (2000). *No Logo: No Space, No Choice, No Jobs*. Knopf, Toronto.

United Nations Environment Programme (1999a). 'Basel Convention on the Control of Transboundary Movements of Hazardous Wastes and Their Disposal'. <www.basel.int/text/con-e.htm> (April 2004).

United Nations Environmental Programme (1999b). 'The Montreal Protocol on Substances that Deplete the Ozone Layer'. <www.unep.org/ozone/pdf/Montreal-Protocol2000.pdf> (April 2004).

United Nations Framework Convention on Climate Change (1992). 'Text of the Convention'. The Kyoto Protocol. <unfccc.int/resource/conv/conv.html> (April 2004).

Chapter 18

Terminating Agricultural Biotechnology? Hard Law, Voluntary Measures, and the Life Sciences Industry

Lisa N. Mills

Investment and trade in agricultural biotechnology products have been facilitated by a number of international hard law instruments created as part of the Marrakesh Agreement establishing the World Trade Organization ([WTO] 1994). The Trade-Related Aspects of Intellectual Property Rights (TRIPS) Agreement, which ensures that intellectual property embodied in genetically modified organisms is protected through domestic patent or plant variety protection regimes, has fostered investment in the industry; the Agreement on Sanitary and Phytosanitary Measures (SPS) ensures that domestic safety standards do not represent a barrier to trade in food derived from these organisms.

It has been argued that hard and soft law should not be seen as binary opposites, but as ends of a continuum of legal measures that vary along three dimensions: obligation, precision, and delegation (Abbott and Snidal 2000, 422). Obligation refers to the giving of a legal commitment; once it has been made, the discourse on the subject of the commitment changes, referring to the law and its interpretation rather than purely to issues of power and interest. Precision refers to the clarity of a rule: the clearer and more specific it is, the less room there is for interpretation. Delegation refers to the extent to which the authority for interpreting and enforcing the law is granted to a third party (Abbott et al. 2000). According to this conceptualisation, hard law refers to 'legally binding obligations that are precise (or can be made precise through adjudication or the issuance of detailed regulations) and that delegate authority for interpreting and implementing the law' (Abbott and Snidal 2000, 421). Soft law is weaker than hard law along one — or more than one — of the three dimensions.

In order to counter the impact of the TRIPS and SPS agreements, nongovernmental organisations (NGOs) and states have campaigned for hard law instruments, as well as soft law ones, to manage potential risks to biodiversity and genetic resources not covered by the two agreements. Two international conventions, the Cartagena Protocol on Biosafety and the International Treaty on Plant Genetic Resources for Food and Agriculture, assert the primacy of domestic interests in biodiversity and food security over the obligations created in the Uruguay Round (Convention on Biological Diversity 2000; Commission on Genetic Resources for Food and Agriculture 2001). However, a lack of coherence among the various agreements and uncertainty about how to resolve conflicts render the future impact of these hard law measures unclear.

The creation of these agreements was preceded by the rise of *ad hoc,* informal soft law measures. Consumer and environmental groups in Europe organised massive boycotts of genetically modified food (see, for example, British Action Against Genetically Engineered Products 1999). Food retailers responded with the introduction of voluntary standards, which in turn resulted in a drop in corporate profits and market capitalisation in the biotechnology industry. Facing criticism from NGOs as well as from prominent members of the scientific community, Monsanto — the world's largest agricultural biotechnology firm — committed to change its practices and repudiated its most controversial technology acquisition. In 1999, the firm was taken over by pharmaceutical giant Pharmacia, and the agricultural biotechnology division was spun off, leading to pronouncements about the demise of the life sciences industry.

Along the hard law–soft law spectrum, voluntary standards and commitments by companies are at the extremely soft end: although they may be precise in the outline of their conduct, their obligation is more moral and normative than legal, and authority to monitor companies' compliance with their own commitments is not delegated to others. Soft law can further political debate, and although monitoring authority is not delegated, activist groups can perform this task and publicise actors' violations of their commitments (Abbott and Snidal 2000). However, although political action — particularly in Europe — has had an impact on the actions of the multinational biotechnology corporations, in the long term soft law measures such as voluntary standards are insufficient to ensure the protection of biodiversity globally; furthermore, these measures cannot ensure that 'technical' questions, which cannot be disentangled from political questions, can be determined through domestic political action. Hard law instruments are also necessary to ensure that the interests of developing countries, as well as developed countries, are taken into consideration in international law. Kenneth Abbott and Duncan Snidal (2000, 452) acknowledge that 'at the 1992 Rio conference, for example, after business interests blocked legally binding agreements on various issues, environmental groups turned their efforts to obtaining nonbinding but highly elaborated documents like the Rio Declaration, Agenda 21, and the Forest Principles'. The legal chronology has been similar in the case of biotechnology; in the wake of hard law agreements protecting corporate (particularly U.S.) interests, NGOs have campaigned for voluntary commitments limiting private actions. Stronger measures have since been sought in the form of binding international agreements. However, whether these agreements can provide a counterweight to the TRIPS and SPS agreements remains to be seen, as negotiations on the issues of biodiversity, protection of traditional knowledge, and equitable sharing of the benefits of biological resources are still underway.

The Codex Alimentarius Commission and the Role of Science in Mediating Conflict over Agricultural Biotechnology

Scientific Analysis and the Legitimacy of Non-scientific Factors

Under the SPS agreement, standards set by the Codex Alimentarius Commission serve as the norm for harmonising food safety measures. Codex standards are based on the principles of sound scientific analysis and evidence. However, controversies about the principles on which scientific analysis is based, and the factors that may be taken into consideration when interpreting the evidence, have made it difficult to arrive at a consensus on the establishment of standards, and to accept their application in international trade disputes. A series of food safety crises around the world have led to increased concerns about the industrial system of agricultural production and the capacity of existing science-based regulatory systems to ensure the safety of food, particularly that derived from genetically modified organisms. These concerns, which are now being reflected in European Union regulatory directives, have led to unresolved conflicts over the role of science in the establishment of Codex safety standards. Paradoxically, it is because current binding international law regulating the safety of agricultural biotechnology products is believed to amount to form of self-regulation that hard law is regarded as inadequate.

The Codex Alimentarius Commission, which was established in 1963, is an international organisation composed of 163 members that set food safety standards in order both to protect consumer health and to facilitate 'fair trade practices in the food trade' (Codex 2003b). Codex formulates food safety standards including, for example, the acceptable levels of pesticides or animal drug residues in food. Standards are established on the basis of recommendations from the relevant Codex scientific committee, which analyses the evidence; these recommendations are then accepted or rejected by consensus or by a majority vote of the Codex members. Residue standards are usually expressed as maximum residue limits (MRLs) — that is, the maximum level of a substance or residue allowable in food — or as an acceptable daily intake (ADI), which is the amount of a substance that can be consumed daily without threat to human health. Because discriminatory trade practices can be justified by food safety standards that are more restrictive than those of a country's trading partners, Codex has aimed to facilitate global trade in food products through the establishment of international standards and the harmonisation of domestic ones (Codex 1999). Under the SPS agreement, standards set by the Codex Alimentarius Commission serve as the norm for harmonising food safety measures and as a reference point for resolving international trade disputes. Although the adoption of Codex standards is voluntary, the standards have acquired the force of hard law when applied in international trade disputes.

Countries can implement standards more restrictive than those promulgated by Codex, but only if they can be justified scientifically (WTO 1995, art. 3(3)). The legality of a trade-restrictive standard is determined by its scientific validity (at least in theory). In 1997, a WTO dispute settlement panel rejected the EU's nine-year ban on the import of U.S. beef treated with (steroid) growth hormones, on the grounds that such a ban was not scientifically justified; Codex had previously found the three hormones in dispute to be safe (Food and Agriculture Organization [FAO] 1997).[1] Although a WTO Appellate Body reversed the conclusion that protective measures must tightly conform to international standards, the panel reviewed the scientific evidence presented to the original panel and found the Appellate Body's opinions were 'highly deferential' to the panel (Wirth 1998, 4). The Appellate Body also rejected the application of the precautionary principle — the principle that under conditions of scientific uncertainty, regulators should err in favour of measures with higher levels of protection. According to David Wirth (1998, 4), the outcome of the case grants the WTO dispute settlement panels 'substantial, virtually unreviewable discretion of questions of science'. The problem with an emphasis on scientific evidence and analysis is that science that serves regulatory purposes is not purely scientific, and nor can it ever be. Regulatory judgement about the implications of a product's introduction cannot be divorced from non-scientific factors, such as assumptions about the context into which the product will be introduced (see Mills 2002).

The difficulty of resolving conflicts about the interpretation of scientific evidence is apparent in the debate at Codex over the first major product of agricultural biotechnology, recombinant bovine growth hormone (rbGH), a genetically engineered product that stimulates milk production in cows. In both North America and Europe, there has been a great deal of concern that the health of humans may be adversely affected by drinking milk produced by cows that have been treated with the drug. It has been widely accepted that the existing evidence shows little likelihood of risk to human health, but some scientists in North America as well as regulatory committees in Europe have noted that there is insufficient evidence on which to base a conclusion because no long-term studies of human health have been conducted. It has also been widely accepted that the use of the drug results in an increased incidence of animal diseases, but the implications for human health have not been extensively examined. Critics of the drug have argued that because the incidence of animal disease results in increased use of antibiotics, potential problems with antibiotic resistance should be examined. Both the critics and the European regulators have argued that because the drug is intended to increase milk production and has no therapeutic benefits, any potential risks should be examined more stringently. The product was approved by the United States Food and Drug Administration (FDA) in 1993 but banned in Europe, in spite of a positive scientific evaluation. In 1998, another evaluation of the product in Europe resulted in a continuation of the ban on animal health grounds, and raised human health concerns. By this time, the European system for assessing pharmaceuticals had been restructured, the committee for veterinary drugs included members who had been critical of the U.S. decision, and further evidence supported questioning the assumptions on which the original human health decision had been based.

A proposed draft standard for rbGH was referred to one of Codex's scientific advisory committees, the Joint Expert Committee on Food Additives (JECFA). In its 1993 evaluation, JECFA found that when the drug was used in compliance with good veterinary practice, the margin of safety was so wide that there was no need to specify MRLs or ADIs (see JECFA1999). The assumption of 'good veterinary practice' was later questioned in Europe. In the U.S., such practice included the use of antibiotics to control outbreaks of disease and allowed those antibiotics to be used in ways that had not previously been approved by the FDA. The extent of pharmaceutical use was consistent with a large-scale, industrial dairy system common in the profitable south and south-west regions of the United States; the drug's approval therefore involved the tacit approval of a particular agricultural system. JECFA's recommendation assumed the appropriate use of the drug — which involved management and agricultural systems that, in its recent concerns for animal welfare, the EU was beginning to reconsider.

JECFA reached the same conclusion when it re-evaluated the drug in 1998. By this time, the EU had raised several additional concerns about the drug's human health implications.[2] JECFA's conclusions were taken by the U.S. and Canada to represent sound scientific analysis, but did not adequately address major concerns of the Europeans, namely the animal health issue and its implications for human health, the interpretation of the human health evidence, and the impact on milk consumption of the drug's introduction.

The Europeans also argued that there were non-scientific issues that should be considered in the establishment of a Codex standard for rbGH. The EU referred to Codex principles, which refer to both scientific and non-scientific factors. Although the first principle concerning the role of science in the decision-making process states that standards and guidelines 'shall be based on the principle of sound scientific analysis and evidence', the second principle allows for the consideration of 'other legitimate factors relevant for the health protection of consumers and the promotion of fair practices in food trade' (Codex 2003a, 181). The inclusion of 'other legitimate factors' potentially represents a method to specify and harmonise the contextual factors on which regulatory judgements are being made, but as yet there is no consensus about which factors might be considered legitimate. Because these other legitimate factors have never been elaborated by Codex, the EU argued that a decision on rbGH should be postponed until a conclusion had been reached regarding them. According to the EU, such legitimate factors included the need to maintain consumers' confidence in the regulatory system, the potential impact of the drug on milk consumption, and animal welfare concerns (Codex 2001a, 11).

The U.S., on the other hand, had consistently argued that risk and analysis and management should be 'science-based' (Codex 2001a, 2–5). This position reflected that of the animal drug industry, which argued that Codex's failure to adopt a standard recommended by JECFA undermined the agency's credibility and thereby threatened world trade and public health (Coalition of the Americas 1997, 30). The U.S. consistently voted for the adoption of a Codex standard for rbGH, and Canada has voted with the U.S. on this issue (Codex 1997; 1999).[3] In 1999, the EU successfully voted to postpone the discussion of rbGH, and the U.S. decided not to pursue the issue further at the Codex meeting in 1999, given that there was no consensus on the issue (Consumers International 1999). Since that meeting, the issue has again been considered by a Codex committee. The EU has proposed that the issue should be taken off the agenda, and the committee 'should consider whether it wants to spend more time on the evaluation of a substance which has not therapeutic or preventive properties and whose use has been proven harmful for animal health and welfare and whose potentially harmful effects on human beings have not yet been fully elucidated' (EU 2000, 2).

Although a Codex ruling would not have forced the EU to approve the use of rbGH, it would not have changed its practices with regard to U.S. imports because the EU had not banned the import of milk or dairy products from the U.S. (Confédération Mondiale de l'Industrie de la Santé Animal 1997, 3). However, the European moratorium on rbGH use may have been increasingly difficult to maintain in the face of a positive Codex decision.

The Principle of Substantial Equivalence

Another contentious issue with regard to the application of scientific principles in the establishment of Codex standards has been the concept of substantial equivalence — that is, that genetically modified food is safe for human consumption if it is the same as its conventionally produced counterpart. Since 1996, joint expert consultations of the World Health Organization (WHO) and the FAO have recommended that the safety assessment of foods derived from biotechnology should be based on this concept. In the view of the experts convened in 2000, the concept embodied 'a science-based approach' (WHO 2000, 1); however, it has been denounced as tautological and meaningless by many NGOs and states. Subsequently, the Royal Society of Canada (the Canadian Academy of the Sciences and Humanities) stated that the ambiguity of the term has meant that its application has effectively become 'a declaration of safety', and argued that product regulation needs to be based on rigorous scientific assessment in order to determine whether the term actually applies (Royal Society of Canada 2001, 177). The Royal Society argued that the concept of substantial equivalence could have two meanings. One, it could indicate that a thorough scientific assessment of the genetically modified product had found that the product is not significantly different from conventional food. Two, the safety assessment could be based on the assumption that because the gene with which the product has been modified is not known to cause harm, and because the product is the same as its conventional counterpart in all other respects, the product will be safe. The problem with this interpretation of the term, according to the Royal Society, is that it is based on an assumption of linearity — that modifying an organism with a particular gene will result only in the expression of that gene, and will not be accompanied by any other potentially adverse changes. However, there may be a number of changes to the organism's genomic structure, protein coding, and metabolic activity.

In order to determine whether a genetically modified organism is really substantially equivalent, the Royal Society argued that regulatory assessment needs to examine the impact of genetic modification on the organism's DNA, gene expression, protein structure, and metabolic patterns. In most regulatory contexts, however, the first meaning of substantial equivalence is applied, and a rigorous scientific examination is not conducted. Consequently, the Royal Society states that 'does not function as a scientific basis for the application of a safety standard, but rather as a decision procedure for facilitating the passage of new products ... through the regulatory process' (Royal Society of Canada 2001, 181–182).

The joint WHO/FAO expert consultations would appear to be amenable to this critique of substantial equivalence. At the consultation in 2000, it was concluded that the application of substantial equivalence should represent a stage in the assessment of hazard, and not its outcome: the 'application of the concept is not a safety assessment in itself; it does not characterize the hazard, rather it is used to structure the safety assessment of a genetically modified food relative to its conventional counterpart' (WHO 2000, 4). However, the consultation report goes on to note that the experts were satisfied with the application of the concept in the assessment of genetically modified foods that have been approved; this acceptance diverges from the Royal Society critique, which argues that the existing application of the term has pre-empted a full safety assessment. The consultation document, on the other hand, concludes that there are simply 'no alternative strategies that would provide a better assurance of safety' (8).

Regulatory agencies' view of the concept of substantial equivalence has also determined their stance on the introduction of guidelines for labelling genetically modified food, which has been debated at Codex for more than eight years without resolution. Consumer, environmental, and other groups that have campaigned against agricultural biotechnology have argued that, in the absence of stringent testing, a mandatory labelling regime should be instituted so that consumers can choose not to consume genetically modified food. They have also argued that labelling may provide a way for consumers to express food preferences based not only on food safety concerns, but also on ethical, socioeconomic, and environmental considerations. Several responses to this position have been articulated at Codex. Six countries, led by Norway, have argued for mandatory labelling of all genetically modified foods, and opposed the application of substantial equivalence to the labelling issue. Ten countries, led by the EU, have argued for thorough safety evaluation before market release, and mandatory labelling if the food may have an impact on health or contains potential allergens or substances not in the conventional food. Seven countries led by the U.S. favour a policy based on substantial equivalence; that is, that genetically modified products that are essentially the same as their conventional counterparts should not be labelled (Phelps 1998).

The emphasis on substantial equivalence in Codex and domestic standards conflicts with the novelty requirement of domestic patent law and the TRIPS agreement. India has pointed out the contradiction between an international regulatory framework for safety predicated on the concept of substantial equivalence and a framework for intellectual property protection predicated on the concept of novelty (Phelps 1998). At the domestic and international levels, agribusiness has lobbied against the introduction of safety regulations that treat biotechnology products differently from their conventionally produced counterparts. Industry exerted a great deal of influence over the final guidelines. Washington insiders stated that 'the control that this nascent industry exerted over its own regulatory destiny … was astonishing … In this area, the U.S. government agencies have done exactly what big agribusiness has asked them to do and told them to do' (Eichenwald 2001). The initial U.S. guidelines for the regulation of biotechnology products, which were introduced in 1986 and subsequently adopted by other regulatory systems, stated that the process by which they had been produced was irrelevant for the purpose of their regulatory assessment. However, biotechnology products needed to be recognised as novel in order to be protected under domestic patent law and the TRIPS agreement.

Hard Law for the Protection of the Environment, Genetic Resources, and Indigenous Intellectual Property

Neither the TRIPS agreement, the SPS agreement, nor the Codex Alimentarius Commission deals with questions of environmental risk or socioeconomic risk. Consequently, NGOs and developing countries have campaigned for the creation of hard law instruments to address the protection of the environment and of the genetic resources that are located mainly in the developing nations of the South. It has increasingly been recognised that programmes of agricultural modernisation, which replace lower-yielding but diverse local crops with higher-yielding, uniform varieties, threaten the existence of diverse genetic resources. Biotechnology furthers this trend toward uniformity (or monoculture). Although biotechnology draws its genetic material from a number of species, its application actually increases genetic uniformity across species (see Shiva 1995). For example, the gene that codes for resistance to Monsanto's pesticide Roundup has been spliced into cotton, canola, and soybean seeds. Three quarters of the area on which genetically modified crops are grown are devoted to the production of plants with the single trait of herbicide resistance (ETC Group 2001, 7). The patenting of genes has also permitted the further control of resources that were originally developed in the South. The Rural Advancement Fund International (RAFI; now the Action Group on Erosion, Technology, and Concentration, or ETC Group) has documented cases of Western organisations making patent claims on plant varieties originally grown in the South that have not been modified, or have had minor modifications introduced (see RAFI 1997).

The Cartagena Protocol on Biosafety, which was formulated as a result of the 1995 Conference of the Parties to the Convention on Biological Diversity (CBD) and adopted in January 2000, recognises that the spread of genetically modified organisms may represent a threat to biological diversity and allows countries to regulate the transboundary movement of genetically modified organisms (GMOs, referred to as living modified organisms, or LMOs, in the document) (CBD 2000). Parties to the protocol can regulate such organisms that 'may have an adverse effect on the conservation and sustainable use of biodiversity' (preamble). The Cartagena Protocol also allows states to take into account human health considerations, and 'socio-economic considerations arising from the impact of living modified organisms on the conservation and sustainable use of biological diversity, especially with regard to the value of biological diversity to indigenous and local communities' (art. 26). Article 11.8 enshrines the precautionary principle: 'lack of scientific certainty due to insufficient relevant scientific information and knowledge regarding the extent of adverse effects ... shall not prevent that Party from taking a decision'. However, in imposing such measures, states must take into consideration their existing international obligations, which leaves open the question of how conflicts between international instruments would be resolved. So far, Codex has not introduced measures to ensure coherence between its standards and guidelines and the provisions of the Cartagena Protocol. In 2001, the Codex Alimentarius Commission refused to endorse its executive committee's recommendation that coherence be achieved between Codex standards and the provisions of the Cartagena Protocol with respect to traceability, labelling, and identification of GMOs (Codex 2001b).

NGOs have argued that the protection and promotion of indigenous genetic resources, as well as of indigenous intellectual resources regarding their development and use, have been threatened by the TRIPS agreement. The agreement does not insist on the application of patent law to plants and animals, but in the absence of patent protection for plant varieties, it states that countries must introduce their own (*sui generis*) systems of plant variety protection. Forms of plant variety protection (PVP), which were first introduced around the end of the nineteenth century, granted breeders of new plant varieties a commercial monopoly on their invention, but historically provided exemptions for farmers and other breeders. The traditional farm practices of saving and exchanging farm-grown seed were not threatened by PVP regimes. However, successive versions of the International Union for the Protection of New Varieties of Plants (UPOV) Convention have resulted in a system that is increasingly similar to the patent regime and correspondingly restrictive of farmers' rights to save, exchange, and sell farm-grown seed (Genetic Resources Action International [GRAIN] 2001). The extension of a more restrictive PVP regime has been facilitated through bilateral agreements. For example, in bilateral negotiations among the U.S., the EU, various development organisations, and Asian countries, the developed country has required the adoption of UPOV as a condition of agreement.

NGOs and states are seeking ways to balance the rights of breeders, or formal innovators (now mainly biotechnology companies that have invested in the production of genetically modified seeds), with those of farmers, or informal innovators, who are located mainly in the developing world and whose on-farm experimentation over millennia has resulted in the development of existing plant genetic resources for food and agriculture. The patenting system currently favours corporations and individuals in the developed world. For example, the Brazilian government estimates that 97 percent of the 4000 patents taken out on natural products in Brazil are held by foreigners (Agencia EFE 2001). Globally, more than 500 000 patents on genes and partial gene sequences in living organisms have been granted or are pending (GRAIN 2001). The complex system of overlapping patents has required research and development of agricultural biotechnology by public organisations. The public research organisation responsible for developing 'golden rice', a grain engineered to express higher levels of vitamin A, handed over the technology to a biotechnology firm because it feared that the cost and time involved in licensing the estimated 70 patents on genes or gene sequences would delay and inhibit the technology's release (GRAIN 2001).

The International Treaty on Plant Genetic Resources for Food and Agriculture attempts to counter the impact of patent law and to codify farmers' right (Commission on Genetic Resources for Food and Agriculture 2001). It enshrines farmers' rights to engage in traditional practices of saving and exchange, but also includes the right to participate in decision making regarding the distribution of benefits from the commercialisation of genetic resources. The treaty advocates the institution of mechanisms to ensure equitable distribution, such as a system of voluntary contributions made by industries profiting from commercialisation, as well as mechanisms for information exchange, technology transfer, and capacity building (see Article 13). NGOs have raised concerns about the voluntary nature of contributions to a benefit-sharing system, but endorse the goals and provisions of the treaty. However, as with the Cartagena Protocol, the parties to the treaty must consider their obligations under other international agreements when fulfilling this one. Again, this raises the question of how conflict among multiple obligations will be resolved. NGOs have requested that the autonomy and integrity of the treaty be clarified and respected ('Statement by Public Interest, Non-Profit Civil Society Organisations to the 31st FAO Conference' 2001). In 2002, China, Brazil, India, and nine other countries formed an alliance to fight for domestic protection of their genetic resources. Calling themselves the Group of Allied Mega-Biodiverse Nations, they seek the creation of a more equitable system for the use of genetic material located predominantly in the global South (Stevenson 2002). The 2001 WTO ministerial meeting at Doha had resulted in a commitment to examine 'the relationship between the TRIPS Agreement and the Convention on Biological Diversity, the protection of traditional knowledge, and other relevant new developments' (Plahe and Nyland 2003, 30). The TRIPS council, which administers the agreement, has also undertaken a review of Article 27.3(b) dealing with the patentability of life forms, plant variety protection, and a review of the entire agreement (WTO 2003).

Threats to traditional farm practices may arise not only from patent protection regimes, but also from technology developments that allow for biological, rather than legal, forms of intellectual property protection. A number of companies have developed what has been dubbed 'Terminator' technology, which results in the production of seeds that cannot germinate beyond the first generation. Because farmers cannot plant seeds from the first generation of crops, there is no point in saving or selling farm-grown seed, and they must make new purchases from their seed supplier each year. The first patent on terminator-type technology was acquired by the U.S. Department of Agriculture and the cotton seed firm Delta and Pine Land. Shortly after the patent had been registered, Monsanto attempted to take over Delta and Pine Land. This merger was later scuttled when U.S. Department of Justice antitrust regulators began investigating the proposed takeover, which would have given Monsanto control of 70 percent of the U.S. cotton seed market. The investigation was later expanded to examine the way that Monsanto's licensing arrangements for patented cotton genes could influence the market (Kilman 1999). The failure to complete the merger resulted in Monsanto paying an US$81 million termination fee ('Monsanto Pays $81m for Cancelled Merger' 2000).

Public Protest, Voluntary Measures, and the Demise of the Life Sciences Model

The development of the biotechnology industry was predicated on what is known as the life sciences model, in which recombinant techniques are applied to a number of fields, from food crops to food products to pharmaceuticals, thereby promoting synergies and increased efficiencies of production among these different areas. There are five major agricultural biotechnology firms: Pharmacia (which owns the Monsanto Company), Dupont, Syngenta, Aventis, and Dow (ETC Group 2001). The recognised pioneer of the life sciences model was Monsanto, and Bob Shapiro, its former chief executive officer. In 2000, Monsanto's genetically modified seed technology accounted for 94 percent of the total area sown to genetically modified crops (ETC Group 2001). The company, originally a manufacturer of plastics, chemicals, pesticides, and other petroleum-based products, transformed itself into the world's largest agricultural biotechnology firm in the mid 1990s. Recognising that the profitability of commodity chemicals had declined and that the success of even its most lucrative product, a pesticide, could not be maintained indefinitely, the company sought to transform its future through a strategy that it considered both financially and environmentally beneficial. Monsanto persisted with its strategy, in spite of the controversy surrounding many of its products, because of its belief in the potential of the life sciences. The company argued that agricultural biotechnology could be consistent with sustainable development in food and agriculture. However, the company's vision remained connected to the production of high yields of single crops; the environmental benefits would be realised because of a reduction in land and chemical use resulting from higher-yielding crops.

Anti-biotechnology groups, on the other hand, have questioned the industrial mode of agricultural production. Sustainability, they claim, cannot be achieved through the high-yield monocultural model, which contributes to pest problems and chemical dependency. In the late 1990s, anti-biotechnology activists co-ordinated a number of campaigns against genetically modified foods that attracted environmental as well as consumer groups. Levidow (2000) has argued that these groups presented definitions of sustainability that challenged corporate definitions and were eventually accepted by EU regulatory bodies and incorporated into their directives regulating the release of GMOs. Since the first EU-wide approval for genetically modified crops was granted in 1996–97, activists have had increasing influence on EU regulatory decision making and earlier assumptions on which regulatory decision making was based have been challenged (Levidow 2000). The activists have widespread support. A UK network campaign for a five-year moratorium on the planting and import of genetically modified food and the patenting of genetic resources comprised 120 organisations representing more than 4 million people (Five Year Freeze 2003). The campaign states that its concerns 'are not limited to the environment or human health, but include the impact [of biotechnology] on small farmers both in the UK and globally, corporate control of the food chain, the threat to consumer choice, ethical dimensions, and the erosion of agricultural diversity. The complexity of the issue of genetic modification in food and farming and its potential to cause massive social, economic and environmental damage drive the continued call for a moratorium. This [call] is now being echoed around the world' (Five Year Freeze 2001, 6).

While seeking to change the regulatory status of biotechnology crops, public protest groups have also sought to change the practices of biotechnology corporations through consumer boycotts of their products, and have been very successful. Although outside Europe hard law has not restricted the planting of GMOs, European consumers' boycott of GMO products has led to the adoption of voluntary measures by major food retailers, which have in turn resulted in pressure on the biotechnology conglomerates. Most major supermarket chains in Europe have rejected genetically modified products. In Asia, the major breweries no longer accept genetically modified ingredients, Japan and Korea have begun labelling certain genetically modified commodities, and Japan's largest flour miller has stopped using them. The largest tortilla maker in Mexico no longer purchases genetically modified corn. There have been a number of ironic outcomes: Novartis, formerly one of the largest producers of genetically modified crops, agreed to withhold genetically modified ingredients from its Gerber brand baby food; the catering firm supplying Monsanto's cafeteria in London also refused to use genetically modified ingredients (Specter 2000), with the result that Monsanto employees and the children of Novartis employees would not be eating their own company's genetically modified produce.

Consumer confidence and the market for genetically modified products were further undermined by a prominent failure of self-regulation in the case of StarLink maize. The maize, produced by Aventis SA, had been genetically engineered to express a protein toxic to corn borers; the U.S. Environmental Protection Agency had not approved the corn for human consumption because of concerns that this protein may cause allergic reactions. Nevertheless, although StarLink corn had been planted on less than 0.02 percent of all U.S. maize cropland in 2000, the offending protein was later discovered in taco shells on supermarket shelves across the country. Farmers had either failed to comply with clauses in their contract with Aventis specifying the distance to be maintained between StarLink and other crops or had not been informed of those clauses. Consequently, cross-pollination led to the contamination of non-StarLink crops, and the collapse of the domestic and international market for U.S. maize (Barboza 2000).

Rejection by European consumers and protests by civil society NGOs and developing countries led to announcements by the major biotechnology firms of changes in practices. Monsanto renounced its approach to the development of biotechnology. The company stated that it would consult with civil society groups it had previously been in conflict with, such as Greenpeace. It also announced that it would not proceed with the development of terminator technology. The latter decision was announced after the president of the Rockefeller Foundation, agricultural ecologist Gordon Conway, reinforced the criticisms levelled by NGOs and publicly urged the company not to develop the technology. The impact of Conway's criticism was all the more damning because Monsanto had expected him to support its position (ETC Group 2003a). Another major pharmaceutical firm, AstraZeneca, also publicly repudiated terminator technology, although the patent owner — Delta and Pine Land — had committed to its commercialisation.

Consumer rejection of GMOs ultimately led to the takeover of the Monsanto by Pharmacia Corporation in 1999. The *Wall Street Journal* reported that the 'crop biotechnology half of the programme has grown so controversial that Monsanto has agreed to a deal that is likely not only to push biotechnology to the back burner, but also to cost Monsanto its independence' (Kilman and Burton 1999). At a cost of US$25 billion, it has been estimated that the price was little more than Monsanto's pharmaceutical division, Searle, would be worth alone (Specter 2000, 68). Upon purchase of Monsanto, Pharmacia spun off the agricultural biotechnology division, under the Monsanto name; in a similar move, Novartis and AstraZeneca spun off their agricultural biotechnology divisions into a new firm, Syngenta (RAFI 2000). Will the new spinoffs will keep the commitments made by their progenitor firms? In 2003, a senior Monsanto employee co-authored a paper with a representative from Delta and Pine Land that advocated terminator technology. Although this is not necessarily indicative of a change in Monsanto's commitment, the ETC Group (2003b) has sounded the alarm.

It may appear that insufficiently stringent regulation under hard or soft law, and consequent consumer rejection, has led to the demise of the life sciences industry. However, consolidation and reinvestment are having contradictory impacts on the industry. Although Pharmacia, Novartis, and AstraZeneca have spun off their agricultural biotechnology divisions, Pharmacia still has 86 percent ownership of Monsanto, and Syngenta is still majority owned by Novartis and AstraZeneca shareholders. BASF and Bayer have also recently announced new investments in agricultural biotechnology. In 2000, BASF announced a US$680 million investment in agricultural research and development, and a year later, Bayer announced its intention to acquire the crops and agricultural chemicals division of Aventis (RAFI 2001).

The latest investments have led RAFI to speculate about the introduction of the third generation of biotechnology crops, which will present new challenges for both hard and soft law. Whereas earlier generations of crops were designed with input characteristics that linked them to the application of particular chemicals, or to output characteristics such as the slower ripening time desired by food processors, third-generation crops will address consumer needs and could conceivably be linked to one's genetic profile. For example, if it were known that an individual had a genetic propensity for a particular disease, that person might be prescribed an appropriate nutraceutical, that is, a food with pharmaceutical properties. The development of the third generation will allow for the increased application of recombinant techniques across scientific and industrial fields, and for the integration of those fields.

The new technological developments are proceeding in the context of increased industry consolidation within and across industries. Since 1993, several major agricultural biotechnology firms have entered into alliances with genomics firms (firms that identify sets of genes in living organisms); the five dominant agricultural biotechnology firms also rank among the worlds top-ten seed corporations, and these firms control 30 percent of the seed market (ETC Group 2001). Human and veterinary pharmaceuticals, agrochemicals, and various aspects of agricultural production and retailing are controlled by a small number of corporations.

Technological integration and industry consolidation will lead to new organisational forms within the industry and to new challenges for the existing protocols to manage environmental risk. The Cartagena Protocol, for example, regulates the transboundary movement of GMOs; yet pharmaceuticals are excluded from its provisions. Will nutraceuticals be defined as a GMO or as a pharmaceutical for regulatory purposes? Generation Three could potentially redefine the nature of biotechnology; rather than splicing genes from different species, new advances in genomics may lead to the modification of the organism's existing genomic profile so that genetic material not currently expressed will be so. Would such an organism be considered modified, or not?

Conclusion

The biotechnology industry and its attendant health and environmental risks present challenges to both hard and soft law instruments. Hard law instruments designed to facilitate trade and investment in biotechnology have resulted in conflict over the role of science and in demands for the explicit inclusion of non-scientific factors in international standard setting. In the absence of effective and binding international law to protect consumers and the environment, consumer pressure and NGO action, along with EU government regulation, have led to the decline in profits and transformation of the destinies of the major biotechnology players and resulted in a number of voluntary commitments to transform business practices and to consult more broadly. However, the application of soft law in the form of voluntary standards and commitments to reject particular technologies has resulted in an uneven distribution of consumer and environmental risk, with companies excluding genetically modified food from markets with high consumer pressure. Ongoing industry consolidation may also represent a threat to corporations' voluntary commitments to change their approach to technology development and public consultation, as companies that have made commitments are merged with others.

Nevertheless, ongoing industry consolidation and new technological trajectories mean that the life sciences era is no longer over, and that further challenges may be in store. As distinctions between categories of GMOs blur in the third generation of biotechnology crops, existing hard law instruments may not be adequate to ensure the protection of biodiversity or the reduction of social and environmental risk. In spite of the challenges to hard law, given the requirements of the TRIPS and SPS agreements only additional hard law instruments can provide universal protection and can allow states to protect their jurisdictions without fear of WTO challenges. Speaking about the International Treaty on Plant Genetic Resources for Food and Agriculture, one NGO representative observed that 'only an international convention of this type can begin to address the deterioration in the flow of genetic resources'. Flawed as it may be, a hard law framework is necessary for the effective regulation of biotechnology.

Notes

1. A consensus was not reached at Codex with regard to either beef hormones or recombinant bovine growth hormone. However, in the beef hormones case, a majority voted in favour of the adoption of a standard; in the case of recombinant bovine growth hormone, a majority voted against adoption.
2. The major human health issue with regard to rbGH is not the hormone itself, but another hormone-like substance, insulin-like growth factor I (IGF-I), which is identical in humans and cows and is produced in response to rbGH. Milk from treated cows has slightly higher levels of IGF-I than normal milk. It has been concluded that because IGF-I breaks down during digestion, it is unlikely to enter the bloodstream in sufficient quantities to cause human health problems. However, only short-term tests have been conducted to examine its potential impact, and there has been some evidence that a protein present in milk (casein) may prevent IGF-I from breaking down during digestion. The Europeans were concerned about this additional evidence, and suggested that it should be investigated further. They were also concerned that drug treatment may lead to an increased expression of retroviruses, such as bovine leukaemia virus. Because there were some data indicating that IGF-I stimulated prion proteins, the infective agents of bovine spongiform encephalopathy (BSE, or mad cow disease), the EU also wanted further consideration of the possibility that rbGH treatment could shorten the incubation period for BSE (JECFA 1999).
3. Although Health Canada eventually rejected the drug on animal health grounds, its human health evaluation had concluded that milk from treated cows would be safe for human consumption; therefore, Canada did not oppose the adoption of a Codex standard.

References

Abbott, Kenneth W., Robert Keohane, Andrew Moravcsik, et al. (2000). 'The Concept of Legalization'. *International Organization* vol. 54, no. 3, pp. 401–420.

Abbott, Kenneth W. and Duncan Snidal (2000). 'Hard and Soft Law in International Governance'. *International Organization* vol. 54, no. 3, pp. 421–456.

Agencia EFE (2001). 'Brazilian Shamans to Discuss Biopiracy, Protection of Practices'. 23 November.

Barboza, David (2000). 'Gene-Altered Corn Changes Dynamics of Grain Industry'. *New York Times*, 11 December.

British Action Against Genetically Engineered Products (1999). 'UK: Status of the Nationwide Frankenfoods Boycott'. 27 June. <www.organicconsumers.org/ge/britact699.cfm> (April 2004).

Coalition of the Americas (1997). 'Food Standard Issues to be Considered by the Codex Alimentarius'. Brochure presented in co-ordination with the Confédération Mondiale de l'Industrie de la Santé Animal.

Codex Alimentarius Commission (1997). 'Agenda Item 7, Codex Committee on Residues of Veterinary Drugs in Food, Comments at Step 8 by the European Community on Bovine Somatotropin'. Alinorm 97/25, Part 10, Addendum 2.

Codex Alimentarius Commission (1999). 'Agenda Item 9, Consideration of Draft Standards and Related Texts, Maximum Residue Limits for Bovine Somatotropin'. Alinorm 99/21, Part 1, Addendum 2. 2 May.

Codex Alimentarius Commission (2001a). 'Report of the Fifteenth Session of the Codex Committee on General Principles'. ALINORM 01/03. <ftp://ftp.fao.org/codex/ALINORM01/Al01_33e.pdf> (April 2004).

Codex Alimentarius Commission (2001b). 'Summary and Conclusions'. 24th Session, Geneva, 2–7 July. <www.codexalimentarius.net/cac24/alinorm0141/summarye.htm> (April 2004).

Codex Alimentarius Commission (2003a). 'Procedural Manual'. <ftp://ftp.fao.org/codex/PM/Manual13e.pdf> (April 2004).

Codex Alimentarius Commission (2003b). 'Welcome'. <www.codexalimentarius.net> (April 2004).

Commission on Genetic Resources for Food and Agriculture (2001). 'International Treaty on Plant Genetic Resources for Food and Agriculture'. <ftp://ext-ftp.fao.org/ag/cgrfa/it/ITPGRe.pdf> (April 2004).

Confédération Mondiale de l'Industrie de la Santé Animal (1997). 'Adopting Standards for the Veterinary Drug BST'. Briefing Document, June.

Consumers International (1999). 'Consumer Representatives Applaud a Decision Which Could Severely Limit the Use of Genetically Engineered Hormone BST around the World'. Press release, 30 June. <www.consumersinternational.org/news> (April 2004).

Convention on Biological Diversity (2000). 'Cartagena Protocol on Biosafety,'. <www.biodiv.org/biosafety/protocol.asp> (April 2004).

Eichenwald, Kurt (2001). 'Redesigning Nature: Hard Lessons Learned'. *New York Times*, 25 January.

ETC Group (2001). 'Concentration in Corporate Power: The Unmentioned Agenda'. *Communiqué*. July/August, Issue 71. <www.etcgroup.org/article.asp?newsid=154> (April 2004).

ETC Group (2003a). 'Broken Promise? Monsanto Promotes Terminator Seeds'. 23 April. <www.etcgroup.org/article.asp?newsid=394> (April 2004).

ETC Group (2003b). 'US Government Forces UPOV to Abandon Terminator Critique'. 17 April. <www.etcgroup.org/article.asp?newsid=393> (April 2004).

European Union (2000). 'Notes for EC Delegates: Agenda Item 3, Matters Referred from the Codex Alimentarius Commission and Other Codex Committees'. CX/RVDF 00/2, CX/RVDF 00/6, and ALINORM 99/31.

Five Year Freeze (2001). 'GM Foods — The Government's Record: Why We Still Need a Moratorium'. <www.fiveyearfreeze.org/gm%20foods.pdf> (April 2004).

Five Year Freeze (2003). 'Who Supports the Five Year Freeze?' <www.fiveyearfreeze.org> (April 2004).

Food and Agriculture Organization (1997). 'European Ban on Hormone-Treated Cattle Rejected by WTO'. Press release, 6 June. <www.fao.org/NEWS/1997/970601-e.htm> (April 2004).

Genetic Resources Action International (2001). 'Intellectual Property Rights: Ultimate Control of Agricultural R&D in Asia'. March. <www.grain.org/publications/asiaipr-en.cfm> (April 2004).

Joint Expert Committee on Food Additives (1999). 'Residues of Some Veterinary Drugs in Animals and Foods'. Food and Agriculture Organization Food and Nutrition Papers No. 41.

Kilman, Scott (1999). 'Antitrust Regulators Are Investigating Monsanto's Control of Cotton Genes'. *Wall Street Journal*, 17 December, p. B12.

Kilman, Scott and Thomas M. Burton (1999). 'Monsanto Boss's Vission of "Life Sciences" Firm Now Confronts Reality'. *Wall Street Journal*, 21 December, p. A1.

Levidow, Les (2000). 'Which Sustainability? Policy Dilemmas over GM Crops'. Paper presented at the POSTI Conference, University of London, 1–3 December. London. <www.esst.uio.no/posti/workshops/levidow.html> (April 2004).

Mills, Lisa (2002). *Science and Social Context: The Regulation of Recombinant Bovine Growth Hormone in North America*. McGill-Queen's University Press, Montreal.

'Monsanto Pays $81m for Cancelled Merger'. (2000). *New York Times*, 4 January, p. C4.

Phelps, Bob (1998). 'International GE Food Labelling Far from Resolved at Codex'. <www.poptel.org.uk/panap/archives/la-codex.htm> (April 2004).

Plahe, Jagjit Kaur and Chris Nyland (2003). 'The WTO and the Patenting of Life Forms: Policy Options for Developing Countries'. *Third World Quarterly* vol. 24, no. 1, pp. 29–45.

Royal Society of Canada (2001). 'Elements of Precaution: Recommendations for the Regulation of Food Biotechnology in Canada'. January. Ottawa. <www.rsc.ca/foodbiotechnology/GMreportEN.pdf> (April 2004).

Rural Advancement Fund International (1997). 'Biopiracy Update: The Inequitable Sharing of Benefits'. 30 September. <www.etcgroup.org/article.asp?newsid=191> (April 2004).

Rural Advancement Fund International (2000). 'Biotech's "Generation 3"'. *Communiqué*. November/December, Issue 67. <www.etcgroup.org/article.asp?newsid=158> (April 2004).

Rural Advancement Fund International (2001). '2001: A Seed Odyssey'. *Communiqué*. January/February, Issue 68. <www.etcgroup.org/article.asp?newsid=157> (April 2004).

Shiva, Vandana (1995). 'Biotechnological Development and the Conservation of Biodiversity'. In V. Shiva and I. Moser, eds., *Biopolitics: A Feminist and Ecological Reader on Biotechnology*. Zed Books, London.

Specter, Michael (2000). 'The Pharmageddon Riddle'. *The New Yorker*. 10 April, pp. 58–71.

'Statement by Public Interest, Non-Profit Civil Society Organisations to the 31st FAO Conference' (2001). 3 November. <www.ukabc.org/iu2.htm> (April 2004).

Stevenson, Mark (2002). 'China, Brazil, India, Nine Other Nations Form Alliance against Biopiracy'. Associated Press, published on Environmental News Network, 19 February. <www.enn.com/news/wire-stories/2002/02/02192002/ap_46427.asp> (April 2004).

Wirth, David A. (1998). 'European Communities — Measures Concerning Meat and Meat Products'. *American Journal of International Law* vol. 92, no. 4, pp. 755–780.

World Health Organization (2000). 'Safety Aspects of Genetically Modified Foods of Plant Origin'. Joint Food and Agriculture Organization/World Health Organization Consultation on Foods Derived from Biotechnology, Geneva, 29 May–2 June. <www.who.int/entity/foodsafety/publications/biotech/en/ec_june2000_en.pdf> (April 2004).

World Trade Organization (1994). 'Marrakesh Agreement Establishing the World Trade Organization'. <www.wto.org/english/docs_e/legal_e/04-wto_e.htm> (April 2004).

World Trade Organization (1995). 'The WTO Agreement on the Application of Sanitary and Phytosanitary Measures'. <www.wto.org/english/tratop_e/sps_e/spsagr_e.htm> (April 2004).

World Trade Organization (2003). 'Reviews, Article 27.3b, Traditional Knowledge, Biodiversity'. <www.wto.org/english/tratop_e/trips_e/art27_3b_e.htm> (April 2004).

Chapter 19

Hard and Soft Law in International Institutions: Complements, Not Alternatives

Nicholas Bayne

This chapter looks beyond the specific topics of this volume — labour standards, corporate social responsibility, and the environment — to examine more broadly the use of hard and soft law in a range of intergovernmental institutions. Some of these institutions are based on hard law, that is, on rule-based systems, and some on soft law, that is, on voluntary co-operation. Two major examples of the first type — the European Union and the World Trade Organization (WTO) — will be compared with two major examples of the second — the Organisation for Economic Co-operation and Development (OECD) and the G7/8 summit. Such a comparison leads to the conclusion that hard law and soft law are not alternatives. Rather, they serve different purposes and complement one another.

The European Union

The European Union, as with the European Community before it, relies heavily on hard law, that is, on formal obligations applying to all members (Woolcock 2003). It is deeply rooted in the statute law tradition of continental Europe, which Britain has found hard to digest. The EU suffers from the drawback of this statute law tradition, in that it tends to over-legislate. Twice in its history the EU has almost ground to a halt under the burden of excessive law making — once in the late 1970s and again in the early 1990s. But there is no doubt that this capacity for collective law making has driven the economic integration of Europe. It is a principal reason for the success of the EU and for its attraction to those who wish to join it.

Because European governments, and European citizens too, are operating within their own hard law system, they tend to favour rule making in wider international contexts. This appears, for example, in environmental issues such as global warming. EU member governments, driven by European public opinion, are the strongest advocates of the rules embodied in the Kyoto Protocol (Rabe 2003). For the same reason, the EU promoted an agenda for a new multilateral trade round that would enlarge the WTO's rule-making capacity, notably in investment and competition.

The EU also makes much use of soft law. Alongside the regulations and directives are recommendations and guidelines. Soft law approaches are used where some or all EU member states are not ready for formal, binding commitments. But in economic, social, and environmental issues, these soft law procedures are seldom seen as ends in themselves. They are more used as a preparatory technique, or even as a delaying tactic, until agreement can be reached on formal hard law measures. Thus the United Kingdom under Prime Minister John Major had the binding 'Social Chapter' of the 1991 Maastricht Treaty converted into a non-binding 'Social Charter'. Similarly, Britain under Prime Minister Tony Blair has been seeking to have aspects of corporate social responsibility handled by voluntary co-operation, rather than through the binding directives that most other member states favour.

In addition, there is a large element of the European Union that is wholly based on soft law. The Maastricht Treaty created a union with three pillars: first, the European Community, narrowly defined, covering economic and social issues; second, political co-operation, mainly in foreign policy; and third, the realm of justice and home affairs. Both the second and third pillars are soft law edifices. But while these may lead to less conflict within the EU, they produce less impressive results. Pascal Lamy, the European Trade Commissioner, has more power to speak and act for Europe than his foreign policy colleagues Christopher Patten or Javier Solana. The divisions over foreign policy toward Iraq, which emerged in early 2003 and were manifested in North Atlantic Treaty Organization (NATO), could never have persisted in international trade and the WTO.

The World Trade Organization

The WTO, as created by the Uruguay Round of trade negotiations, is also very largely based on hard law — more so than its predecessor, the General Agreement on Tariffs and Trade (GATT). The WTO has created a worldwide trade regime covering not only manufactures, but also agriculture and services. Its dispute settlement regime has legal force that cannot easily be evaded (Jackson 1998; Carden 2003).

The hard law of the WTO, with 147 members, naturally differs from the EU with only 25. Agreements reached at Geneva do not apply automatically, like regulations agreed at Brussels. They must be ratified by the national legislatures of all the WTO members. But once in force, they have the effect of law. That makes the WTO appear to be a very strong regime — stronger, in some respects, than its members realised or than the world is ready for (Abbott et al. 2000). The WTO, like the EU, has fallen into the error of over-legislation.

The WTO is an institution equipped to set rules for trade and to remove trade barriers. It is not equipped to regulate other subjects. But because the WTO system looks so strong, in relation to other multilateral regimes, it has got drawn in where it does not really belong. In retrospect, it was a mistake to include intellectual property and food safety in the Uruguay Round agreements. But the damage has now been done. Yet it would be wrong to have the WTO regulate labour or environmental standards. This particularly aggravates developing countries, which see new protectionist barriers being raised against them. Developing countries have already concluded that the Uruguay Round results did not help them as much as they had hoped. They have come to feel that the WTO did not treat all its members equally. It was this more than anything that destroyed the Seattle WTO meeting in 1999 (Bhagwati 2001).

The November 2001 WTO ministerial meeting in Doha was a second attempt to launch a new round of trade negotiations. This time it was successful. The declaration agreed to by ministers includes many points of importance to developing countries, especially the poorest (Laird 2002). Although the non-trade issues have not gone away entirely, largely because of EU pressure on the environment, they have much less salience. Some of the earlier problems, for example in intellectual property, are being addressed (Abbott 2002). Thus the prospects for the WTO look much more hopeful.

Soft law has its place in the WTO, as in the GATT before it. The voluntary regime for industrial subsidies under the GATT, dating from the late 1970s, had little effect on its own. But it prepared the way for the formal rules agreed during the Uruguay Round. More important, the WTO's Trade Policy Review Mechanism (TPRM), through which countries explain their trade policies to their peers and respond to criticisms, leads only to non-binding recommendations. This part of the WTO works more like the OECD.

The Organisation for Economic Co-operation and Development

In contrast to the EU and WTO, the OECD works almost wholly through soft law, that is, through voluntary co-operation (Henderson 1993; Bayne 2000). The OECD's method could be summed up in the two slogans inscribed in the temple of Apollo at Delphi in classical Greece: 'Know Thyself' and 'Nothing in Excess'. The institution's value to its members is, first of all, as a source of information and comparative analysis across the whole range of economic and social policies. Second, it provides a place where policies are discussed and modified by persuasion and peer pressure, to make them more internationally coherent and compatible. Without the requirement to produce legally binding results, the OECD is free to be more innovative and experimental. Countries can get good ideas from the OECD, which they can reflect in their national policy and legislation — even though often they will claim to have thought of the ideas themselves.

The OECD has the power to conclude legally binding agreements, if its members so decide. But it seldom tries to do so, and when it does, it often does not succeed. This is partly because the persuasion and peer pressure technique of the OECD is not very transparent, partly because it has no compliance machinery, and partly because rules agreed to in the OECD will inevitably have an impact on nonmember countries, which resent it. This explains why the OECD's Multilateral Agreement on Investment (MAI) did not work, and also why the OECD's rules on tax havens have provoked such resistance (Henderson 1999; Rugman, Kirton, and Soloway 1999; Persaud 2001).

But if the OECD does not often make rules itself, its soft law activities profoundly influence the making of hard law elsewhere. The OECD makes a vital input into the rules drawn up by wider multilateral institutions. The main content of the Uruguay Round agreements on agriculture and on services — two of the areas of greatest advance — were worked out in the less demanding, more intellectual, and more experimental context of the OECD.

The G7/8 Summit

The G7/8 summit, with its supporting apparatus, relies entirely on voluntary co-operation and soft law (Bayne 2000). It has no capacity to make hard law and there is no demand that it should. It would be quite unacceptable for this powerful and self-selected group to dictate to the rest of the world. As with the OECD, G7/8 discussions can be innovative and experimental. The G7/8 does not have the intellectual resources of the OECD Secretariat. But because it has so few members, it can be very flexible and move surprisingly fast (Budd 2003).

Very little that is agreed to among the G7 or G8 applies to the members alone. Nearly everything affects wider international decision making. Here the G8 has a double contribution to make. The first is its capacity to resolve differences among its own members. If the G8 members do not agree on how to act, there will be no agreement on what to do more widely. Second, the G8, if its members can agree, can then provide leadership to the multilateral institutions. This is not provided, as sometimes in the past, on a take-it-or-leave-it basis, but as the foundation for wider consensus.

A good illustration of this comes from the field of money laundering and freezing terrorist assets. In 1989, the G7 Paris Summit launched the Financial Action Task Force (FATF) to combat money laundering. This became a free-standing body of wider membership. Starting in 2000, the G7 brought international attention back to the FATF, so as to discipline certain countries providing havens for criminal funds (Wechsler 2001; Williams 2000). So the terrorist attacks of 11 September 2001 found the G7 well positioned both to extend the remit of the FATF rapidly to the pursuit of terrorist assets and later to involve the International Monetary Fund (IMF) in the process.

Conclusion

Some broad conclusions on hard law and soft law can be drawn from this comparison of the EU and WTO with the OECD and G7/8. First, hard law has the greatest potential and produces the most far-reaching results. It is more durable, more predictable, and, usually, more transparent. But it is very demanding, and, if it is abused or otherwise goes wrong, putting things right is very difficult.

Second, soft law can do some things that hard law cannot. It is less demanding and more flexible. It facilitates innovation and experiment. It can provide the essential preparatory phase before people are ready for hard law commitments. But it can also be used as an excuse for avoiding firm commitments and it can come apart under pressure.

Third, it is not easy to mix hard law and soft law in the same institution. There are some examples of successful combination, but it is rare — none of the institutions considered in this chapter does it very well. Normally very different disciplines are required to produce results in rule making and in voluntary co-operation.

Finally, soft law stimuli are often essential to get the hard law moving. Thus the OECD and G7/8, soft law institutions themselves, make things happen in wider rule making.

To sum it up in a single image: hard law is like marriage; soft law is like courtship or cohabitation.

References

Abbott, Frederick M. (2002). 'The Doha Declaration on the TRIPs Agreement and Public Health: Lighting a Dark Corner at the WTO'. *Journal of International Economic Law* vol. 5, no. 2, pp. 469–505.

Abbott, Kenneth W., Robert Keohane, Andrew Moravcsik, et al. (2000). 'The Concept of Legalization'. *International Organization* vol. 54, no. 3, pp. 401–420.

Bayne, Nicholas (2000). *Hanging In There: The G7 and G8 Summit in Maturity and Renewal.* Ashgate, Aldershot.

Bhagwati, Jagdish (2001). 'After Seattle: Free Trade and the WTO'. *International Affairs* vol. 77, no. 1, pp. 15–30.

Budd, Colin (2003). 'G8 Summits and Their Preparation'. In N. Bayne and S. Woolcock, *The New Economic Diplomacy: Decision-Making and Negotiation in International Economic Relations*, pp. 139–146. Ashgate, Aldershot.

Carden, Richard (2003). 'The World Trading System'. In N. Bayne and S. Woolcock, *The New Economic Diplomacy: Decision-Making and Negotiation in International Economic Relations*, pp. 275–283. Ashgate, Aldershot.

Henderson, David (1993). 'International Economic Co-operation Revisited'. *Government and Opposition* vol. 28, no. 1, pp. 11–35.

Henderson, David (1999). *The MAI Affair: A Story and Its Lessons.* Royal Institute of International Affairs, London.

Jackson, John (1998). *The World Trade Organization: Constitution and Jurisprudence.* Royal Institute of International Affairs, London.

Laird, Sam (2002). 'A Round by Any Other Name'. *Development Policy Review* vol. 20, no. 1, pp. 41–62.

Persaud, Bishnodat (2001). 'OECD Curbs on Offshore Financial Centres: A Major Issue for Small States'. *The Round Table* no. 359, pp. 199–212.

Rabe, Patrick (2003). 'Making EU International Environmental Policy'. In N. Bayne and S. Woolcock, *The New Economic Diplomacy: Decision-Making and Negotiation in International Economic Relations*, pp. 215–228. Ashgate, Aldershot.

Rugman, Alan M., John J. Kirton, and Julie A. Soloway (1999). *Environmental Regulations and Corporate Strategy: A NAFTA Perspective*. Oxford University Press, Oxford.
Wechsler, William F. (2001). 'Follow the Money'. *Foreign Affairs* vol. 80, no. 4, pp. 40–57.
Williams, Frances (2000). 'Poorest Countries Attack WTO Aid Package'. *Financial Times*, 11 April, p. 14.
Woolcock, Stephen (2003). 'The Regional Dimension: European Economic Diplomacy'. In N. Bayne and S. Woolcock, *The New Economic Diplomacy: Decision-Making and Negotiation in International Economic Relations*, pp. 197–214. Ashgate, Aldershot.

Index